AIP Conference Proceedings

Series Editor: Hugh C. Wolfe

No. 19

Physics and the Energy Problem–1974

(APS Chicago)

Editors

M.D. Fiske

General Electric Corporation, Research and Development

and

W.W. Havens, Jr.

Columbia University

American Institute of Physics

New York   1974

D
621.4
TOP

Physics and the Energy Problem—1974

(APS Chicago)

The Proceedings of

The American Physical Society

Topical Conference
on
Energy

Held in Chicago, Illinois, February 4-7, 1974

Edited by

M.D. Fiske
General Electric Corporation, Research and Development

W.W. Havens, Jr.
Columbia University

Conference Organization

These are the Proceedings of the first Topical Conference of the American Physical Society to be held as an integral part of its regular meetings. This conference mode was deemed by the Society's Committee on Meetings to be appropriate to those topics which should have a broad appeal to the membership and which as a matter of policy should be brought to the attention of members in as wide a format as possible.

Such a concurrent topical conference has the advantage that most of the administrative machinery for it - for a meeting place, local arrangements, lodging, etc. - exists as part of the Regular Meeting. The conference committee can then focus its efforts on the program alone.

The program planning for this conference, started in early 1973, was given substantial help by the August meeting at Los Alamos of the APS Energy Study Planning Committee.

The members of the program committee for this topical conference have been:

S.J. Buchsbaum (Bell Telephone Laboratories)
B.L. Cohen (University of Pittsburgh)
M.D. Fiske (chairman) (General Electric Corporate
        Research and Development)
W.C. Griffith (North Carolina State University)
W.W. Havens, Jr. (American Physical Society)
J.K. Hulm (Westinghouse Research Laboratories)
F.E. Jamerson (General Motors Research Laboratories)
B. Weinstock (Ford Scientific Laboratories)

Acknowledgement for contributions to the planning of the conference are due to:

A.J. Bennett (General Electric Corporate Research and
        Development)
S. Gratch (Ford Scientific Laboratories)
E.F. Hammel (Los Alamos Scientific Laboratories)

PREFACE

The application of physics to new technology has been the concern in recent times principally of the physicist in industry and of his engineering colleagues, except during times of national emergency. The focus of the academic physicist -- beyond teaching -- has been on the discovery and elaboration of new physics. The application and consequences of his research he has left to others.

The separation between "academic" and "industrial" physics which is now regarded as the norm was not so pronounced before World War II. Many of the academic physicists were employed to teach engineers. The physicists were required to know how physics was used in the development of new processes and products in order to arouse the interest of their students and if possible persuade the best of them to embark on a career in physics. Since the livelihood of many of the physicists depended on their usefulness either directly or indirectly to the industrial community, the gap between "academic" and "industrial" physics was almost nonexistent. At that time, many industrial physicists were the leaders in the physics community and a new Ph.D. regarded a position in industry at least as desirable as an academic position.

Another factor which has changed significantly since World War War II is the education of engineers to the doctorate level, many of whom are equipped to perform the work which was once the province of the industrial physicist. In the 1930's there were very few opportunities for advanced graduate work in Engineering. Many of the leaders of the physics community received their undergraduate education in engineering and then did graduate work in physics. In the 1930's the number of doctorate degrees awarded in the U.S. in engineering was considerably less than the number in physics, whereas at the present time, the number of annual doctorates in engineering is about three times the number of doctorates in physics. Many of the engineering doctorate programs concentrate on applied physics and many of them are taught by physicists. The engineering doctorate has become the "applied physicist". In view of these changes in circumstances it should not be surprising that the younger physicists came to view physics in industry as the domain of the engineer, with the norm of the academic physicist being to develop new physics without regard for its application.

In the last few years, we have witnessed a rather remarkable departure from that norm, with words like "relevancy" and "social responsibility" taking on new force. The new APS Forum on Physics and Society and its activities and this conference itself are measures of a sort of that departure, and it will remain for some time an interesting question as to whether these represent an excursion or a lasting new direction. Even before energy problems assumed the crisis proportions some of them have today, many of the members of

this Society -- particularly the younger ones -- had been turning
their attention away from more conventional physics areas toward
pressing social problems: pollution, transportation, the cities,
the environment. Part of the driving force in this new direction
was due, to be sure, to the drying up of jobs in the traditional
sectors of physics employment. But a significant part of it, without doubt, has been a growing sense of social responsibility, a
wish or a need to see physics used for social good. This sense has
been expressed most sharply in the universities and in civic action
groups, but those in industry have seen it also especially in the
young people who have gone there in the last few years.

Now, one can do a great deal of fruitless threshing about if
one's desire to do something socially useful is not coupled with a
good sense of what really needs to be done -- and here is where
this conference comes in. A year ago, when the planning for this
conference began, it was abundantly clear that energy problems were
going to be with us for a long time and that, of the large problem
areas, energy probably offered the greatest opportunities for contribution by physicists. Indeed, as a rough estimate there were
already 2-3 thousand physicists in the United States concerned
with production and delivery of energy. Best known of these efforts
to the physics community has no doubt been the research on fusion.
Less glamorous than this has been work ranging from fission reactors
to high temperature turbine materials to switchgear to catalysis.
Indeed the normal business enterprise has seen to it that technical
developments were pressed wherever they could be seen to create a
new business opportunity. And where a new technology had to be
created or an old one improved, there also were physicists lurking
in the background or in the thick of the fray. For it is here that
physicists have had their unique contribution to make in the innovative organization of technical complexity. New understanding,
discovery, invention are aspects of this contribution already well-
known to the academic physicist. Not so well-known is the milieu
in which these are applied where costs, product safety, public acceptance, etc., may be much more important than technical achievement. It is a world which can be both arcane and frustrating to
the novitiate. But willy-nilly, it is -- we believe -- a world
physicists must come to know much better if physics is to grow as
a profession.

What the American Physical Society has tried to do in the 7
sessions of this conference is to present to physicists a set of
problems in energy-related fields to which they may be able to contribute. The set is obviously but a sampling and many important
areas have been left out. But within this sampling, each speaker
has been asked to present first some of the context of his topic
so that those of the audience not acquainted with the area may
appreciate its significance. And then he has been urged to delineate the technical problems where a lack of physical understanding

blocks desired developments. It is our hope that physicists looking for new areas to enter may find some at this conference.

- M.D. Fiske

W.W. Havens, Jr.

February 1974

PROGRAM AND TABLE OF CONTENTS

The Energy Problem in Perspective, M.D. Fiske, presiding
Monday, 4 February, 9:00 A.M.

Resource Development, W.C. Griffith, presiding
Monday, 4 February, 2:00 P.M.

Mobile Powerplants, F.E. Jamerson, presiding
Tuesday, 5 February, 9:00 A.M.

Energy Transmission and Storage, B. Weinstock, presiding
Wednesday, 6 February, 9:00 A.M.

THE FUEL SITUATION

J. C. Fisher
General Electric Company
345 Park Ave., New York, N. Y. 10022

ABSTRACT

The United States has an abundance of energy resources; fossil fuels (mostly coal and oil shale) adequate for centuries, fissionable nuclear fuels adequate for millenia, and solar energy that will last indefinitely. Current fuel shortages reflect a shortage of productive capacity, not the depletion of resources. Shortages can be alleviated by the development of additional mineral reserves and the construction of additional facilities for refining and transporting natural and synthetic fuels. The cost of energy, measured in constant dollars to strip away the effects of inflation, has declined as long as records have been kept, largely through economies of scale and new technology. These cost-reducing forces are still effective. The constant-dollar cost of environmentally acceptable clean fossil fuel (refined oil and natural gas now, followed by synthetic oil and gas later) may continue its historic downward trend for several centuries, and the cost of nuclear fuel may follow a downward trend for millenia, before economies of scale and new technology are finally overcome by the effects of depletion.

ENERGY CONSUMPTION

Nearly three-quarters of the consumption of energy in the United States is related in one way or another to people working in industry, commerce, and government, including the associated transportation of goods and of people to and from work which I estimate to aggregate about 3/4 of all transportation.

Industrial	39.9 percent
Commercial and Government	13.9
3/4 Transportation	19.2
	73.0 percent

The rest of the nation's energy consumption is related to residential and recreational uses.

Residential	20.6 percent
1/4 Transportation	6.4
	27.0 percent

Job-related energy usage has been growing rapidly in recent years, in large part because an increasing proportion of the adult population has been finding employment in factories and offices. People are still leaving farms and going to work in cities, but much more significantly, women are leaving their homes and joining the labor force. As a result of these changes in the employment patterns of Americans, the number of non-farm workers has been growing more than twice as fast as the population in recent years, and job-related energy usage has had to grow more rapidly than the population to keep pace.

Overall affluence has increased as the number of wage-earners per family has increased. Some of the increased affluence has allowed the purchase of a second automobile so the second wage-earner could drive to and from work, and some has enabled residential and recreational energy usage to grow more rapidly. Most of this increased energy usage has been for hot water, comfort heating and cooling, and home appliances.

Although growth of the non-farm work force is currently high, as the last of the farm workers leave the farms and the influx of women into the work force is at its maximum, this rapid growth must come to an end fairly soon when migration from the farms is complete and when women achieve equal participation in the work force. And the rapid growth of energy consumption must slow down at the same time, partly because the need for additional factory, office, and transportation facilities for workers will decline, and partly because the general level of affluence will grow more slowly as the number of workers per family stabilizes.

An estimate based on the anticipated future pattern of employment and the anticipated future level of affluence suggests that energy consumption in the United States may level off at about 40 percent more, per capita, than it is today.[1] Relatively little of this increase is due to the consumption of energy (whether at work, travelling to or from work, at home, or at play) by families where both husband and wife already have non-farm employment. Most of the increase is due to the increased consumption of energy (at work, travelling to or from work, at home, or at play) associated with families adding a second non-farm wage-earner and joining the more-affluent majority.

Viewed in this light, we see that the current rapid growth in energy consumption is a measure of the pace at which Americans are coming to participate in the material benefits of our society; and we can anticipate a level-

ling-off in energy consumption as full participation is
achieved. It would be unfair to the minority who have
not yet joined the non-farm work force if, by stopping
the growth of energy consumption prematurely, we were to
deny them jobs and economic independence.

SOURCES OF ENERGY

A number of different sources have provided significant energy inputs to the United States at one time or
another. In approximate order of their historic development, they can be classified under the headings of solar
energy, fossil fuels, and nuclear fuels:

> SOLAR ENERGY: conversion via
> > Fuel wood
> > Work animal feed
> > Windpower
> > Direct waterpower
> > Hydroelectricity
>
> FOSSIL FUELS: combustion of
> > Coal
> > Petroleum
> > Natural gas
>
> NUCLEAR FUELS: fission of
> > Uranium
> > Thorium

Solar energy is dilute, but large in magnitude and
unlimited in time. Fossil fuels are concentrated and
cheap, but are exhaustible and can become exhausted after
several centuries. Nuclear fuels are practically in-
exhaustible, particularly if breeder reactors are able to
utilize the common isotopes of uranium and thorium.
Broadly speaking, for the industrialized societies
the years of significance for fuel wood, work animal feed,
windpower, and direct waterpower have passed; and the
years of significance for nuclear fuels are just beginning. The energy sources of current significance are
hydroelectricity and the fossil fuels. In the United
States in 1970, 94 percent of the energy consumed came
from fossil fuels, 4 percent from the fuel equivalent of
hydroelectric power, and the remaining 2 percent from
various minor sources, including a rapidly growing one-
third of 1 percent from nuclear fuels. These figures are
summarized in Table 1.

TABLE 1.   SOURCES OF ENERGY FOR
THE UNITED STATES IN 1970

Source	Percent
Fossil fuel	
Coal	20
Petroleum	41
Natural gas	33
Solar energy	
Hydroelectricity	4
Other	
Miscellaneous	2
	$\overline{100}$

Other sources of potential significance for large-
scale energy production include

> SOLAR ENERGY: conversion via
> Fuel crops

> FOSSIL FUELS: combustion of hydrocarbons from
> Oil shale
> Tar sand

Sources of energy are judged to be potentially significant where the available quantities are large and where
technological and economic considerations show that the
costs are competitive or close to competitive.

Heat and other forms of energy can be measured in
terms of British thermal units, or Btu for short.   One
Btu is the amount of heat it takes to warm up 1 pound of
water (approximately 1 pint of water) $1^{o}F$.   The United
States consumes so much energy that the annual number of
Btu is very large.   To bring such large numbers down to
size, I define a C-unit as

$$1 \ C \ = \ 10^{16} \ Btu.$$

One C of heat is approximately the heat that would be
generated by burning 400 million tons of coal, roughly
the amount of coal consumed in the United States each
year for the past half century.   (Hence the use of the
letter C for the unit.)   It is also approximately the
heat required to warm up Lake Michigan $1^{o}F$, for there are
just about 10^{16} pints of water in the lake.

Now it is possible to quantify the various energy
inputs as shown in Table 2.

TABLE 2. SOURCES OF ENERGY FOR THE UNITED STATES IN 1970

Source	Conventional Quantity	Energy Content C $(=10^{16}$ Btu)
Fossil fuel		
Coal	525 million tons	1.28
Petroleum	5.36 billion barrels	2.65
Natural gas	21.4 trillion cubic feet	2.13
Solar energy		
Hydroelectricity	253 billion kilowatt-hours	0.26
Other		
Miscellaneous		0.13
		6.45

Overall in 1970, the energy input to the United States amounted to 6.45 C. (Subtracting out 0.26 C of hydro-electricity and adding in 0.39 C of fossil fuels used for nonenergy purposes, the overall fuel input to the United States in 1970 was 6.58 C.)

World energy consumption in the same year amounted to about 21.5 C including 19.3 C of mineral fuels and hydroelectricity and an estimated 2.2 C of traditional fuels as shown in Table 3.

TABLE 3. SOURCES OF ENERGY FOR THE WORLD IN 1970

Source	Energy Content C $(=10^{16}$ Btu)	Percent
Fossil fuel		
Coal	6.5	30
Petroleum	7.7	36
Natural Gas	3.8	18
Solar energy		
Hydroelectricity	1.3	6
Wood, waste, feed	2.2	10
	21.5	100

ENERGY RESOURCES AND RESERVES

We must distinguish between total resources and re-serves. A lack of clear thinking in this regard has led to much misunderstanding of the current energy crisis. At any given time, some of the known mineral deposits have been well mapped and have been developed for produc-tion, while others remain undeveloped. A deposit is de-

6

veloped when the necessary investment has been made in
whatever capital equipment is required to produce the
mineral from the deposit and move it to market. For example in the coal industry this means acquisition of mining rights, investment in mining equipment, and installation of transportation equipment to move the coal to the
nearest existing railroad or waterway. Investments of
this nature are made in anticipation and expectation of
profitable production. The minerals that have been developed for profitable production are called "reserves."
This definition of reserves is followed most closely in
the oil, gas and uranium industries, and less closely in
the coal industry where well-known easily developable deposits are often counted as reserves.

Reserves amount to current inventory of minerals in
the ground. As mineral production proceeds, material is
withdrawn from inventory and reserves are diminished. At
the same time, as investments are made in developing additional deposits, new inventory is created and reserves
are increased. Current reserves at any given time reflect
the interplay of these opposing tendencies. Reserves can
be measured in terms of the reserve-to-annual-production
ratio, which equals the number of years that the inventory would last if production continued at its present
rate (and if no new inventory were developed). Economic
forces keep the reserve-to-production ratio for many minerals at 10 to 20 years.

Should the reserve-to-production ratio for some mineral drop from 12 years supply to 10 years supply over
some period of time, we need not necessarily assume that
ultimate mineral depletion is approaching. It may reflect only prudent trimming of inventory at a time of
rising interest rates, or perhaps it may reflect growing
uncertainty about impending price regulation or a potential flood of low-cost imports.

Known undeveloped deposits are often called "submarginal," because they cannot be developed to produce minerals at a profit with today's technology, today's costs,
and today's prices. Yet as the economy of scale reduces
costs, and as new technology reduces costs, submarginal
deposits tend progressively to be developed and are transformed into reserves. As an example, when oil is withdrawn from an oilfield, it flows easily at first, then
less easily, then must be pumped, and finally, with whatever state of technology exists at the time, the cost of
getting it out exceeds the price that it will bring. The
inventory of recoverable oil has been exhausted. The reserves are gone. Yet 70 percent or so of the original
oil in place in the field still remains there as a submarginal resource. As new recovery technology is developed, a time usually comes when it pays to redevelop the
same field for additional production by waterflood, fire-

sweep or some other technology.

New reserves are created whenever an old oil field or coal mine is redeveloped, and in principle, the process can be continued time after time until the oil and coal are completely developed.

In addition to reserves and known submarginal deposits, there are additional deposits not yet discovered. And (particularly for oil and gas) there is substantial worldwide activity devoted to discovering them. Once discovered, some will prove to be easily developable for low-cost production, and these may be classified as unknown deposits of economically recoverable minerals. Others will prove not to be profitably developable, and these may be classified as unknown submarginal deposits. The quantities of undiscovered resources clearly are the most difficult to determine, but geological and geophysical experts have learned to make respectable estimates.

RESOURCE AND RESERVE ESTIMATES

U.S. Geological Survey specialists have compiled estimates of United States fossil-fuel resources[2] including coal, petroleum liquids, natural gas, and oil from shale. Their estimates are generally made on geological projections of favorable rocks and on anticipated frequency of the energy resource in the favorable rocks. The estimates of submarginal resources of oil from shale include only relatively rich deposits that might be recoverable with today's technology at less than two or three times 1970's United States oil prices, and exclude much larger quantities of lower-grade shale. I believe these Geological Survey estimates are the best objective estimates available for the United States.

Total cumulative United States consumption of fossil fuels has amounted to about 200 C through 1970, and 1970 consumption amounted to about 6.5 C. Projections of future consumption suggest that an additional 300 C may be required between 1970 and 2000. The data in Table 4 show that the United States has abundant fossil-fuel reserves and resources. Present fossil-fuel reserves are more than adequate to supply the 300 C of energy required for the balance of this century.

If we accept the projections that annual per capita energy consumption will level off at 450 million Btu per capita and that annual per capita fossil-fuel consumption for nonenergy purposes such as asphalt tiles, road oil and petrochemical feedstocks will level off at 50 million Btu per year, then, in an economy totally energized by fossil fuel, the annual per capita fossil-fuel consumption would level off at 500 million Btu.[1] If we assume in addition that the population of the United States will stabilize at about a billion people within the next few

TABLE 4. U.S. FOSSIL & NUCLEAR FUEL RESERVES & RESOURCES

Energy Source	Reserves*	Additional Resources*
Fossil fuels		
Coal	900	6,600
Petroleum	30	1,640
Natural gas	30	640
Shale oil	-	15,000
Total	960	24,000
Nuclear fuels		
Uranium	22	22,000,000
Thorium	-	34,000,000
Total	22	56,000,000

*Units of $C = 10^{16}$ Btu

centuries, it follows that annual fossil-fuel consumption
(energy uses plus non-energy uses) would level off at
about 50 C per year. Under these conditions, in an economy where all energy was derived from fossil fuels, our
fossil-fuel reserves and additional resources would be
enough to last about 500 years.

$$\frac{25{,}000 \text{ C reserves and resources}}{50 \text{ C per year}} \approx 500 \text{ years.}$$

Uranium and thorium contain so much energy per pound
that it makes good sense to consider very low-grade ores
in estimating resources. The most comprehensive estimates have been made by an Interdepartmental Study[3] with
participation by nine Federal departments and agencies.
The uranium reserves shown in Table 4 correspond to utilization of 1.5 percent of the potential energy of the uranium, as is appropriate for today's light-water reactors.
The nuclear resources correspond to utilization of 80 percent of the potential energy of uranium and thorium as
may become possible with the new technologies of breeder
reactors. Although nuclear fuel reserves available with
current technology are not particularly large in relationship to projected energy consumption, domestic nuclear fuel resources are adequate for a million years.
So far our consideration of fossil and nuclear fuels
has focussed on the United States. The Interdepartmental
Study considered world resources as well as United States
resources, and using data available in 1962, they estimated that in terms of total resources "the United States
is endowed with approximately one-fourth of the coal, one-

seventh of the oil, possibly one-tenth of the natural gas, one-twelfth of shale oil, and one-seventeenth of uranium and thorium." Since the United States has about one-seventeenth of the world's land area, it appears it may have somewhat more than its share of fossil fuels, particularly coal. I expect that as better data become available, the overall energy resources of the world will prove to be more or less uniformly distributed. Most major land areas should have their proportionate shares, with a few exceptions such as Japan having none, the Middle East having more than its share of oil, and the United States having more than its share of coal.

In summary, we find that fossil-fuel resources alone are adequate for about 500 years, and nuclear resources for a million years; provided, of course, that these resources can be utilized within the bounds of environmental acceptability. Solar energy is even more abundant; the energy reaching ground level in the United States amounts to about 10,000 C annually and is assured indefinitely. World energy resources are substantially larger, in approximate ratio of the world's land area to the United States' land area.

Although it is true that nuclear and solar resources exceed fossil-fuel resources by a wide margin, there is no immediate threat of fossil-fuel depletion to cause us to shift to nuclear fuel or solar energy. The United States supply of coal and shale oil is sufficient to last for many centuries if we choose to use it, and the same is true for the rest of the world. It is the overall sociology, politics, economics, and technology of energy supply and utilization that will determine which fuels are actually used and on what timetable.

FUEL COST ESTIMATES

Mineral recovery costs in general tend to rise, because recovery tends first to exhaust rich deposits, then to move on to leaner deposits with higher unit costs. But at the same time the economies of scale and of new technology tend to reduce costs, and the overall cost of a mineral at the mine or well can either rise or fall depending on the relative significance of these opposing tendencies. For crude oil production in the United States, the opposing influences of depletion and of new technology (economy of scale is relatively unimportant for crude oil recovery) have been in approximate balance for many decades. During the period 1935-1971, while cumulative production of crude oil increased nearly six-fold and the depletion of total oil resources grew from about 1/2 percent to about 3 percent, the price of crude oil at the well fluctuated around $3.50 per barrel (measured in 1970 dollars to strip away the effects of inflation), with fluctuations amounting to about 20 percent each way.

Over the same time span, the price (in 1970 dollars) of a gallon of gasoline was cut in two, as the transportation, refining, and distribution segments of the oil industry all responded to significant economies of scale.

Estimates of future crude oil recovery costs, as affected by increasing depletion and further economies of new technology, suggest that an approximate balance may continue until depletion reaches about 10-15 percent, after which the effects of depletion will become progressively stronger and costs will rise at an accelerating rate.[1] Before the costs begin to rise significantly, an approximate additional 200 C of crude oil may be recoverable from the United States resource base. This quantity of oil, supplemented by other fossil and nuclear fuels and by hydroelectric power, would be adequate for several decades. Estimates of future transportation, refining, and distribution costs of gasoline and other refined products suggest that these elements of cost will continue to decline as a result of further economies of scale and new technology.

It is important to keep in mind that natural crude oil is not the only potential source of refined products. The full spectrum of refined hydrocarbon products can be produced from synthetic crude oil derived from solid fossil fuels such as coal, oil shale, and tar sand. Because mining responds to the economy of scale in addition to the economy of new technology, synthetic oil costs are expected to decline (in constant dollars) as synthetic oil production grows. Even if synthetic oil costs are somewhat higher than natural oil costs at the beginning of the synthetic industry, they are likely to drop below as the industry matures.

The cost of oil in the future will depend in part on the timing and extent of synthetic oil production. The sooner and more rapidly synthetic oil facilities are brought into production, the more rapidly the cost of oil will fall. We may anticipate that, from time to time, demonstration synthetic oil plants will be built and the current economics will be tested. When production costs are favorable, a synthetic oil industry will take off. I anticipate that political, economic, and social pressures will cause a relatively smooth shift from natural to synthetic oil, so that refined hydrocarbon products may continue to flow on their historic downward (constant dollar) price trajectory.

ORIGINS OF THE ENERGY CRISIS

The United States' energy resources are adequate for centuries to come, yet we find ourselves in the midst of

an energy shortage of crisis proportions. Many objective
factors have contributed to the crisis, and in aggregate
I believe they are entirely adequate to account for it.

In summary, the energy crisis is the result of a
lack of adequate productive capacity; capacity for recovering fuels from the ground and capacity for refining
raw fuels.

First, on the demand side, the growth in energy consumption has been unusually great as farm workers and
women move into the non-farm work force. Since most energy consumption is associated with factory and office work
and related transportation, and since the rest of energy
consumption is accelerated by the affluence that comes
from two jobs per family, we find that the changing life-
style of Americans, with women entering the labor force
in the same manner as men, approximately doubles our per
capita energy consumption. Much of this rapid growth is
behind us; we have just passed through the maximum growth
rate and can anticipate a slowing-down. But the growth
in consumption over the past decade has been much higher
than normal, and it put a substantial strain on the industrial energy infrastructure in its efforts to keep pace.

Against this background of unprecedented growth in
overall demand for energy, the environmental movement un-
expectedly shifted demand more heavily toward refined
fuels, with the consequence that a shortage of refinery
capacity was created.

Because of the limitations on sulfur oxide emissions,
many coal-burning electric power plants, particularly in
the Northeast, found that they could no longer continue
legally to burn the high-sulfur coal that was mined in
their geographical area and that they had been burning
for years. In principle they could have installed sulfur-
removal equipment to cleanse their stack gases of sulfur,
but reliable equipment of this type was not available,
the need for such equipment having become apparent too
late for technology to respond in time. They could have
brought in low-sulfur coal from the western states, but
the transportation cost would have been high, and the
western mines and railroads were not ready to meet such
stepped-up demand. Thus many of them chose to switch
from coal to low-sulfur oil. The switch by northeast
utilities increased the demand for petroleum and for refining capacity beyond that which had been anticipated by
the petroleum industry who---along with most others in
industry and government---were taken unaware by the environmental movement.

At the same time, construction of nuclear plants was
being delayed, partly because of the usual growing pains
of a new technology, but partly and significantly because
of changing AEC regulations and environmental legislation
(much of it after plant design and construction had be-

gun). In addition there have been intervention and delaying legal actions by environmental groups. Delays aver-
aging several years have been experienced.

At the beginning of 1967, thirty-one full-scale nuclear plants were scheduled by the electric utility industry to be producing electricity by the end of 1971. The
industry was confident of meeting this timetable because
of their long history of bringing conventional power
plants into commercial operation on schedule. Yet of the
31 plants scheduled to be in commercial operation by the
end of 1971, only 10 made it. Six more went into commercial operation in 1972 and one in the first quarter of
1973. One plant was cancelled, and thirteen were re-
scheduled for 1973-1975. Unanticipated construction and
licensing delays ruined the 1967 schedule, the average
plant being delayed about 2 years (neglecting the cancelled plant and assuming no further delay for the 13
plants not yet in commercial operation).

Two significant consequences arose from these nuclear
plant delays. First, such delays are expensive and add
to the cost of nuclear power. Second, by the end of 1971,
the electric utility industry found itself short 21 nuclear plants that it had been counting on for 16,000 megawatts of generating capacity. (Meanwhile, other environ-
mentalists were delaying the construction of pumped storage hydroelectric plants, so that the total deficit in
generating capacity was somewhat greater, amounting to
about 19,000 megawatts.)

It takes a long time, after the initial decision by
an electric utility to add nuclear capacity, before the
new plant goes into commercial operation. As much as 12
to 13 years are now required to go through all the steps
from the initial search for a suitable site through site
acquisition, plant design, construction, licensing, testing, and finally commercial operation. When the utilities
found themselves with a generation capacity deficit because of nuclear and pumped storage delays, they had
little choice of where to turn to get the generation
capacity they needed to fill the gap. Only oil-burning
gas turbines, similar in many respects to the turbines
that power jet aircraft, could be manufactured and installed rapidly enough to meet the remaining very short
timetable. The utilities turned to them for 22 percent
of the new generating capacity installed in the United
States in the three-year period 1970-1972, just managing
to bridge the nuclear generation gap, and the threat of
widespread blackout was averted.

But the utilities had not planned the introduction
of gas turbines on anywhere near this scale, and the impact on the demand for fuels was dramatic. Instead of
powering new capacity with uranium in nuclear plants as
anticipated, the electric utilities were burning distil-

late oil in gas turbines. Here was a second unanticipated demand for petroleum and for refinery capacity.

Still another unanticipated demand for petroleum products and refining capacity is making itself felt as new safety and exhaust pollution control equipment is being added to automobiles. More gasoline is being required per mile driven, and more refinery capacity is being required to refine it.

In the coincidence of these various unanticipated demands for petroleum products and refinery capacity, we can identify one cause of the current energy crisis in the United States. These coincident demands were not forseen because the environmental movement was not forseen, and capacity was not provided to meet them. The demands could and would have been met if anticipated, and they can and will be met within two or three years as refinery capacity adjusts to the new situation. In the meantime we can expect shortages of refinery capacity.

At the same time, the productive capacity for recovery of fossil fuels---and thereby the supply of fossil fuels---was cut back from anticipated levels. The oil pipeline across Alaska was delayed because of environmental considerations, reducing crude oil productive capacity by about 2 million barrels per day, or perhaps 10 percent of current consumption. More effective mine health and safety laws cut into coal mining productivity, substantially diminishing the productive capacity for coal recovery.

Natural gas price regulation has been another contributing factor. Undertaken to help keep the price of gas low for residential consumers, regulation of the wellhead price of natural gas intended for interstate shipment interfered gradually but progressively with competitive interfuel substitution and pricing. In 1970 the owner of a gas well got only about a third of the money from selling gas as did the owner of an oil well from selling oil with the same heating value. The motivation for developing new gas reserves eroded, and the supply began to decline. At the same time the increasing desirability of gas, relative to other fuels, caused demand to soar. With supply and demand decoupled from their customary meeting ground in the marketplace, they moved so far out of balance that rationing, imports, and synthetic gas have become necessary. Hence we find ourselves short of gas, and the demand for refined oil to substitute for gas is still another pressure for increased refining capacity that must be added to those already listed.

In recent months, of course, planned imports from some Arab nations have been cut off, adding to what were already problems of crisis proportions.

As a result of these many causes we find ourselves today with a shortage of productive capacity. We lack the full complement of refineries, mines, wells, and transportation facilities needed to supply our new pattern of consumption---a pattern with increased energy usage as women and farm workers flow into the non-farm labor force, with demand shifted toward refined fuels to protect the environment, and with need for greater reliance on our own capabilities. Fortunately, the various elements of productive capacity can be expanded, each with its own characteristic time-constant. New refinery capacity can be built in about three years. New pipelines can be laid in about the same length of time. New oil and gas reserves can be developed in from one to ten years depending on the nature of the resource base, the longest time corresponding to new discoveries in inhospitable offshore locations. New coal mines can be opened within several years.

Hence we can anticipate that today's shortage of productive capacity will be temporary, and that the current energy crisis will ease as new capacity is put in place. When the crisis has passed, we can look forward to a long future in which solar energy, fossil fuels, and nuclear fuels compete again as they have in the past, under the ever-present sociological, political, economic, and technological forces that shape our energy infrastructure.

REFERENCES

1. J. C. Fisher, Energy Crises in Perspective, John Wiley & Sons, New York, 1974.
2. P. K. Theobald, S. P. Schweinfurth, and D. C. Duncan, Energy Resources of the United States, Geological Survey Circular 650, U.S. Geological Survey, Washington, D.C., 1972.
3. A. B. Cambel, et al, Energy R&D and National Progress: Findings and Conclusions, U.S. Government Printing Office, Washington, D.C., 1966.

Physics and Energy Conservation

John H. Gibbons

U.S. Department of the Interior

ABSTRACT

Opportunities abound to conserve energy. Losses, many of them avoidable or potentially capable of large reduction, occur all the way from energy resource extraction to ultimate use. Other presentations in this conference touch on ways to conserve energy by increased efficiencies in the energy cycle up to the point of use. Some of the greatest scientific challenges lie in this area. Another presentation in this conference (Mobile Powerplants) covers the important conservation impacts and technological challenges of more efficient engines. In this presentation we therefore focus on energy conservation, other than propulsion engine technology, at end-use. About half of total U.S. energy is used for heating or cooling buildings or industrial processes. Therefore, any improvement in the technology of both high and low heat transfer materials and processes will help conserve energy. Associated engines (e.g., heat pumps, air conditioners) also offer opportunities for large improvement. Control systems for industrial processes, heating, ventilating, and air conditioning can be improved and have large payoff. Half of our communications, especially lower costs, could enable decreased physical travel. But physicists have a habit of going where the problem takes them, even when it leads seemingly far afield. Thus, a leading figure in energy input-output economics is a physicist. So too are experts on energy impacts of recycling, modes of transportation, and pollution controls. Physics, as a way of thinking about things, seems ubiquitous in energy conservation activities and has already benefitted from physicists' contributions even though most of them have been in seemingly "non-traditional" ways. The pragmatist could claim that this is a measure of the power and relevance of the discipline.

RESEARCH PRIORITIES FOR THE ELECTRIC UTILITY INDUSTRY

Chauncey Starr, President
Electric Power Research Institute
Los Angeles, California

Research and development in the electric utility industry is the key to meeting the continuously increasing demand for electricity under a continuously decreasing availability of energy resources. Fuel resources are constrained by nature. They are non-uniform geographically, and are only partially accessible to man. They exist in two forms: the depletable fuels which represent nature's endowment to mankind, and the continuous or non-depletable resources originating in the sun or the central heat of the earth. It is our responsibility as a technological society to utilize these resources as effectively as possible, to preserve their value for future generations to the maximum possible degree, and to provide our energy services with the minimum of environmental impact to the biosphere.

Until recently fossil fuels have been relatively abundant, easy to obtain from the confines of the earth and have not presented problems of availability whenever needed. However, during the past few decades the exponential increase in demand has gradually revealed the future implications of a continuous drain upon the depletable resources. The economic pressure to utilize these natural fuels more efficiently has grown steadily, and this has been further compounded by a parallel growth in our awareness of the environmental impact of their use. Thus, we now have accepted a spectrum of national objectives for technological development. These include the efficient utilization of depletable resources, increased emphasis on their replacement by continuous energy sources such as solar radiation wherever possible, the use of these resources in a non-polluting fashion, and finally, improvement of the efficiency of the end-point use of various energy forms, and in particular electricity, so that needed functions can be accomplished with a minimum demand on the original sources.

It is now generally recognized that the mix of social amenities and economic welfare of a large industrial society, which we commonly describe as the quality of life, is heavily dependent on the continuous availability of energy. Human labor and energy machines are the basic ingredients for modifying nature's raw materials into those forms needed for human existence and for the material needs of a social structure. In a modern industrial society energy is almost as essential as a food supply. In fact, our food supply is itself dependent upon an intensive man-created energy input. To provide an overview perspective of the role of energy in our society, Figure 1 lists the major uses of energy resources in the U.S. Clearly, production of material goods and transportation represent a combined use of roughly two-thirds of all our energy. It is in these areas that the largest impact is felt from either energy constraints or from technological improvements.

The total social cost of energy supply is a key component of the

U.S. Energy — How it's Used

	Percentage of All Energy Used in U.S.
Running Industry	37.3%
Powering Transportation	24.8%
Heating Homes and Offices	17.9%
Providing Raw Materials for Chemicals, Plastics	5.5%
Heating Water for Homes and Offices	4.0%
Air conditioning Homes and Offices	2.5%
Refrigerating Food	2.2%
Lighting Homes and Offices	1.5%
Cooking Food	1.3%
Other Uses	3.0%

Fig. 1

final cost of all the end products resulting from this combination of machines and labor. In a society in which all sectors have been encouraged to expect steadily improving quality of life, the cost of energy may well determine the social distribution of the amenities which our productive systems produce. It is thus a function of research and development to ensure, first, an ample availability of energy in forms most suitable for energy machines, and, second, to make this energy available at a minimum social cost. We now recognize that the national accounting necessary to determine these costs must include not only the direct investment of capital goods and labor in the production of energy, but also the indirect cost of the environmental impacts on both the generation using the energy and on future generations. These long-term burdens include the future consequences of the depletion of limited natural resources, irreversible or slowly reversible environmental changes, and the consequent adjustment in future life styles resulting from increased costs of energy.

Because the technical, social and economic frameworks in which future decisions will be made are not clearly apparent to anyone today, the most useful function for research and development is to provide a spectrum of technical options for meeting the most probable future problems of energy systems. From the experience of the past half century it appears that for the indefinite future the most desirable and versatile form of energy for widespread public use is electricity. On the assumption that the present growth in electricity use will, therefore, continue at the expense of other energy forms, the research and development conducted in the electrical utility industry may be a vital factor in determining the availability of optional future energy systems. Fortunately, electricity can be created by conversion of every known energy source, and thus the growth of a national electricity distribution system does not inhibit the development of alternative generation systems.

Most of us have been exposed to some of the stages through which a new technological concept must pass before becoming commonplace in public use. It is useful to enumerate these stages as shown in Figure 2. The first is that of scientific feasibility, involving both the analytic development of the concept and its experimental verification. The second is that of engineering feasibility, which moves from component development to subsystem development and to the total integrated performance demonstration needed to establish the interactions of a complete system. Commercial development then tests the concept against competitive alternatives. Finally, the technical system, although proven as a commercially desirable alternative, must now face the difficulties of integration both physically and operationally into ongoing electric utility service systems. If this hurdle is successfully achieved, we have reached the final growth stage leading to substantial public use. The figure also lists some of the criteria which have to be met at each stage. It is certainly clear to all of us that a sequence of this nature is very time-consuming. Occasionally, scientific discoveries, new technological insights, and new economic conditions will speed up this process. Such discontinuous events are certainly not predictable, even though history has shown that they have a finite probability of occurring.

R & D Development Phase

Development Phase	Controlling Parameter
Scientific Feasibility	Natural Laws
Analytic	
Experimental	
Engineering Feasibility	
Small-scale Pilot Plant	Sub-system Performance
Intermediate Demo	Operational Integration of Commercial Size Components
Commercial Plant	Cost, Performance, and Reliability
Utility Integration	Costs of phasing in multiple plants, including supporting systems
Significant Use	Competitive Alternatives

Fig. 2

I have tried to illustrate in the next two figures the probable time sequence for technical systems with which we have some familiarity. Figure 3 shows the various stages for two of the hydrocarbon alternatives now under commercial development. The first is the production of low Btu gas from coal for direct use in utility boilers. The function of coal gasification is to provide a process for removing coal impurities. This system faces technological competition with improved combustion methods in the boiler and with stack gas cleanup technologies. A second illustration is the development of shale oil as a fuel resource. Both of these have been under study for several decades. The sharp acceleration in their development in the past year or two is, of course, a direct result of the rapid increase in cost of the competitive natural oil and gas which they are intended to replace. While such extrapolations into the future are certainly speculative, I believe that curves illustrate the typical lead-time for such available technologies before they can make substantial impact in public use.

Figure 4 illustrates in the same fashion the probable rate of development of advanced generation alternatives, including conventional nuclear power, the fast breeder, solar photovoltaic sources and fusion. I would certainly agree that all of these can have their rate of development accelerated by changes in national allocation of resources, by technological break-throughs, and by changes of national policy. I doubt, however, whether the rate at which these would become significant in end-use is likely to be importantly altered by such changes.

Of particular interest to this audience of physicists may be the estimated twenty-year differential lead-time between the fast breeder and fusion. This has been a subject of much public discussion, and certainly the projection presented here represents a judgment which may not be shared by others. Nevertheless, I believe the history of development of new technologies would support this estimate.

Because of these very long lead-times required to bring new technologies into use, if we expect to make a contribution to society's welfare some decades hence, their development should be started now. It is obvious that the more speculative developments will primarily benefit the next or succeeding generations. It is always possible, however, that performance which might be completely out of reach today may suddenly turn out to be feasible much sooner, due to new technological developments, new scientific insights and new discoveries. If research is going to be called upon at some future date to "pull a rabbit out of the hat" to meet some future unforeseen need, that "rabbit" will have to have been placed there by previous decades of development. I don't believe we should be uncomfortable about the fact that these long-range projects do not represent technological fixes for today's problems. They may represent such fixes for the crisis problems of the future, some of which we will be creating by our national actions today. As a major example, there is no question in my mind that the national commitment to develop the fusion process is justified in terms of the potential values to future generations.

21

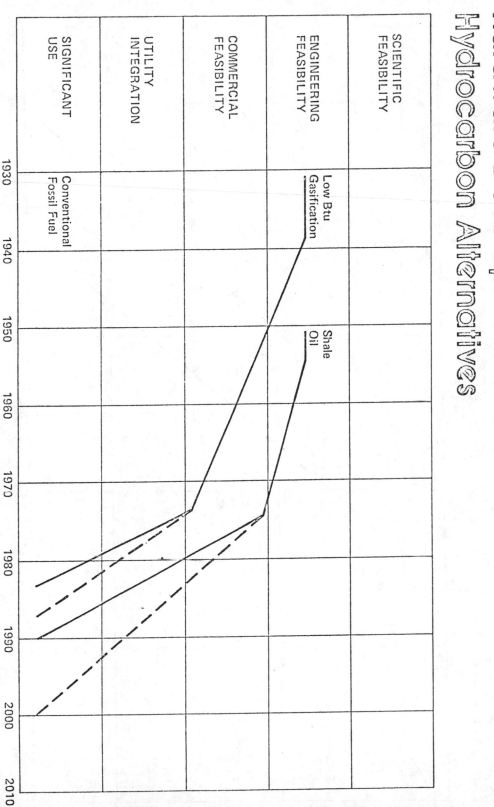

Hardware Development Phases for Hydrocarbon Alternatives

22

Hardware Development Phases for
Advanced Generation Alternatives

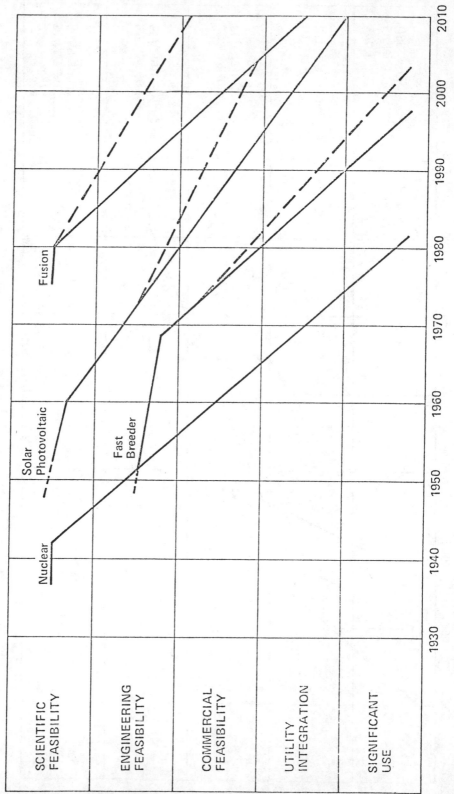

Fig. 4

To this audience of physicists, whose interests may be primarily focused in areas of applied science, I would like to suggest some typical generic research areas which may be very important for the successful engineering development of our advanced concepts. I have illustrated on Figure 5 some such examples. This tabulation is obviously not meant to be all inclusive --- rather it illustrates some of the scientific limitations to successful engineering development of advanced concepts for electrical energy systems. While the research areas are quite fundamental in nature, the scientific barriers which they address affect all aspects of energy systems, from the efficiency of the conversion of fuel resources to the ability to utilize these in an energy conserving manner. I believe the tabulation is self-explanatory and I will not belabour the subject. It is, however, very important to point out that practically every aspect of engineering development of the many potential systems, is faced with performance limitations of an inherently scientific nature. In only a few of our engineering systems is the practical performance of the system near to the theoretical limit of performance. In fact, the difference between how the system should perform theoretically and how it actually performs in engineering practice is a good measure of the inadequacy of the basic scientific understanding of the phenomena involved in the technical concept.

The program of this week's meeting of the American Physical Society will provide you with a comprehensive review of most of the technological developments now being actively considered in energy system research planning. I hope that, in listening to these presentations, you will give consideration to the common problem associated with moving these developments into substantial public use, which I have described. I believe that it would be particularly useful if you were to extract from these presentations those key scientific barriers to substantial progress in the performance of these various technologies. The success of the national program for developing a large variety of usable energy options, acceptable both environmentally and economically, depends greatly on improving our scientific understanding of the phenomena which now limit technological development.

Some Examples of Areas Requiring Scientific Investigation

Generic R & D Area	Typical Application	Typical Results
Materials	High-temperature turbines, reactors, MHD channels, fusion containers.	Increased Carnot efficiency of energy conversion.
	Low-temperature materials and super-conductors.	Increased power transmission capabilities.
	Solar cells	Improved economics of solar energy conversion.
Surface Phenomena	Catalysis	Improved energy recovery in coal gasification and liquefaction.
	Electrode Process	Long life electrochemical cells.
	Corrosion	Improved equipment life and reliability.
	Coolant-fuel interaction in nuclear reactors.	Increased safety factor and lower cost.
Laser Development	Greater power and pulsing capabilities.	Laser fusion, isotope separation, new molecular processes.
Fracture Mechanics	Piping, Vessels	Increased reliability, safety, and lower cost.
	Rock	Improved oil and gas recovery, hot rock heat sources.
Fluid Dynamics	Two-phase flow	Geothermal development, liquid metal MHD system, higher reliability of nuclear reactor cooling, improved coal conversion.
Drying, coating, and melting techniques	Paper, paint, materials processing	Conservation of energy.

Fig. 5

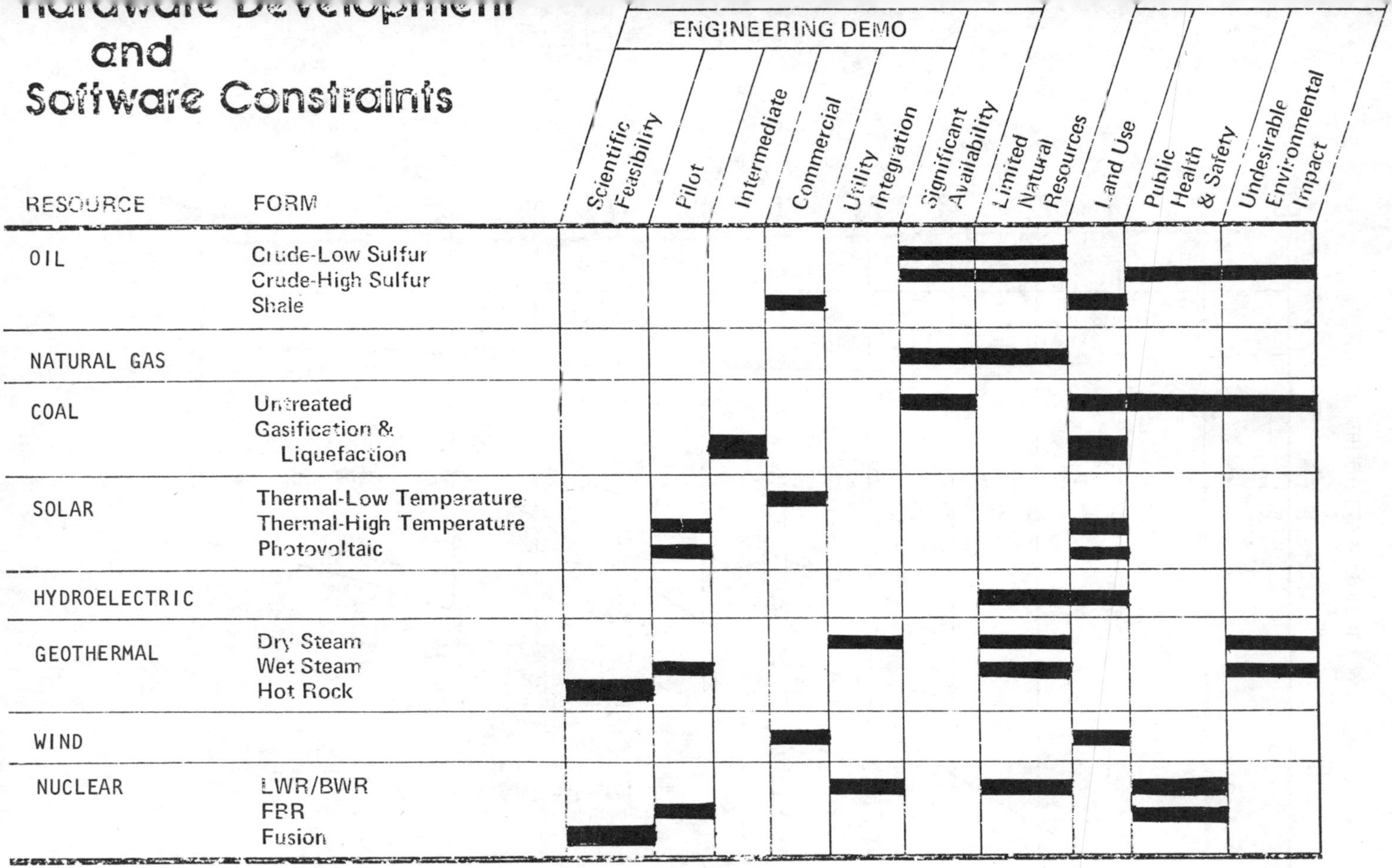

Hardware Development and Software Constraints

ENGINEERING DEMO

Column headings:
- Undesirable Environmental Impact
- Public Health & Safety
- Land Use
- Limited Natural Resources
- Significant Availability
- Utility Integration
- Commercial
- Intermediate
- Pilot
- Scientific Feasibility

RESOURCE	FORM
OIL	Crude-Low Sulfur
	Crude-High Sulfur
	Shale
NATURAL GAS	Untreated
COAL	Gasification & Liquefaction
SOLAR	Thermal-Low Temperature
	Thermal-High Temperature
	Photovoltaic
HYDROELECTRIC	
GEOTHERMAL	Dry Steam
	Wet Steam
	Hot Rock
WIND	
NUCLEAR	LWR/BWR
	FBR
	Fusion

Fig. 6

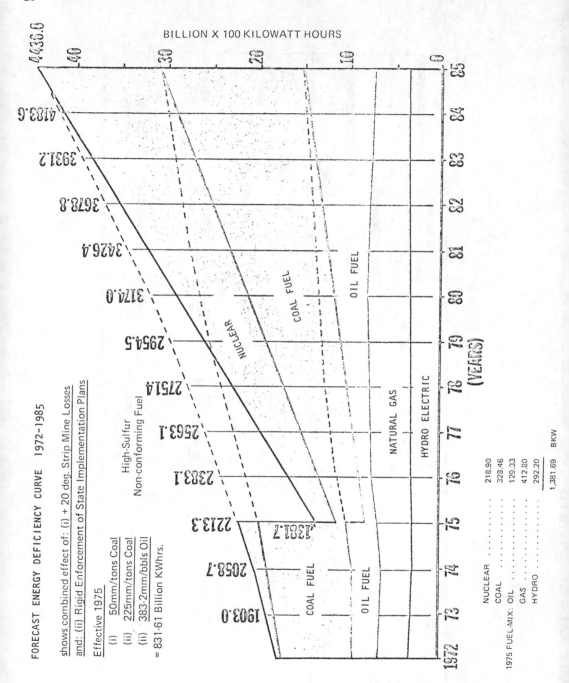

The Energy Problem in Perspective

Paul F. Donovan

National Science Foundation

ABSTRACT

The Energy Problem in this country today is a very complex issue
which underlies many facets of the operation of our entire society.
Energy is the source of this nation's economic growth ... that
growth over the past decades has been very rapid, and as a consequence, our ability to supply new sources of energy -- growing
sources of energy -- has not been able to keep pace with the demand.
At this time, there is a very broad spectrum of research programs
underway, both supported by the Federal government and by private
industry. The three main targets for these research areas are to
increase energy supply, to control energy demand, and to improve
energy utilization, particularly in minimizing the effect of energy
utilization on health and on the environment. A brief overview of
the role of R & D in the evolving energy posture of the United
States is presented. Among the topics discussed are:

°The present funding situation in Federal and private sector
  energy R & D,
°The present and evolving role of Federal agencies in energy
  R & D,
°The formulation and implementation of Federal energy R & D
  policy,
°Plans for increased Federal R & D support,
°Long-range energy R & D priorities, and
°A national systems approach to energy R & D.

CONTRIBUTIONS OF PHYSICISTS
TO THE EXPLORATION FOR HYDROCARBONS

Franklyn K. Levin
Esso Production Research Company, Houston, Texas 77001

ABSTRACT

Exploration for gas or oil involves either detecting changes in some physical parameter which is affected by the hydrocarbons or, more commonly, mapping the configuration of rock layers making up the first thirty thousand feet of the subsurface. Among the quantities measured during the search for hydrocarbons are variations in the earth's magnetic and gravity fields. However, the predominant method of exploration geophysicists is reflection seismology. Only the seismic method combines great depth penetration with reasonable resolution.

In reflection seismology, elastic waves are introduced into the earth; and reflections from impedance changes in the subsurface, detected and recorded at the surface. Common sources on land are explosives or large vibrators; marine exploration involves air bursts or gases exploded in an expandable sleeve. 24 to 96 detector stations separated by up to 200 m are occupied simultaneously, each station consisting of a linear or areal array of 12 to 144 detectors. Signals are recorded in digital form on magnetic tape for 5 to 15 seconds. The useful frequency range of 10 to 100 Hz is limited on the low end by surface waves and on the high end by attenuation within earth materials. Data may be accumulated at a daily rate of 100 reels of tape.

Processing the data requires computers of the 370/165 size. Among the routine processes are band-pass filtering, correction for variations in near-surface material and for non-zero source-detector separations, computation of velocity, summing of traces around a common surface point, collapse of sinusoidal events, and return of reflections to their point of origin. Processed data are displayed in variable density or area form as two-dimensional distance-depth or distance-time surfaces. Interpretation by geologists involves theory and experience.

Limitations of reflection seismology include unsatisfactory depth resolution, an unknown attenuation mechanism, and interfering scattered energy. Work on going from data to subsurface (the inverse problem) has barely begun. To the solution of these and numerous processing problems physicists can expect to contribute.

INTRODUCTION

There is some irony in a discussion which lays out for physicists the techniques of exploration geophysics, for, with few exceptions, the founders of the field were physicists. The early workers on electrical and potential methods and the members of the first seismic crew were trained in physics. Indeed, we count among the pioneers of exploration geophysics men who subsequently earned substantial reputations in more conventional areas, including at least one Nobel prize winner.

As I describe methods presently used to explore for hydrocarbons, it will be clear why we have become, if not the pariahs, at least unknown members of the physics community. However, we do not need a detailed description to understand broadly the gulf separating exploration geophysics from modern physics. Exploration geophysics is classical physics: very little of what we do would surprise the great physicists of the nineteenth century. Exploration geophysics is concerned with the earth as it exists: we cannot isolate our system from its surroundings, for the surroundings are the system. Because of the dimensions involved, exploration geophysics is not concerned at all with the quantum phenomena. Finally, exploration geophysics is an interdisciplinary activity, one which calls upon sciences unknown to most physicists.

So, exploration geophysics differs from areas that concern most of you. What then is this subject which combines in its name the familiar term "physics" with a prefix meaning "earth"? What do geophysicists do? How are probable locations of hydrocarbon accumulations found? An exploration geophysicist measures any physical parameter which either is affected by the presence of hydrocarbons in the subsurface or which can be used to delineate rock layers in the subsurface. The importance of geophysical methods has increased as the search for prospective reservoirs has extended to greater depths and into deeper water. Many of the remaining accumulations of gas and oil occur at 20 to 30 thousand feet and are sought beneath 1000 or more feet of water. Hence, whatever method is used must be long range in the sense of a physicist. For this reason, some effort is devoted to mapping areal variations in the Earth's gravitational and magnetic fields. However, when we add to the requirement of great penetration that of reasonable resolution, only one method remains—reflection seismology. In 1972, 97 percent of the 828 million dollars used to acquire and process geophysical data for petroleum exploration in the Free World was spent on the seismic method. In the remainder of this paper, I shall restrict myself to seismology.

Before beginning, I'd best clear up a misapprehension some of you may have. Contrary to what is shown in TV cartoons Saturday mornings, oil does not occur as black liquid lakes in which trilobites swim. Oil occurs as drops which fill the pores—usually microscopic pores—in sedimentary rocks. The rocks themselves lie layer on layer from the surface to the base of the sedimentary column—sandstone, shale, and limestone layers ranging in thickness from a fraction of an inch to hundreds of feet. Fluids other than oil, notably salt water, fill pores in sedimentary rocks. A porous rock may contain 30 percent gas or liquid by volume but lower porosities are usual. With this said, let us return to the seismic reflection method.

CURRENT SEISMIC TECHNIQUES[1]

In its simplest version, the seismic method takes the form shown in Fig. 1. An impulse is introduced into the earth by detonating an explosive charge in a hole. Elastic waves travel out from the source, are reflected at changes of acoustic impedance in the subsurface, and return to the surface, where the waves are detected and recorded as a function of time. In subsequent steps, this entire setup is moved along the surface until a picture of the subsurface emerges.

30

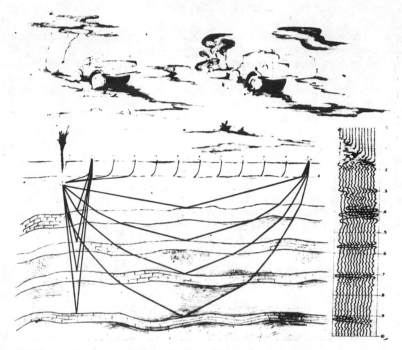

Fig. 1. The basic form of the seismic reflection method as implemented in the late 1930's and early 1940's.

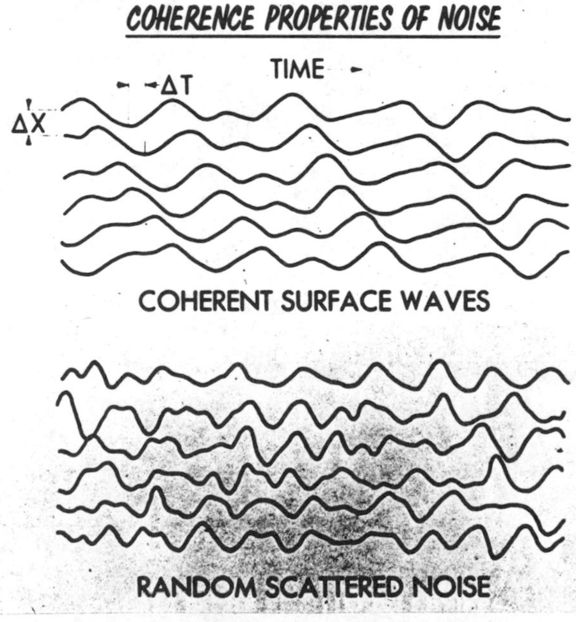

COHERENCE PROPERTIES OF NOISE

TIME

ΔT

ΔX

COHERENT SURFACE WAVES

RANDOM SCATTERED NOISE

Fig. 2. Source-generated coherent and random noise.

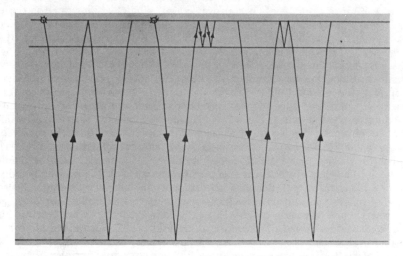

Fig. 3. Examples of ray paths for multiple reflections.

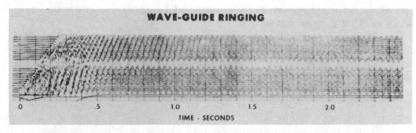

Fig. 4 Sinusoidal data recorded in Lake Maracaibo, Venezuela.

Fig. 5. A truck-mounted vibrator for introducing elastic waves into the earth.

If seismic exploration were as easy as Fig. 1 indicates, there'd be little point in my taking more of your time. Of course it is not. What I've illustrated is the technique of the 1930's, when much of the world had to be written off as non-reflection or NR areas. Let me now indicate those changes which have reduced the percentage of NR areas to very nearly zero for offshore exploration and to a small value on land. For purposes of discussion, I'll break the system into three parts—the source, the detectors, and the medium. Since the medium sets the characteristics of the source and detectors, we'll consider the medium first. The elastic field is a tensor field. Even for an infinite isotropic solid, two types of waves, compressional and shear, can propagate. When the medium is layered, additional waves become possible, the simplest of which are named for the physicists Rayleigh and Love. For seismic exploration, the signal consists of the primary or first-order compressional wave energy reflected from interfaces in the subsurface. All other types of energy are classified as noise. Some of the noise is really noise in the sense of electrical engineering—energy which is present whether or not the source is activated. It originates in distant earthquakes, in energy called microseisms, and, primarily, in cultural activities—machinery, automobiles, airplanes, and the like. However, most of what a geophysicist calls noise is energy he has generated but can't use for exploration. Some of the noise is coherent, i.e., can be traced from detector to detector (Fig. 2). These events are surface waves, the exact analogue of waves which have received much attention in the solid state literature because of their use in delay lines. Early delay line papers published in the Journal of Applied Physics referred to a text by the geophysicists Ewing, Jardetzky, and Press[2]. Now, of course, there is abundant literature in AIP publications, and I suspect geophysicists could benefit from what solid state physicists have learned.

A second type of source-generated noise cannot be tracked from detector to detector. We call this random noise, although it is reproducible (Fig. 2). Because we deal with functions of space and time, we can have a function random in one coordinate but not in others. As best we can tell, random noise is due to scattering by near-surface inhomogenieties. We don't know scattering is the cause; we just haven't found any other. Theoretical guidance has been skimpy; nor has anyone thought of, let alone performed, a definitive experiment.

Other types of unwanted energy plague geophysicists. An annoying type is multiple reflections, particularly in water-covered areas. Multiple reflections have bounced more than once from an interface (Fig. 3). In extreme cases, the multiples add to produce sinusoidal signals nearly as pure as those from an oscillator. Fig. 4 shows data recorded in Venzuela's Lake Maracaibo. Energy has bounced between the surface and the bottom, which contains gas and has a reflection coefficient of -0.7 to -0.8. Although the data look as if they contain no information, processing will collapse the sinusoids into discrete pulses.

Turning from the noise, which we wish to eliminate, to the reflections, which we want to keep, we find attenuation in the earth sets an upper usable frequency of about 100 Hz. Most experiments yield an attenuation constant proportional to the first power of frequency; most theories predict a square law. A large but unknown fraction of the attenuation results from transmission losses and concomitant multiple reflections in layered subsurfaces. Part of the attenuation is probably intrinsic, particularly for that energy traveling in weathered material near the surface. Because coherent noise consists of waves bound to the surface, the high frequency components of coherent noise are absorbed preferentially, leaving frequencies below 10 to 15 Hz. Limited on one end by attenuation within the material being explored and on the other end by the need to avoid coherent noise, the usable reflection band is roughly 10 to 100 Hz.

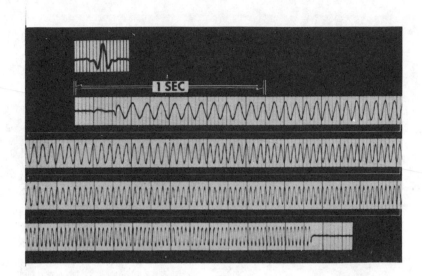

Fig. 6. A chirp signal of the type emitted by the vibrator of Fig. 5 and the pulse resulting from crosscorrelation of detected chirps with a pilot signal.

Fig. 7. An elastic sleeve marine seismic source.

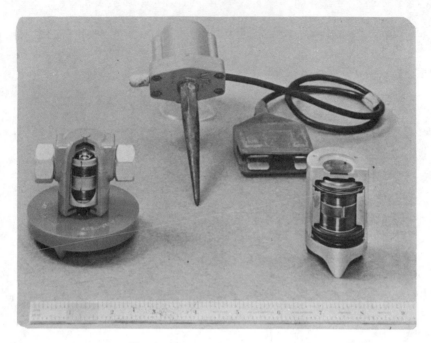

Fig. 8. Three typical geophones.

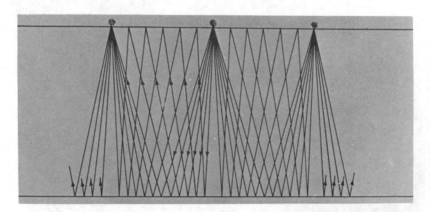

Fig. 9. Diagram of the arrangement for split-spread seismic shooting. A series of detector stations stretch out symmetrically in opposite directions from the source.

I said earlier that gas, oil, and salt water fill the pores of rocks. The elastic properties and densities of oil and water don't differ greatly. This fact, coupled with the occurrence of the liquids as a relatively minor phase of the total medium, means that direct detection of oil is rarely possible. Because gas has a much lower density and higher compressibility than either oil or water, gas in unconsolidated material can be detected under favorable conditions. Direct detection of gas has received much attention recently.

With this brief brush of media properties, let us consider sources. A high explosive charge detonated in a borehole drilled through the weathering was the first type of seismic source and is still a favorite on land. In recent years, explosives have had the competition of large vibrators[3] (Fig. 5) which emit swept frequency signals of the chirp radar type. The wave trains (Fig. 6), usually 10 to 20 seconds long, are collapsed to short time functions during data processing. Common sweeps are 10 Hz to 40 Hz. Vibrators can be used where dynamite cannot; exploration through the streets of Los Angeles was a triumph of the method. For marine exploration, high explosives are no longer competitive nor, in many waters, allowed. Instead, seismic ships drag either sources which suddenly release large bubbles of air into the water (air guns) or elastic sleeves in which gaseous mixtures are detonated (Fig. 7). In either case, the sources are used in groups, the air guns having different volumes to broaden the input frequency band and the sleeves doubled or quadrupled to ensure adequate energy. The four sources I've mentioned—dynamite, vibrator, air gun, and sleeve exploder—are popular at present but they by no means exhause the list. Nearly every device that emits a useful amount of energy in the seismic frequency band has been tried at one time or another.

Only two types of detectors are in common use. For land work, the detector, called a geophone (Fig. 8), consists of a coil suspended in a magnetic field by springs whose resonant frequency is lower than that of the lowest frequency of exploration interest. The output of a geophone is proportional to the vertical component of the velocity of the surface. Geophones are rugged enough to be thrown from a moving truck, cost less than $25 each, and have sensitivities of about 200 mv/cm/sec. They are among the most inexpensive of all sensitive transducers. For marine exploration, the detector is called a hydrophone and usually consists of polarized lead zirconate titanate units which respond to pressure changes. Both geophones and hydrophones are used in arrays of 12 and 144 units with 18 to 40 units being common. Arrays are designed with considerable care to cancel coherent noise. They also reduce random noise, although at a slower rate. The requirements that both kinds of noise be reduced simultaneously are not always compatible and compromises must be made. Separations between the centers of successive arrays increase with increasing depth of exploration interest; separations of 30m to 200m are typical.

Various arrangements of sources and detectors have been used. Fifteen years ago split spreads (Fig. 9) were the rule but most crews today record in common depth point mode (Fig. 10). Here the center point is held fixed and each reflection point receives multiple coverage of 3 to 48 with 24-fold coverage common. The number of detector stations, i.e., array centers, stretching out from a source may be anywhere from 26 to 60. For marine work, all of the detector arrays are in a cable pulled behind the same ship which carries the sources. Such a vessel, towing a 3000m or even 4000m cable, is an awesome beast. Whether on land or water, exploration is along a grid of lines. Thanks to recent advances in navigation, which include satellite positioning and doppler sonar, marine exploration 24 hours a day at distances several hundred miles from shore is practical.

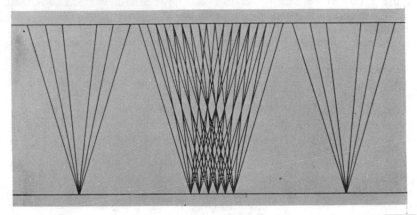

Fig. 10. Diagram of the arrangement for common depth point (CDP) shooting. Three-fold coverage is illustrated here.

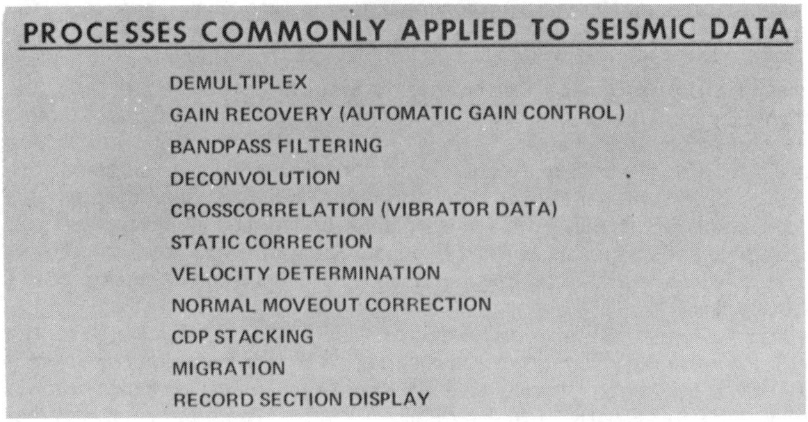

PROCESSES COMMONLY APPLIED TO SEISMIC DATA

DEMULTIPLEX

GAIN RECOVERY (AUTOMATIC GAIN CONTROL)

BANDPASS FILTERING

DECONVOLUTION

CROSSCORRELATION (VIBRATOR DATA)

STATIC CORRECTION

VELOCITY DETERMINATION

NORMAL MOVEOUT CORRECTION

CDP STACKING

MIGRATION

RECORD SECTION DISPLAY

Fig. 11. A list of processes commonly applied to seismic field data.

Data, recorded in digital and multiplexed form on tape at sampling rates of 1 to 4 msec for 5 to 15 seconds, accumulate in unbelievable quantities. By intermingling samples from different stations, multiplexing permits multichannel data to be recorded on half-inch or one-inch magnetic tape. A marine crew easily collects 100 reels of tape in a working day and does this week after week. Reel after reel pour onto shore. It is no surprise that geophysicists are major users of digital computers. Some data processing may be carried out with truck or shipboard minicomputers, but the major processing occurs in a center with one, two, or more computers of the IBM 370/165 or CDC 6600 or 7600 size.

We'll take a quick look at processes commonly applied to seismic field data (Fig. 11). Here we have a brief list of principal processes. I've omitted all editing which, in practice, eats up large chunks of time. First the data are demultiplexed and returned to a trace sequential form. A trace is the data from one detector station played out as a function of time. Filters which maximize the signal-to-noise ratio are applied. A process called deconvolution collapses into simple pulses sinusoidal energy of the type you've seen earlier. Data from vibrator sources are crosscorrelated with the input signal, i.e., the chirp signals are shortened as if the source had emitted an impulse. If the data come from a land survey, the individual traces may need to be shifted in time to correct for variations in thickness or velocity of the weathered material. Velocities with which the waves have traveled are determined from the data themselves. In Fig. 12 we have a typical variation of velocity with depth. Interval velocity is the velocity with which compressional waves travel within a layer. The RMS velocity is determined directly from the data and is slightly greater than the quotient of the depth to a reflector divided by half the traveltime for the corresponding reflection. Note that these velocities, together with the frequency range discussed earlier, imply wavelengths in the earth of about 60m. Once the velocity structure is known, a correction which in effect brings the source and detector into coincidence is applied (Fig. 13). This correction lowers the frequency content of the pulses in an undesirable manner but is essential for the next process—common depth point (CDP) stacking. In CDP stacking, traces having a common center point half way between the source and detector stations are added together, a process which increases the signal-to-noise ratio and reduces the amplitudes of multiple reflections relative to those of primary reflections. Finally traces from successive surface positions are plotted side by side (Fig. 14). For the plots, the abscissa is distance from a selected starting point; the ordinate is traveltime; and the amplitude is shown in variable area or variable density form. The result is called a record section. To a first order, a record section reproduces the subsurface below the line along which data have been collected.

For many reasons, the resemblance may be no more than superficial. Resolution is limited by the 60m wavelength of the probing pulses. The energy on the section includes arrivals other than primary compressional wave reflections. Multiple reflections abound on records from deep water. Diffractions from sharp discontinuities and the various types of noise discussed earlier are rarely eliminated completely. A tacit assumption of the method places the reflecting points directly below surface points. For dipping interfaces or folded structures this assumption is invalid and a process called migration is applied to move a reflection back to its origin in the subsurface. Fig. 15 shows a section before migration and Fig. 16 the same data after migration. The structure has been clarified.

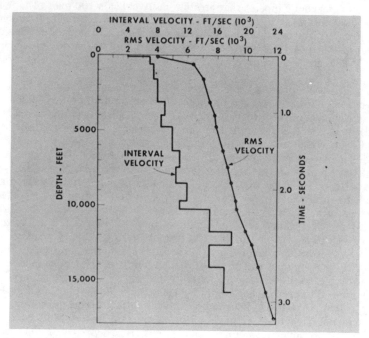

Fig. 12. The variation of seismic velocity with depth. Although the variation shown here was constructed, it is typical of that found for earth columns. English units are used by geophysists in this country; elsewhere metric units are usual.

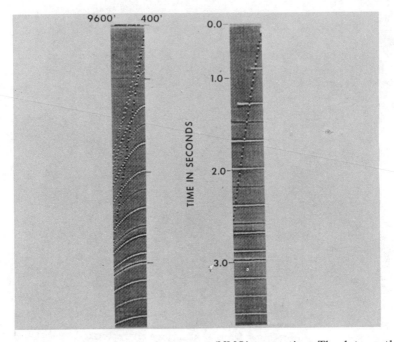

Fig. 13. Illustration of normal moveout (NMO) correction. The data on the left are those that would result from the layering of Fig. 12 for 24 detectors 400 ft. to 9600 ft. from the source. NMO correction flattens the reflections as shown on the right. Pulses that were stretched more that 100 ,percent by the process have been eliminated, resulting in a loss of data at small reflection times.

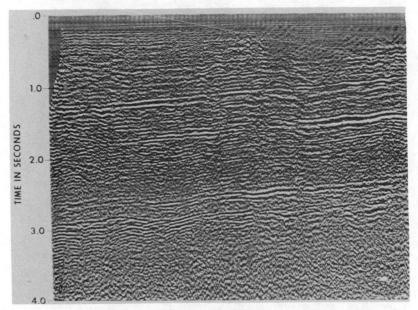

Fig. 14. A typical display of processed seismic data. A time of 3 seconds corresponds to a reflection from an interface at perhaps 5000m. Data from a line about 16 km long are displayed in this figure.

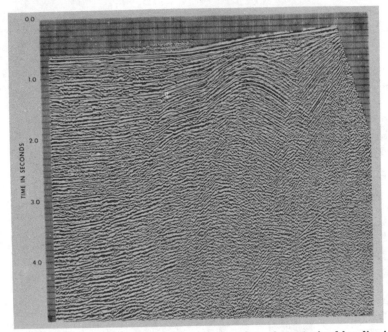

Fig. 15. Seismic data recorded from a subsurface characterized by dipping interfaces and sharp folds.

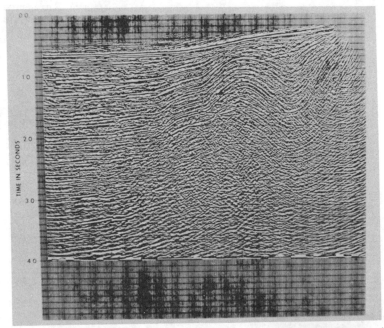

Fig. 16. The data of Fig. 15 after migration to return the reflections to their proper subsurface position. The crossed cycles on the upper left of Fig. 15 are seen to arise in a concave fold (syncline) and convex events (anticlines) have been narrowed.

Because of time limitations, I have brushed quickly over the complicated processes applied to seismic data. All the processes have some theoretical basis and many lean on sophisticated communication theory considerations. Masses of data must be handled efficiently. For this reason, the last three physicists who joined our research group came with degrees earned for work with bubble chamber photographs or spark counter data at large universities. Processes applied to particle and seismic data differ but the experience gained with the former let the young men begin to contribute without the usual delay for reeducation.

I was to speak on exploration for hydrocarbons and ended with record sections. What happens to the sections? After a geophysicist is certain he has obtained the best picture of the subsurface he can hope for, either alone or working with other geophysicists and geologists he interprets the record section in terms of subsurface geology. Years of experience and geologic theories indicate what types of geology may have hydrocarbons associated with them.

Once the subsurface configuration is known or gas inferred, a drilling location is selected and the prospective hydrocarbon accumulation tested. Nine times out of ten, there is no accumulation of commercial size. However, without the geophysicist's guidance, the odds may be 50 to one. In deep water or at great depths, the drilling costs demand the assurance of geophysical exploration.

Although my purpose in this discussion was not to detail the seismic reflection method, a review was necessary if I hoped to indicate where physicists can contribute toward solving our remaining problems. Even partial solutions would be welcome as we hunt the hydrocarbons needed to ease the energy crisis. Before pointing to those areas where help is wanted, I'll mount a hobbyhorse and make a brutal, general observation. If you're a physicist who believes there was no physics before Heisenberg and Schrodinger, you'd best find another way to make your contribution. The physicists of the 20th century are not our household gods. On the other hand, if you're interested in classical physics, are intellectually flexible, and are prepared to learn enough geology to feel comfortable with the subject, exploration geophysics can challenge you. When physicists turned from the classical areas, men trained in other disciplines moved in. Many of us believe the broad training which has characterized physicists has been missed.

I'll dismount now and turn to the remaining problems. In spite of the success of the seismic method, there are still parts of the earth with great economic potential where usable data can't be recorded. West Texas is such an area. There random noise and low signal level stop us. Some of the Arctic areas are seismically difficult because of variations in the permafrost layer. In all areas, geologists would like greater resolution: we need to generate, transmit to great depths, and record a broader band of frequencies. Before we can hope either to subdue the noise or obtain higher resolution data, we must understand better the way elastic waves travel in earth materials rather than in the perfectly elastic materials with which theoreticians love to play. We don't understand and have great difficulty measuring attenuation in a sedimentary earth section.

Turning to other directions, we note designers of filters assume a seismic trace is a piece of a stationary time series. It isn't. We use the methods of communication theory whether or not they apply. Many of the processes we apply to our data are based on unrealistically simple models of the subsurface—liquid layers bounded by planes. Although much research has involved solution of the elastic wave equation, for all except the simplest cases we resort to ray theory. Finally, the output from the seismic system looks like the subsurface but isn't. We are in the position of a man deducing the details of a Victorian parlor from the echoes recorded after a firecracker was exploded in the room. It's the details we want, not the echoes. Thus we face an inverse problem, a challenge familiar to physicists.

42

In this short account I've tried to give you the flavor of a field founded by physicists and abandoned by them. It is a mature field, one in which spectacular advances are unlikely. It is also a field of great practical importance in the present energy crisis. Physicists have contributed in the past and can expect to contribute in the future.

REFERENCES

1. Since this account is a survey of art common to geophysical practice, no attempt will be made to reference original papers. Contributions of exploration geophysicists normally are published in Geophysics or Geophysical Prospecting, the journals of the Society of Exploration Geophysicists and of the European Association of Exploration Geophysicists. Geophysics also publishes review papers and, every few years, a survey of advances in the field.
2. W. M. Ewing, W. S. Jardetzky, and F. Press, Elastic Waves in Layered Media (McGraw-Hill, New York, 1957).
3. Figs. 5 and 6 were furnished by S. W. Schoellhorn (Seismograph Service Corporation) to whom I should like to express my appreciation.

COAL LIQUEFACTION

Harry Perry
Resources for the Future, Inc.
Washington, D. C.

For presentation at meeting of the American Physical Society,
February 4, 1974

Abstract

Two different processes were used by the Germans during World
War II to produce liquid fuels in industrial quantities from coal.
One such plant is in operation in South Africa. Discovery in the
early 1950s of large Mid-East reserves that can yield oil at very low
costs reduced interest in coal processes.

The current shortage of domestic clean energy sources and the very
large reserves of indigenous coal have revived interest in the production
of synthetic fuels from coal. Several oil companies have conducted
sporadic research on proprietary processes in their own laboratories
and have supported a joint program on the H-Coal process of Hydrocarbon
Research, Inc. The government-supported coal liquefaction programs of
the Office of Coal Research have concentrated on Project Gasoline and
Solvent Refined Coal. The Bureau of Mines has conducted several small-
scale, in-house coal liquefaction experiments in an attempt to discover
lower cost methods for producing low-sulfur boiler fuel.

No recent cost estimates have been published for producing oil
from coal by either of the processes used during World War II by the
Germans but by adjusting previous estimates inflation, costs of over
$15 per barrel of oil would be expected. This compared to a crude oil
price that had remained relatively stable (until early 1973) at $3.30
to $3.60 per barrel for a number of years. The National Petroleum Coun-
cil estimates costs of a refinery feedstock at $6.50 to $7.50 per barrel
for western coals and $7.50 per barrel using eastern coals. With improve-
ments expected in second generation plants these costs might be reduced
to $6.00 per barrel with western coals. Other estimates for refinery
feedstocks from coals are higher.

One of the most important current needs is to produce a low-sulfur
boiler fuel from high-sulfur coal. Since the hydrogen requirements and
processing conditions for producing a low-sulfur boiler fuel from coal
are less severe than for producing a refinery feedstock, lower product
costs should be possible. Unfortunately, less experimentation has been
carried out on methods to produce a low-sulfur boiler fuel than for refinery feedstocks. However, the Bureau of Mines is currently concentrating on development of the Synthoil process to convert coal into a very
low-sulfur boiler fuel and has successfully operated a half ton per day
unit (1.5 barrels/day). Estimates of cost of a low-sulfur boiler fuel
are $4.50 to $6.75 per barrel.

I. INTRODUCTION

In the past, the degree of interest in the United States in coal liquefaction technology at any time has largely depended upon the then current view of the adequacy of proved reserves of oil--both domestically and worldwide. In Germany extensive research was conducted during the 1920s and 1930s on methods of producing liquid fuels from coal. By then it had become obvious that adequate and secure supplies of liquid fuels were vital to the conduct of modern war with the heavy emphasis on mobility of operation and the greatly expanded use of tanks and planes operating on liquid fuels. In the absence of indigenous liquid fuel supplies, Germany could only become a major power by either conquering lands with oil reserves (Roumania and the U.S.S.R.) or producing synthetics from other indigenous fossil fuel resources, which for Germany meant using its large coal resources.

Research on coal liquefaction in the United States has been characterized by periods of both intensive interest and of almost complete neglect. The activities of the German scientists in pursuing various coal liquefaction technologies prompted the U.S. Bureau of Mines to conduct small-scale laboratory studies in the 1930s to determine the feasibility of the German processes for U.S. coals and the problems that still remained.

There had been concern expressed periodically over the adequacy of domestic oil supplies starting as early as the turn of the century. However, the discovery of very large deposits of oil in 1930 in East Texas, just as the depression of the 1930s began, depressed oil prices to the point that production rates were placed under controls (through state prorationing and the use of the Connolly "Hot Oil" Act) to protect the oil industry from economic chaos. As a result, activity on coal liquefaction remained at a low level in the United States from the mid-1930s until World War II.

The success of Germany in producing industrial quantities of synthetic oil from coal with which to operate its military forces during World War II again stimulated interest in coal conversion in the United States. A Synthetic Liquid Fuels Act was passed by Congress in 1944 which provided funds for construction and operation of pilot and demonstration plants for gasification and liquefaction of coal, for shale oil production, and for the manufacture of liquid fuels from farm products and farm wastes. The Act was extended several times and the program had an eleven-year life. During this period $60 million was spent on coal conversion, $22 million on shale oil, and $0.5 million on agricultural materials. The program was concluded in 1955 since the discovery and development of very large oil resources in the Middle East that could be produced at very low costs (as little as 10 cents per barrel) made coal conversion to liquids uneconomic under normal political and commercial conditions.

The availability of large low-cost Middle East oil deposits caused Western Europe to shift from a coal-based economy to one which is now largely dependent on Middle East oil. In the United States, oil import quotas controlled the amount that could be imported so as to protect the domestic petroleum industry. This, however, provided no protection for the coal industry, and coal production declined from

687 million tons in 1947 to 420 million tons in 1954. In order to revitalize the domestic coal industry, the Congress created the Office of Coal Research (OCR) whose responsibility it was to develop new markets for coal. Since the only market that would be large enough to have any significant impact on the economic health of the coal industry was its use as a fuel, the OCR directed much of its research effort toward converting coal to the more convenient and cleaner fuel forms--oil and gas.

While production of synthetic fuels included coal gasification to either low-, medium- or high-Btu gas, coal liquefaction to either a refinery feedstock or to a clean liquid boiler fuel, production of shale oil from oil shale, and the production of clean liquid and gaseous fuels from agricultural products, agricultural wastes and other types of organic wastes, this review will concentrate on coal conversion to refinery feedstocks and to a clean, low-sulfur boiler fuel. Nothing will be said about the complex set of economic, social, environmental and technical problems that are involved in the determination of whether to concentrate on making oil from coal or from oil shale.

Until the most recent revival of interest in coal liquefaction, brought about by a decline in domestic oil production capacity and the actions of the OPEC countries on that part of the oil which we previously imported to make up domestic shortfalls, all coal liquefaction research had been directed toward making a substitute refinery feedstock from which the usual range of petroleum products could be produced. As a result, the processes that will be described have been mainly aimed at producing refinery feedstocks. The need for additional research for developing processes to make a low-sulfur, low-ash fuel suitable for boiler use will then be discussed.

II. INDUSTRIAL PROCESS--COAL TO REFINERY FEEDSTOCKS

A. The Bergius Process

Figure 1 is a schematic diagram of the Bergius process, one of the two processes used industrially during World War II. The Bergius process produced ⌐85 percent of all synthetic liquid fuels. The coal is first ground to a fine size and mixed with a hydrocarbon liquid, which has been produced by the process itself, and with a catalyst. This mixture is reacted with hydrogen (produced by the gasification of coal) at a pressure of 10,000 pounds per square inch (psi) at a temperature of 850°F. The products from the first reactor are separated into a light fraction, a middle fraction, and a bottom fraction. The middle fraction is further treated over a catalyst in a vapor phase and under relatively mild conditions to produce petroleum-like products. The bottom fraction is filtered to remove solids (unreacted coal, catalyst, ash) and the remaining liquid is used to mix with the fresh coal that is processed in the first reactor.

The extreme operating conditions in the primary reactor (resulting from the poor quality of the catalysts available) and the large volume of hydrogen consumed, which is expensive to produce from coal using existing technology, made the liquids produced very expensive.

FIGURE 1

BERGIUS HIGH PRESSURE HYDROGENATION

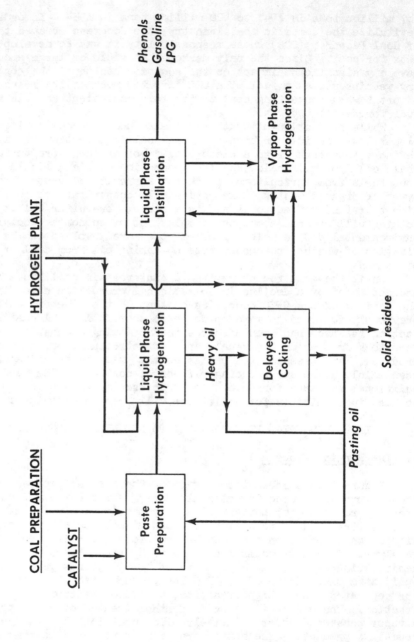

Overall process efficiency was about 55 percent. No commercial Bergius plants are currently operating.

Detailed cost estimates for producing gasoline made by the Bureau of Mines in the early 1950s differed sharply from those made by industry. The Bureau estimates indicated that gasoline could be produced at about 19 to 20 cents per gallon, but this included large by-product credits from the chemicals that were produced. The National Petroleum Council, using the same basic information but not allowing for by-product credits and including other components in the capital cost (e.g., company housing), estimated that gasoline would be produced for about 35 cents per gallon.*

Even if one assumes that the Bureau of Mines estimates were correct, after allowing for escalation in capital and labor costs, the price of gasoline (1972 dollars) made by the Bergius process would be about 40 cents per gallon, with coal at $10 per ton (40 cents per million Btu).

B. The Fischer-Tropsch Process (F-T)

The F-T process was the other one used in World War II by the Germans to produce industrial quantities of synthetic hydrocarbon liquids. The major difference in the products resulting from the Bergius and F-T processes was that the coal hydrogenation oil was aromatic and the F-T was paraffinic (with cobalt) or higly olefinic (with iron catalysts).

In the F-T process (Figure 2) coal is first gasified completely to form a gaseous mixture which, after purification, consists of a mixture of carbon monoxide and hydrogen (synthesis gas). The ratio of the hydrogen to carbon monoxide can be adjusted to produce different final products and for use with different types of catalysts. The purified gas is passed over a catalyst at temperatures of 570°F to 640°F and at pressures of about 450 psi, and a mixture of straight chain paraffinic and olefinic products are made that can be further refined to give the types of liquid fuels that are needed.

While the F-T process avoids the extreme processing conditions of the Bergius method, the complete gasification of coal and its reconstitution to liquid products gives an overall conversion efficiency of only 38 percent. One small industrial F-T plant has been in operation in South Africa since about 1955. The plant uses fixed-bed Lurgi gasifiers to make the synthesis gas and two different methods of contacting the purified gas with catalyst: a modified fixed-bed German process and an entrained-bed catalyst. The major engineering problem is the removal of the large volumes of heat that are released when the gas is converted to a liquid by the catalyst.

Although only one F-T plant is in operation and no commercial Bergius plants are, every cost estimate that has been made which compares the costs of liquids produced by these two methods indicates that fuels made by the F-T process will be more expensive than those made in the Bergius process. The F-T plant in operation in South

*A 6 percent return on the total investment after a 50 percent income tax.

48

FIGURE 2

FISCHER-TROPSCH PROCESS

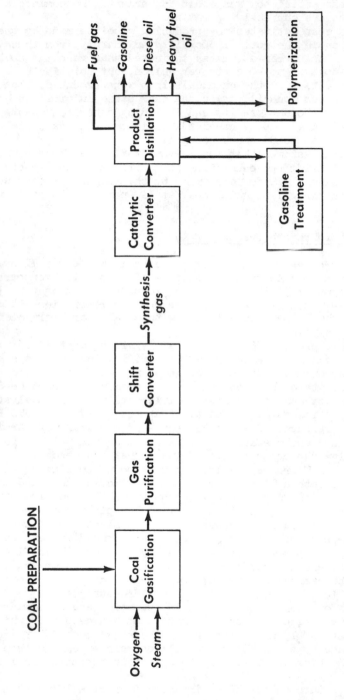

Africa was justified on the basis of the special values that could be assigned to some of the by-products and on social and political grounds.

III. RESEARCH AND DEVELOPMENT PROGRAMS--
COAL TO REFINERY FEEDSTOCKS

Because of the high costs associated with the industrial processes that are available for coal liquefaction some research has been underway to develop improved and lower cost methods for producing liquid fuels. Of these, three have received the greatest amounts of funding over the past ten years. These are the COED process (developed by the FMC Corp.), Project Gasoline (developed by Consolidation Coal Co.), and the H-Coal process (developed by Hydrocarbon Research, Inc.). The first two processes, after their initial conception, have been largely government supported. The HRI process, while it received limited government support, has been largely a private effort.

In addition to these three major efforts, the Synthoil process of the Bureau of Mines has recently received increased attention, and experimental work is said to be underway on coal liquefaction by several private oil companies using their own funds.

A. The COED Process

While the liquid produced by this process can be used as a refinery feedstock, the process is basically aimed at removing valuable liquids from the coal before it is burned. The amount of liquid and char produced will vary with both the coal used and processing conditions but in all cases the char is the major product. As a result, COED must be viewed as a way to supplement supplies of liquids from coal and not as a primary coal liquefaction process unless methods to liquefy the char are found.

The process has been under investigation for a number of years. The process consists of a series of fluidized, low-temperature carbonization reactors. The number of reactors and the processing conditions necessary to optimize liquid yields will vary for different coals. Some medium-Btu gas is produced during the carbonization, but the two major products are a hot char and liquids.

A number of coals have been tested in a small unit (PDU) with a capacity of 100 pounds per hour. A few coals have been tested in a larger pilot plant with a capacity of 1-1/2 tons per hour, and there are plans to test other coals in this pilot plant. The usefulness of the COED process as a source of liquids from coal will depend on how successful is the development of a method for using the char in an environmentally acceptable way. If the overall economics justify it, there should be no major problems in engineering and constructing a large commercial plant.

B. Project Gasoline

This process was to have been tested on a 1 ton per hour scale, following successful laboratory tests, but the plant constructed at

Cresap, West Virginia, has been plagued with a large number of engineering difficulties, only some of which are directly related to the hydrogenation process itself. Moreover, operations were handicapped by labor unrest so that only a limited amount of data was obtained from the pilot plant before work was terminated.

In this process, the coal is partially converted to liquid in a first step by a hydrogen transfer recycle solvent (which is a product made in the process) and solids are separated and then the liquid is treated with hydrogen in a fluid-bed catalyst reactor. The solvent must then be separated from the product which is a satisfactory refinery feedstock. The residuals solids separation has been one of the most persistent engineering problems encountered, with the hot filters originally used not ever operating satisfactorily. When cyclones were used they gave a product somewhat higher in ash and sulfur than desired.

The current plans of OCR apparently intend to convert Cresap to a unit operations testing center where engineering problems common to many of the coal conversion schemes could be tested on a large scale. In the meantime, Standard Oil of Ohio, which still retains some interest in the background patents of Project Gasoline, has proposed that a 900-ton/day pilot plant be erected, supported by fifteen energy companies. Total capital costs are estimated at $70 million.

C. H-Coal Process

This process has been under study by Hydrocarbon Research, Inc. for a number of years. It represents a modification of their H-oil process, which is being used commercially, so as to be able to handle coal. After initial development work by HRI it was supported by OCR for a limited period. In recent years the research has been funded by a consortium of oil companies, the membership of which has changed from time to time.

In this process, coal is mixed with a recycle oil and fed into a hydrogenerator containing a granular catalyst which is maintained in an ebullient state by the flow of the liquid-solid mixture and hydrogen. A number of coals have been tested in an 8-inch diameter reactor that can handle 200 pounds of coal per hour. The catalyst can be added or removed continuously as required. A high yield of low-sulfur liquids are produced that are suitable for a refinery feedstock. Because of its high cost, the catalyst that is attrited must almost all be recovered and this means separating it from the unconverted coal and coal ash. Construction of a larger plant is expected to start in 1974.

D. Synthoil

This particular process has been under development by the Bureau of Mines for only the past three years, but they have drawn upon their experience in this field over the past several decades to invent this method.

The key feature of this one step hydrodesulfurization process is the use of rapid, turbulent flow of hydrogen to propel coal slurry through an immobilized bed of catalyst pellets in a reactor. The slurry vehicle for conveying the coal is a recycled portion of its own product oil. The combined effect of the hydrogen, turbulence, and catalyst is to liquefy and desulfurize the coal at high yields and high throughput. Sulfur is removed as H_2S, which is easily converted into inert elemental sulfur for industry or storage.

Both a low-sulfur liquid suitable for boiler fuel use and a refinery grade feedstock have been produced from five different high-sulfur, high-ash coals. The nature of the product, whether it is to be a boiler fuel oil or refinery feedstock, depends upon the pressure and residence time to which the coal is exposed.

Operations with 100 pound/day and 1/2 ton/day process development units have proven long-term operability (seven week runs). The smaller reactor was 5/16 inch in diameter and 68 feet long; the larger is 1 inch in diameter and 14 feet long, both containing a fixed bed of 1/8 inch catalyst pellets. Coal reaction residence time is only 2 minutes.

The process can be operated to produce a gasoline at much less severe operating conditions (4,000 vs. 10,000 psi) than the Bergius process because of improved catalysts (Co, Mo on alumina) that have been developed and the benefits in rapid reaction rates obtained through the use of a highly turbulent regime. Thus it should be much less costly.

E. Other Research Programs

Although no details have been revealed, it has been reported that at least three oil companies now have coal liquefaction processes currently under study and other companies were known to have had programs in the past. Gulf Oil Company is said to be investigating on a relatively small scale at least one, and possibly two, different coal hydrogenation technologies. Exxon has said that it has at least one process under study, and Standard Oil of Ohio is said to be considering still another coal hydrogenation scheme (see above).

A process similar to the COED method, which produces a low-sulfur oil, char, and a gas, is being investigated by Occidental Petroleum Company at its Garrett Research Laboratories. The rapid pyrolysis which the process uses is said to produce more oil that is low in sulfur than when the pyrolysis is carried out more slowly. It, like COED, must be viewed as a supplement method to liquid fuel supplies and not as a primary liquefaction process.

Until these processes reach a stage where these companies have secured adequate protection for their proprietary interests, or are willing to reveal their processes in seeking government support for additional or larger scale tests, the details of their investigations will remain unknown. It is further evidence, however, of a revival of interest in coal liquefaction technology.

IV. RESEARCH AND DEVELOPMENT--COAL TO LOW-SULFUR,
LOW-ASH SOLID OR LIQUID FUELS

The reasons for the past emphasis on coal liquefaction to make
refinery feedstocks were noted earlier. With the current energy situation in which the nation is seeking to reduce oil imports and use its
large indigenous coal deposits in an environmentally acceptable way,
the emphasis on coal liquefaction research should be for the production of a low-sulfur, high-Btu, low-ash solid or liquid boiler fuel.
The development of a successful low-cost process would free large
volumes of natural gas and liquid fuels now used at electric generating and large industrial plants for other uses in which such substitution is not possible. Moreover, a low-sulfur liquid fuel that
is entirely acceptable for use under boilers would be lower cost than
a refinery feedstock since it would use less hydrogen and should be
able to be produced under milder processing conditions.
A low-sulfur liquid from coal is a preferred boiler fuel over a
low-Btu gas from coal since it can be more easily stored and transported. These factors permit the fuel to be produced at a central
plant which can be base loaded (and not have to follow the daily and
hourly fluctuations of the fuel required for the process) and can be
large enough to serve several plants, thus achieving economies of
scale. The ability to store the fuel at low cost permits the manufacturing process or the electric generating plant to continue to
operate if the coal conversion plant has operating difficulties. Low-
sulfur liquids are preferable to low-sulfur solids because they can
be handled and transported at lower costs. Low-sulfur liquids also
require less chemical transformation and hydrogenation, less severe
processing conditions, and conversion efficiencies are higher.
While all of the processes discussed above may be able, by modifying the processing conditions, to produce a low-sulfur boiler fuel
from coal at lower cost than a refinery feedstock, only the Synthoil
process has reported the results of such experiments. In addition,
extensive experimental work has been carried out on the PAMCO process
that makes a low-sulfur boiler fuel that solidifies at ambient temperatures.

A. PAMCO

As originally conceived, fine coal is dissolved in a recycled
stream, hydrogenated with small quantities of hydrogen at modest
pressures (1,000 psi) and filtered to remove the solids. These
process steps eliminated part of the sulfur and added some hydrogen
to the coal. The product is a relatively low-sulfur, low-ash material, solid at room temperature but a liquid when heated to 350°F.
It could be burned in a boiler, either as a solid or as a liquid, if
it is first heated. Unless more stringent processing conditions could
be used, the product would not be a satisfactory refinery feedstock.

B. Synthoil to Low-Sulfur Boiler Fuel

The process described earlier for producing a liquid fuel from coal in a fixed-bed turbulent reactor has also been operated so as to make a low-sulfur, low-ash boiler fuel. Coals with a sulfur content of as high as 5.5 percent have been treated at pressures of 2,000 psi at temperatures of 840°F to produce oils with sulfur contents in the range of 0.2 to 0.4 percent. Production of 3 barrels of low-sulfur oil per ton of coal has been achieved using only 3,000 cubic feet of hydrogen per barrel of product oil so that overall energy conversion efficiencies of 75 to 78 percent have been obtained.

V. ANTICIPATED PRODUCT COSTS USING
ADVANCED PROCESSES UNDER STUDY

The high costs of liquids made by either the Bergius or F-T processes were given earlier. Recent estimates of the costs of making a refinery feedstock by coal liquefaction using one of the processes now under study were made by the National Petroleum Council.[1] In view of the relatively small experimental effort and the lack of success of some of the programs that were undertaken, the NPC figures appear to be optimistic. This same optimism characterized the early estimates of the costs of high-Btu gas from coal, and these costs now appear to have been less than one-half of what the costs are now thought to be.

The NPC estimated that a refinery feedstock could be produced from western coal ($2 per ton, 10 cents per million Btu) for $6.50 to $7.50 per barrel, and from eastern coal ($6.50 per ton, 27 cents per million Btu) for $7.00 to $8.00 per barrel (all in 1970 dollars and 10 percent discounted cash flow). In a survey made for a study for the Senate Interior Committee,[2] estimates (in 1973 dollars) by experts were in the range of $7.50 per barrel, but only after considerable additional successful research was concluded. Estimated costs made by the NPC for a low-sulfur boiler fuel were $4.50 per barrel and those in the Senate study were $6.75 per barrel, but in both cases the processes are yet to be proved. The large differences in cost estimates will not be resolved until either large demonstration scale or commercial plants have been constructed and operated.

Until early in 1973 crude oil prices had remained relatively constant at about $3.20 to $3.60 per barrel, so that no matter which of the estimates of synthetic crude oil prices were correct, they were still too high to be economic. With the recent actions by OPEC in repeatedly escalating world oil prices and the use of an oil embargo to serve political purposes, future oil prices are impossible to predict. Spot prices in late 1973 appear to indicate that imported crude landed in the United States will be about $10 or more per barrel, but whether these high prices will continue is unclear because the cost of producing and transporting much of that oil is in the range of $1.50 and $3.00 per barrel. At $10 or more per barrel, production of a low-sulfur boiler fuel from coal would probably be economic but it is doubtful that a refinery feedstock could now be produced at that cost until additional process improvements are made.

Material costs alone contribute a significant share of final product costs. For a refinery feedstock, if an overall coal utilization efficiency of 65 percent and a coal cost of $10 per ton are assumed, then coal costs are about $3.50 per barrel. If hydrogen could be produced from coal at 60 cents per 1,000 cubic feet, then hydrogen costs would be another $5.40 per barrel for a total of material costs of $8.90 per barrel. For the same set of assumptions, material costs for a low-sulfur boiler fuel would only be $5.50 per barrel.

REFERENCES

1. National Petroleum Council, U.S. Energy Outlook: Coal Availability (Washington, 1973), p. 250.
2. U. S. Senate Committee on Interior and Insular Affairs, Energy Research and Development--Problems and Prospects, Committee Print 93-21 (1973).

CATALYSIS

T. E. Fischer
Esso Research and Engineering Co.
Corporate Research Laboratories
Linden, New Jersey

ABSTRACT

Catalysis presents the physicist with an excellent opportunity to contribute to the technology of energy supply by doing basic research. Recent developments in the experimental techniques for the preparation and investigation of solid surfaces and adsorbates and in the electronic theory of surfaces and of chemisorption quite naturally will provide insights into the fundamental processes of catalysis that were not accessible before and hold the promise of making catalysis an exact science in the foreseeable future. Such knowledge can be expected to promote in a significant way the discovery of new catalysts that will be necessary in the technology of energy production. An analysis of our natural resources reveals that for the next 50 years the major part of our energy supply will have to be provided by fossil fuels. The depletion of oil reserves makes it necessary to transform coal, tar sands and shale oil to convenient fuels that can be burnt without unacceptable pollution of the environment. Chemical processes exist but much progress in efficiency and price could be achieved by the development of novel catalysts. Solid catalysts are generally prepared with high specific surface area. Their catalytic activity correlates roughly with their electronic properties. Metals are mainly used to promote reactions involving hydrogen; semiconductors catalyze charge transfer and oxidation-reduction; insulators possess surface sites with strongly acid or basic properties and promote skeletal rearrangements of hydrocarbons through the formation of ionic intermediates. Progress in metal catalysis is characterized by the introduction of multimetallic systems. Development in this field will profit from the establishment of a relationship between the physical and electronic and chemisorption properties of alloys.

INTRODUCTION

The purpose of this paper is to investigate whether there are opportunities for physical research to contribute to progress in catalysis and whether such progress would help solve the problems of energy supply in the future. We shall find that recent developments in physical methods have provided us with novel insight into phenomena occurring at solid surfaces that quite naturally leads to progress in catalysis. We shall also conclude from a rapid survey of

the technologies and resources that catalysis must be expected to play
a major role in the supply of energy, at least in the intermediate
future, that is, as long as we shall be dependent on fossil fuels.

THE ROLE OF CATALYSIS IN THE SUPPLY OF ENERGY

The energy shortage that we experience at this time reflects the
fact that our rapidly expanding demand for fuel has exceeded the
annual yield of the resources that were available to us with the
existing technology. We shall not address ourselves here to the
question of the need for slowing down the rate of increase in energy
consumption, for any non-catastropic change will still leave us with
a problem of energy supply. In the long run, non-exhaustible sources
of energy will have to be found and exploited. Solar and geothermal
energy are being explored. Breeding of nuclear fuels is being envisioned. There is little doubt that they will provide the bulk of
energy in the future, but their stage of development does not allow
us to expect them to play a major role in the next 50 years. Nuclear energy from the fission of uranium is ready. If, as expected,
their utilization will be chiefly in base load electric power generation, nuclear reactors will contribute less than 25% of total consumed energy by the end of the century. Fossil fuels are quite
plentiful in the North American continent. There is enough off-shore
oil, coal, tar sand and oil shale to satisfy our needs until other
technologies are developed. These resources present us with a series
of challenges. The first is to get these fuels out of the ground
without causing irreparable damage to the environment. A second
challenge is to give them a form that allows their convenient and
efficient use. A third challenge is presented by the necessity of
avoiding excessive pollution.

Figure 1 shows a hypothetical coal molecule that assembles the
various constituents that have been experimentally identified. We
note in passing that the structure of various coals is still a subject of controversy. An alternate model proposes a large fraction of
carbon with sp^3 bonding and tetrahedral coordination, as in diamond.
Although their shapes and sizes are different, molecules of coal,
sand tar, shale oil and petroleum residua (which are left after gasoline, jet fuel and residential heating oil have been extracted from
crude oil by distillation) share a number of undesirable characteristics. Their large molecular weight gives them the form of solids or
viscous liquids. They contain relatively large amounts of impurities,
especially sulfur and nitrogen the combustion of which forms toxic
pollutants. They contain various amounts of minerals (ashes) whose
combustion products are solid (see Table I). Large scale industrial
burners can be designed to use these fuels directly and to extract
the pollutants from the effluents. Transportation and residential
burners require a fuel that is easily transported and handled and
burns efficiently with minimal pollution. We are thus faced with the
challenge of upgrading the various hydrocarbons described above to
fuels that allow clean and efficient combustion. Solutions envisioned
now are the gasification or liquefaction of coal with simultaneous removal of sulfur and nitrogen and/or the development of effective

58

Table I Major Constituents of Typical Hydrocarbon Sources

	Crude Petroleum	Petroleum Residua	Tar Sands Bitumen	Shale Oil	Dry Bituminous Coal
C/H, Wt Ratio	7.0	8.0	8.0	8.0	14-20
Sulfur, Wt%	1.4	2-4	4.0	1.0	1-4
Nitrogen, Wt%	0.2	0.5	0.4	2.0	1-2
Oxygen, Wt%	--	1.2	1.0	1.5	5-10
Ash, Wt%	<0.1	0.2	0.6	<0.1	10-13

Figure 1 A Model of the Molecular Structure of Coal

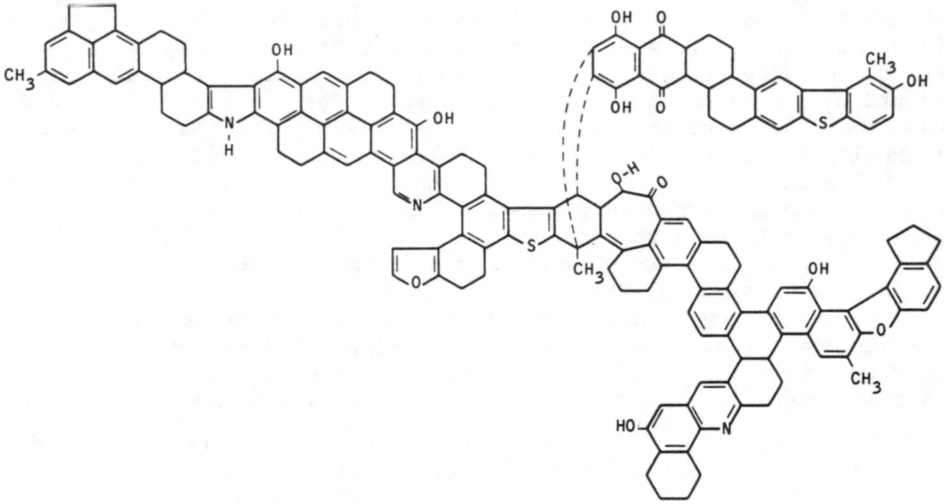

methods to remove oxides of sulfur and nitrogen and small particles
from the flue gases. For oil shale, tar sands and heavy residua, the
problem is chiefly the removal of sulfur and nitrogen and of metals,
such as vanadium and nickel, from the molecules. Methods for the desulfurization of hydrocarbons exist and are used on an industrial
scale. In fact, oil desulfurization is a major contributor to the
world's supply of sulfur. These methods work well on the smaller
hydrocarbons. With the large molecules the process becomes very expensive because of large hydrogen consumption and low rates of
reaction. There is a considerable economic incentive for a catalyst
which would allow a rapid reaction that breaks the strong sulfur-
carbon and nitrogen-carbon bonds with less consumption of hydrogen.

It is well known that the rate of a chemical reaction at a given temperature can be increased remarkably by bringing the reactants in contact with a suitable substance which is itself not consumed in the reaction. This phenomenon is called catalysis, and the rate-increasing substance is the catalyst. The activity of a catalyst measures the reaction rate achieved, usually in moles of reactant consumed per unit time and unit weight of catalyst. The selectivity of a catalyst is the ratio of its activity for a given reaction over that of an alternate reaction. In view of our dependence on the chemical transformation of fossil fuels, it is quite obvious that catalysis will play a major role in the satisfaction of our needs for energy.

Most industrial chemical processes are promoted by catalysts. It is not surprising that this phenomenon has been the object of extensive scientific investigations and that a large body of know-ledge and experience has been accumulated. Yet, the rich diversity of phenomena observed and their complex nature has prevented so far a detailed understanding of the fundamental processes involved. Up-grading of heavy hydrocarbons has received little attention so far since oil was plentiful until now. The following are some of the most outstanding challenges.

Coal Gasification

Several processes for the gasification of coal exist, either in the industrial or the experimental stage. Most of them make use of the so-called steam gasification reaction

$$C + H_2O \longrightarrow CO + H_2 \tag{1}$$

This reaction is endothermic (33 kcal/mole) and occurs at 930°C. Further reaction between CO and H_2 to produce hydrocarbons occur between 480°C and 540°C over various catalysts and are exothermic. For example:

$$CO + 3H_2 \longrightarrow CH_4 + H_2O \qquad \text{Methane}$$

$$6CO + 9H_2 \longrightarrow C_6H_6 + 6H_2O \qquad \text{Benzene} \tag{2}$$

$$CO + 2H_2 \longrightarrow CH_3OH \qquad \text{Methanol}$$

The synthesis reactions (2) cannot be run at higher temperatures than 550°C because equilibrium would favor the left (decomposed) side of the reaction too heavily.

There is thus a great incentive to discover a catalyst that would allow the gasification reaction (1) to proceed at useful rates at a temperature below 550°C. Then most of the energy released by reaction (2) could be used in reaction (1) instead of being wasted. The biggest gains would be realized if one discovered a catalyst that would allow to combine reactions (1) and (2) in the same reactor.

Room for improvement also exists in the catalysts for synthesis of hydrocarbons. We have shown above only a few examples of the many synthesis reactions that occur. The catalysts known to date lack selectivity so that it is difficult to obtain a specific product with reasonable purity.

Coal Liquefaction

Industrial processes for the production of liquid hydrocarbons from coal have existed for some time. During the second world war, an important fraction of Germany's gasoline supply came from coal liquefaction. These processes are presently in use in South Africa; they are costly and relatively inefficient and had no economic justification as long as oil was plentiful.

Modern coal liquefaction processes are not directly catalyzed. They involve the use of liquids that act as solvents and hydrogen donors. The preparation of these liquids requires large amounts of hydrogen. The production of the latter consumes so much energy that liquefaction and purification of coal already consumes about 1/3 of the original energy content of this resource. It is obvious that more efficient processes for producing environmentally acceptable fuels from coal must be discovered.

Desulfurization and Denitrogenation

Desulfurization is now accomplished industrially by hydrogenation over cobalt-molybdate catalysts. It is relatively efficient for the small molecules found in gasoline and residential heating oil. However, sulfur and nitrogen removal becomes a problem in the case of the large refractory molecules of residua and of tar and shale oil. The catalysts become rapidly inactive due to severe coking and poisoning by the metallic impurities. The process, to be operative, requires high pressures of hydrogen (60-200 atmospheres) and temperature (400-450°C). It is expensive and produces objectionable amounts of light gases. The search for a more efficient process for the desulfurization of heavy molecules is centered on the exploration for a novel catalyst capable of breaking the sulfur-carbon and nitrogen-carbon bond, with a minimum consumption of hydrogen.

We have mentioned above a few unsolved problems of energy production where the discovery of a suitable catalyst could unlock the technology. It is unlikely that research with physical methods in these specific reactions will lead directly to the desired results. However, as we shall see below, recent developments in surface physics are more than likely to shed new light on the fundamental processes in catalysis and thereby will be able to give the catalytic chemist novel tools with which he will attack his problems. Let us first describe some of the new methods of surface physics that have emerged in the recent years. Then we shall briefly describe various kinds of catalysts and the reactions they promote; on the way we shall point out some advances that can already be credited to surface science. These examples will show specifically the kind of contributions to be expected from physical investigations.

THE METHODS OF SURFACE SCIENCE

During the last 10-15 years, the physics of solid surfaces has developed into a mature and quantitative science. This development was based on the progress in vacuum technology and was supported, in its early phases, mainly by the successes and problems of the semi-conductor industry.

Today, the solid surface is one of the most reproducibly pre-pared and well characterized research objects. Clean surfaces can be prepared by evaporation or sputtering onto substrates, or, for bulk samples, by a combination of ion bombardment and annealing; a limited number of clean surfaces can be prepared by cleavage. Very sensitive (10^{-13} torr partial pressure) mass spectrometers (parti-cularly the versatile quadrupole and monopole spectrometers) deter-mine the composition of the gases in the experimental chamber. These instruments have made it possible to observe the catalytic activity of surfaces as small as a fraction of a cm^2. The introduction, in rapid succession, of many measuring techniques based on the absorp-tion and emission of photons and electrons have allowed the surface scientist to determine most of the properties of surfaces that are relevant to chemisorption and catalysis. These methods make use of the variation of the free mean path of electrons in a solid with their energy; this mean free path, which characterizes the penetra-tion or escape depth of electrons beneath the surface has a minimum of 2-10Å, depending on the material, at around 30 to 60 eV.

The crystallographic structure of surfaces is investigated by Low Energy Electron Diffraction. The latter technique is much more difficult to interpret than, say, x-ray diffraction because of the large interaction cross section of the low-energy electrons with atoms. However, the theory of LEED has recently reached the stage where it becomes practical to deduct the positions of atoms at the surface from measurements of the angle and intensity of diffracted beams. Limited information can also be obtained about the morphology of the surface, in particular the presence of steps.

Auger spectroscopy permits the identification of atoms on and near a surface with a sensitivity of 10^{-2} monolayer in the best cases. It measures the small peaks that appear in the energy dis-tributions of secondary electrons because of the replacement, by an Auger process, of a core electron ejected by the impact of a primary electron. Auger peaks appear at several characteristic energies for every element. Often, with the help of the energy dependence of the escape depth, it is possible to differentiate between atoms at the surface or beneath it. Other methods for identification of surface atoms are secondary ion mass spectroscopy which analyzes the ions ejected by ion bombardment and helium ion scattering. This latter technique measures the energy transfer of helium ions to surface atoms and thus determines the mass of the latter. Auger spectro-scopy is by far the most widely used; it has the great advantage of being performed with the same equipment as used for LEED or for electron spectroscopy measurements.

Photoelectric emission is performed in various ranges of the optical spectrum. It was first used in the visible and near ultra-violet for the measurement of work functions. Then its usefulness was discovered successively for studying the band structure of solids, and for determining surface properties of semiconductors. Photoemission by x-rays (also called ESCA) has been widely applied for the spectroscopy of core electrons. At photon energies capable of exciting electrons to energies corresponding to their shortest escape depth, photoemission has recently been used to measure the energy of valence electrons of adsorbed molecules and to observe directly the emission from surface states on clean surfaces. Auger spectroscopy is also beginning to find some use as a spectroscopic tool because the shape and exact energy of the Auger peaks depend on the valence state of the atom. Interpretation of these latter measurements is still controversial. Energy distributions in field emission have been used successfully to measure the energy of valence electrons in adsorbates and to observe surface states on semiconductors. Other spectrometric methods are as follows: Appearance Potential spectroscopy measures the various thresholds in the energy of bombarding electrons to cause step increases in the production of x-rays. Energy Loss Spectrometry again concentrates on certain peaks in a secondary emission experiments; it analyzes the energy of electrons that are inelastically reflected from the solid after having excited a well defined interband transition of an electron in the solid near the surface. Incoming electrons are also loosing well defined amounts of energy by exciting surface and bulk plasmons. With special care to achieve high energy resolution of the electrons, it is possible to observe energy losses due to the excitation of phonons.

Ion neutralization spectroscopy measures the energy of electrons ejected after having received, by an Auger process, the energy released by another valence electron from the solid in transferring to the low lying empty level of a noble gas ion. It is particularly sensitive to the electronic structure at the surface and thus is well suited for the study of adsorption.

Most experiments combine several of these techniques. A typical vacuum chamber will contain an ion gun, provision for sample heating, LEED, Auger and mass spectrometry facilities and perhaps one electron spectroscopy experiment.

The introduction of experimental techniques have been followed quite closely by the development of theory. Early theories described the various electron emission phenomena on the basis of the free electron theory of metals. Qualitative theories of the work function of metals, surface states on semiconductors, and chemisorption were not at a stage which allowed comparison with experiment. In recent years however, much progress has been made in theory of Low Energy Electron Diffraction (LEED) of the electronic density near the surface, the work function of metals, surface states, as well as in the theory of photoemission.

The electronic treatment of chemisorption represents such an active field at this time that one expects, in the near future, quite realistic calculations of the energies and charge densities in molecules as they are modified by adsorption onto a surface.

CATALYSIS

The chemical science of catalysis has a much longer history, and its methods are in sharp contrast with those of surface science. The catalysis researcher works usually at 10^{-2} atmospheres or higher pressures, rather than in ultrahigh vacuum. He prepares his samples with as high a surface area as possible rather than flat surfaces of single crystals. His attention is focused on the energetics and kinetics of reactions. The catalysis researcher's methods permit rapid screening of many catalysts, his experimental situations are quite close to the reality of the applications, but are often too complicated to permit detailed knowledge of what is occurring. By contrast, experiments in surface physics are slow and elaborate, their object is simplified in the extreme, but the information they provide is fundamental and usually detailed. It is this difference in outlook and method that forms the basis of our optimism concerning the chances of a meaningful contribution of surface science to catalysis. Surface science offers a completely novel approach, it focuses its attention on the fundamental processes at the surface. Its contribution to the design or discovery of novel catalysts obviously depends on the combination of the two approaches and therefore, on the successful cooperation between surface scientists and catalytic chemists.

A short survey of some broad concepts in catalysis will show a few specific opportunities.

One distinguishes between homogeneous catalysis, where the catalyst is in the same phase (usually a fluid) as the reagents and heterogeneous, or contact, catalysis where the catalyst is a solid. We shall concern ourselves only with the latter because it is the most widely used and presents the most opportunities for progress to be achieved with the help of physical investigations. Practical heterogeneous catalysts are usually solids prepared in such a way as to present a high specific surface area. Metal oxides, such as silica or alumina, for example, can be obtained by precipitation from a very concentrated solution, with specific surfaces in excess of 200 m^2/gr. Metallic catalysts have the form of very small particles supported on high surface-area insulators. They are prepared by soaking the support in a solution of a salt of the metal, drying, calcining in air and reduction at elevated temperature in hydrogen.

Solid catalysts can be divided roughly into three classes, according to their electrical properties. The distinctions so made only represent trends, they do not include all chemical reactions and are riddled with exceptions. Metal catalysts are predominately used in reactions involving hydrogen. Some examples are the synthesis of ammonia catalyzed by iron, the oxidation of ammonia to nitric acid over platinum and the various hydrogenations and dehydrogenation reactions of hydrocarbons over platinum in the manufacture of gasoline. Semiconductors are most active for reactions involving charge transfer, such as oxidation and reduction. Insulators, in particular metal oxides, catalyze the skeletal rearrangement (i.e. cracking and isomerization) of hydrocarbons. Their activity is similar to that of liquid acids which catalyze isomerization and cracking by the formation of carbonium ions as reaction intermediates.

Solid Acid Catalysts

It is an experimental fact that many insulating solids are strong acids. The surfaces of Al_2O_3, $SiO_2 \cdot Al_2O_3$, $Al_2O_3 \cdot B_2O_3$, to name a few, contain in the order of 10^{13} cm^{-2} Lewis or Bronsted acid sites. A Lewis acid is defined as a species that can accept a pair of electrons from a base. A Bronsted acid can donate a proton to another molecule.

The acidity of these solids is determined by the high adsorption energy of bases on them and by the fact that these adsorbed bases show changes in their infrared absorption spectra very similar to those caused by reaction with liquid acids. It has also been observed that certain solids possess basic, i.e. electron donor properties.

The precise nature and structure of acid sites has not been studied very extensively. Several interesting models have been proposed, but the problem is far from being considered close. In all these models, the acid sites are related to some defect on or near the surface. On alumina, for instance, acidity is only observed after the solid has been heated in air or in vacuo at temperatures above 300°C. Adsorption of water covers every aluminum ion with an OH group. When the alumina is heated in vacuo, evaporation of water proceeds randomly at first, in a sense nucleating small domains of perfectly dry silica. These domains will meet at boundaries where they mismatch and contain such defects as adjacent oxygen ions, adjacent uncovered aluminum sites and single OH groups in different (oxygen-rich, oxygen-deficient or "perfect") environments. These defects thus present a range of electron affinities and therefore constitute acidic and basic sites of various strengths.

Silica-alumina is a much stronger acid than alumina, i.e. the electron affinity of its acid sites is larger. By contrast, silica itself is not acidic. Maximum acid strength is obtained with silica-rich compounds, namely with 70 wt% silica and 30 wt% alumina. The explanation is reminiscent of semiconductor physics. Strong acid sites owe their existence to an isomorphic substitution of aluminum for the tetravalent silicon in the silica lattice. Like in the silicon crystal, the aluminum forms an electron acceptor site.

When water is adsorbed on silica-alumina, Lewis acid sites are transformed into Bronsted sites. The strong Lewis acid site abstracts an electron from the water molecule and transforms it into an adsorbed OH group and a loosely bound proton.

Numerous experiments have related the catalytic activity of oxides to their acid or base properties. A direct correlation, often linear, has been found between the rates of polymerization of small molecules or cracking of large hydrocarbons with the number of acid sites. In general, it is found that Bronsted acid sites are responsible for the polymerization, isomerization and cracking, in other words, for the skeletal rearrangement of molecules. It is generally thought that the skeletal rearrangements occur via the formation of carbonium ions (positively charged hydrocarbons) as intermediates.

In effect, it has been known for a long time that the addition of acids such as H_2SO_4 to hydrocarbons would produce carbonium ions and skeletal rearrangements. It must be borne in mind, of course that the carbonium ion intermediates are not free in the gas phase, but generally bound to the (negatively charged) acid site.

Semiconductor Catalysts

As for the case of electrical conduction, the difference between insulators and semiconductors is not essential, but a matter of degree, since it depends on the width of the conduction band. Most practical semiconductor catalysts, again, are metal oxides, although some halides find applications as well. One distinguishes between cooperative and localized surface reactions. Cooperative reactions are those in which electron exchange occurs between the adsorbate and the conduction or valence band of the catalyst. The development of the theories of cooperative catalysis by semiconductors occurred in parallel to the quantitative description of the electron concentration at semiconductor surfaces. The theories were first based on equilibrium statistics and later, more realistically, made use of the analogies with electron-hole recombination kinetics.

Localized interactions of adsorbed molecules, by contrast, are those which involve either the catalytic activity of a defect on the surface or the chemical binding forces of the admolecule to specific surface atoms. A theoretical treatment of this kind of interaction is obviously much more difficult, since it is not a simple matter of statistics. Such a theory does not exist at this time; its elaboration will probably follow a more complete theory of chemisorption on metals.

Metal Catalysts

Metal catalysts have been known and used in the chemical industry for a long time. Understandably enough, a large amount of work has been devoted to the investigation of metallic catalysts, systematic trends in catalytic activity for specific reactions with the position of the elements in the periodic table were established. The fact that most elements showing measurable catalytic activity are transition metals of the Groups VIII and IB leads to the often successful search for a correlation between catalytic activity and Pauling's percentage d character of the metallic bond. A striking example of such a correlation is presented by the hydrogenolysis of ethane ($H_3C - CH_3 + H_2 \longrightarrow 2CH_4$) where the catalytic activity of the metals of Group VIII varies by eight orders of magnitude and correlates well with the d-character of the metallic bond. A similar trend is observed in the hydrogenolysis (i.e. the breaking up by the action of hydrogen) of a larger paraffin (i.e. straight hydrocarbon saturated with hydrogen). These reactions proceed at a measurable rate at temperatures above 200°C only.

By contrast to hydrogenolysis, the simple addition of subtraction of hydrogen to or from hydrocarbons occurs readily at room temperature on all Group VIII metals. The catalytic activity for

this type of reaction varies much less with the metals than for
hydrogenolysis and does not correlate at all with the per cent d
character of the metal bond. The distinction between these two
types of reactions is also related to the way they respond to the
state of dispersion of the catalyst.

The rate of the hydrogenation-dehydrogenation reactions discussed above which does not correlate with the per cent d character
of the metal, is also independent of the state of dispersion. In
other words, the reaction rate per <u>surface area</u> of metallic catalyst
is independent of the size of the crystallite. Dispersion has
merely the effect of increasing the overall reaction rate by increasing the available area. By contrast, the hydrogenolysis
reactions are influenced in a major way by the size of the metal
particles. The catalytic activity per surface area of small
(12-40Å) particles can be as much as 30 times greater than for
particles larger than 200Å.

It is a valid question to ask to what extent catalytic activity
depends directly on the size, i.e. the small number of atoms in the
particle and to what extent it reflects the radius of curvature and
thereby the density of atomic steps on its surface.

An answer to this question was provided with the help of the
methods of surface science by Somorjai and his students [1] who prepared macroscopically flat surfaces of platinum by cutting a single
crystal in such directions as to contain atomic steps of various
lengths. The structure and cleanliness of the surfaces were established by LEED and Auger spectroscopy. They measured the catalytic activity of these surfaces with the help of sensitive mass
spectrometers and compared it with the activity of a (111) surface of
platinum carefully prepared to minimize the number of steps and
other surface defects. They found the step free surface to be inactive. The stepped surfaces showed catalytic activity and even
selectivity to different hydrocarbon reactions, depending on the
predominant step length.

Computer simulations of the growth of very small particles by
Burton [2] and by Hoare and Pal [3] indicate that the structure of these
clusters is not identical to the crystal structure of bulk matter.
The most stable particle of 55 atoms, for example, has the shape of
an icosahedron which has five fold symmetry. An important characteristic of these particles is that it exposes only close-packed, smooth
(111) surfaces. The regular fcc structure of 55 atoms is bounded by
100 and 111 surfaces. Clearly, the question of the structure sensitivity of the dispersed metallic catalysts has been brought into
new light by physical investigations, but it is far from being resolved.

The present thrust of progress in research and industrial utilization of metallic catalysts is in the direction of alloys, or more
properly, considering the small size of particles, of multi-metallic
clusters. The new reforming catalyst KX 130 developed by Sinfelt
and coworkers at the Esso Research and Engineering Company is based
on this concept. It possesses much higher activity and remains
active much longer in the refinery than the previously used platinum

catalysts. Together with the prospect of new technological advances, the introduction of multi-metallic catalysts present the researcher with a whole new set of questions and problems. To begin with, there is no reason for the surface of an alloy to have the same composition as the bulk. Williams and Boudart [4] for instance, have shown by Auger spectrometry that the surface of a 5% gold and 95% nickel alloy is completely enriched with gold when heated in vacuo; subsequent heating in oxygen produces a NiO surface. Quantitative theory of surface segregation, of the more volatile element in vacuo and of the more reactive in gas ambient was given by Meijering [5] and by Williams. [6] Burton, Hyman and Fedak [7] extended the treatment to the surface segregation in small particles.

The catalytic activity of alloys varies very rapidly with composition. In copper nickel alloys, for instance, Sinfelt and co-workers [8] observed a 1000 fold decrease in the rate of ethane hydrogenolysis when the copper content is increased from 1 to 5 per cent. In the fact of such variations, it is necessary to separate concentration effects from a modification of the chemical properties of surface atoms by the proximity of a different element. The theory of Burton et. al., for instance, was able to confirm Sinfelt's interpretation [8] in terms of surface segregation of copper.

We have noted earlier that we do not yet understand how catalytic activity for a given reaction varies from one material to another. In the case of elements and simple compounds, a catalogue based on experience can be established to guide the chemical engineer. Alloys, by contrast, represent a multidimensional continuum.

Catalysts are no longer selected, they are designed. Guidelines for such designs will have to be established and should be based on scientific understanding. Theories of chemisorption, assisted and tested by measurements of the electronic spectra of the bare and the adsorbate covered surface, will in due time provide us with the necessary insight. It should then be possible to identify those properties of solid surfaces relevant to catalysis. Parallel to these developments, experimental explorations of chemisorption and the simplest cases of catalysis by the methods of surface science are likely to uncover some unexpected facts. For instance, Somorjai and his coworkers discovered, [1] in their studies of stepped surfaces described above, that the bare platinum surface is catalytically inactive. Only when adsorption of hydrocarbons has allowed the formation of an ordered, thin, carbonaceous layer on platinum is catalytic activity observed.

Often a catalyst will lose its activity during use. The deactivation can be due to coalescence or sintering of the dispersed catalyst, to coking, where the surface is covered with fragments of the reactant or of the product, or it can be due to poisoning. Catalysis requires the adsorption of the reactants with strong enough forces to affect the chemical binding in the adsorbed molecules and make them more susceptible to change, it requires also that the heat of adsorption be weak enough for the reaction products to desorb readily. It becomes immediately apparent that a molecule that is strongly adsorbed on the active sites will in time stop the catalytic activity. These strongly adsorbed molecules can be impurities in the reactant or they can be reaction products, in which case the reaction is self poisoning.

Metal catalysts can be poisoned by nonmetallic species, of which oxygen, sulfur and nitrogen atoms (not N_2) are the most prominent, or by the adsorption of other metals. Poisons can decrease the overall activity or change the selectivity. By use of Auger spectroscopy, LEED and mass spectrometry, Bonzel and Ku[9] investigated the poisoning by H_2S of a platinum catalyst used for the oxidation of CO. They found that H_2S forms a strongly bound sulfur layer on platinum by dissociative adsorption. Sulfur-covered platinum does not adsorb CO. When a clean area of platinum surface is available for oxygen adsorption, the latter combines with the sulfur to form a weakly bound SO_2 which desorbs at low temperature. This result is in keeping with the experience on automotive exhaust catalysts where it is found that sulfur poisoning is no serious problem because the sulfur is burned off.

CONCLUSIONS

We have seen that the recent developments in the experimental techniques for the investigation of surfaces and in the electronic theory of surfaces and chemisorption quite naturally tend towards catalysis. We have also found that catalysts play an important role in the exploitation of our natural resources in energy since that depends so heavily on the chemical transformation of fossil fuels. New, as yet undiscovered catalysts will be needed. This confluence of circumstances presents the physicist with an opportunity to contribute significantly to society by doing fundamental research. Success, however, will depend on recognizing the limits of that contribution. The strength and limit of surface science resides in the fact that its methods and its results are entirely complementary to those of traditional catalysis research. Surface science will provide insights that were not accessible otherwise and will help make catalysis an exact science. The development of the novel catalysts needed in building the energy technology of tomorrow will be done mostly by the catalytic chemist. This development can be much accelerated by an efficient flow of ideas, back and forth, between the two disciplines.

ACKNOWLEDGMENTS

The author is indebted to his colleagues at the Esso Research and Engineering Company Corporate Research Laboratories for many illuminating discussions and excellent advice, particularly to R. B. Long, J. P. Longwell and F. J. Wright.

REFERENCES

Fields as active as surface science and catalysis have a very extensive literature. It would be impossible to give a correctly representative list of references of manageable length. References were restricted to the few examples explicitly described in the text.

1. G. A. Somorjai, Catalysis Reviews 7, 87 (1972).
   K. Baron, D. W. Blakely and G. A. Somorjai, Surface Science 41, 45 (1974).
2. J. J. Burton, Catalysis Reviews, in print.
3. M. R. Hoare and P. Pal, J. Crystal. Growth 17, 77 (1972).
4. F. L. Williams and M. Boudart, J. Catalysis, in press.
5. J. L. Meijering, Acta Metallurgica 14, 259 (1966).
6. F. L. Williams, 3rd North American Meeting of the Catalysis Soc., San Francisco, February 4-6, 1974.
7. J. J. Burton, E. Hyman and D. G. Fedak, 34th Physical Electronics Conference, Murray Hill, February 25-27, 1974.
8. J. H. Sinfelt, J. L. Carter and D. J. C. Yates, J. Catal. 24, 283 (1972).
9. H. P. Bonzel and R. Ku, J. Chem. Phys. 59, 1641 (1973).

TO DRIVE OR BREATHE?

P. S. Myers
University of Wisconsin, Madison, Wisconsin 53706

ABSTRACT

All types of energy systems for transportation produce pollutants. The different types of
energy systems available for transportation use
are reviewed and the types of pollutants each system puts out and the effects of that pollutant are
indicated.

The effect of pollutants on individuals is
via the respiratory tract. The characteristics of
the respiratory tract are discussed. The effect
of the individual pollutants on the respiratory
tract and on the individual are discussed. Data
on air quality standards for a particular pollutant are presented together with typical concentrations of pollutants found in urban atmospheres.
The effect of the different pollutants on individuals at different levels are presented.

The automobile is next looked at as a source
of pollutants and the different places where pollutants are formed and emitted are discussed. The
advantages and problems of devices that have been
used and are proposed for use to reduce these pollutants are discussed. A variety of energy sources
have been proposed for use in transportation. The
advantages and disadvantages and the emission characteristics of these different types of energy
sources are presented.

Emission controls have been accused of contributing to the energy shortage. The effect of present and proposed emission controls on fuel economy is discussed. Comparisons are made of changes
in fuel economy due to emission controls with
changes in fuel economy due to vehicle weight,
type of transmission, air conditioning, etc.
Trends of engines and fuels for the future as affected by the conflicting need for energy conservation and clean air are presented.

INTRODUCTION

Both air pollution and public concern with air pollution have increased dramatically in the last 10 years.
The increase in air pollution is a result of too many polluters[1] and too much pollution per person while the increased concern is the result of increased pollution,

increased affluence so that we can be concerned with pollution and the distracting but inevitable entry of politics
into the problem.

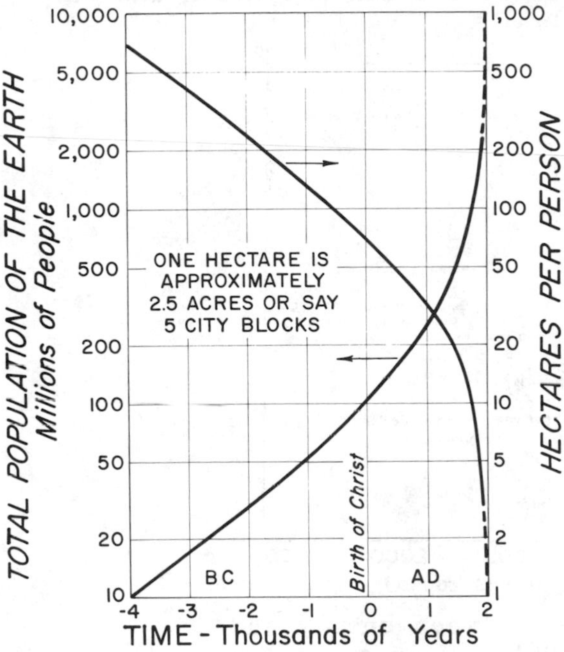

Figure 1 shows
the rapid increase
with time in the num-
ber of polluters.
While not completely
satisfactory probably
the best measure of
pollution per person
is energy consumption
per person. Conver-
sion of energy to use-
ful work produces both
air and thermal pollu-
tion. Use of this use-
ful work in producing
material things pro-
duces pollution during
the birth, life and
death of the product.
Figure 2 shows the in-
crease with time in in-
dividual energy con-
sumption in the United
States.

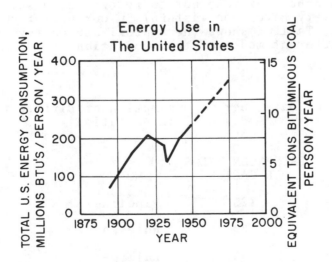

Fig. 1.   World Population growth

Fig. 2.   Energy use per capita in
          United States.

However, there is a relationship between individual energy consumption and material standard of living[2] (Fig. 3). Thus to a certain extent there is an inevitable conflict and necessary compromise between pollution and our material standard of living.

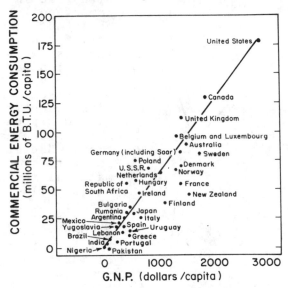

Fig. 3.   Relation between energy and
          GNP - 1961.

It would be a mistake not to recognize, however, that we as a society have been wasteful in our use of energy. For example, Table I shows the approximate energy consumption for different modes of transportation.

TABLE I

Approximate Energy Consumption of Different
Modes of Transportation

MODE	PASSENGER MILES PER GALLON	MODE	TON MILES PER GALLON
Buses	125	Pipeline	300
Railroads	80	Waterways	250
Automobiles	32	Railroad Trucks	200 58
Airplanes	14	Airways	4

It is no secret that the highest energy consumptive modes of transportation also show the most rapid rates of growth.

At the same time it should be clear that there is no pollution free energy converter. This fact is illustrated in Table II which shows the type of pollutants put out by different energy converters and the effect of these pollutants.

TABLE II

Pollutants Emitted by Different Powerplants

Energy Converter	Pollutant	Affect			
		Toxic	PC	Smog	Visibility
Spark Ignition	Lead Compounds	X			X
Diesel	Particulates	X			X
Gas Turbine	Carbon Monoxide	X			
Stirling	Nitric Oxide	X		X	X
Steam-Petroleum	Unburned Hydrocarbons			X	X
Electric-Coal	Sulfur Dioxide			X	X

In summary, even if our material standard of living does increase more slowly in the future the standard of living and thus the energy consumption of the rest of the world will increase which will inevitably give rise to more pollution. Thus our best judgment and efforts will be required to balance desired material standards of living for all people with desired low levels of pollutants.

TYPE AND EFFECT OF POLLUTANTS

As indicated in the introduction even with our best efforts there will be inevitable compromises between numbers of polluters, material standard of living of the polluters and the extent of pollution that exists. In making the inevitable compromises between these factors a rudimentary understanding of the health effects of the pollutants is necessary. It should be understood that there are also material and aesthetic effects due to pollutants but space limits dictate that we concern ourselves here only with health effects.

Air pollution primarily affects the human respiratory system[3]. The main function of the respiratory system is to supply oxygen to the blood and take away the carbon dioxide resulting from the food oxidation process - all through the same intake and exhaust pipe! Air first enters the nose (or the mouth) where some 2-3% of the basic metabolism (idling fuel rate in engine terminology) is spent in humidifying and conditioning the air. From here the air goes to the lungs which have the characteristics shown in Table III.

TABLE III

Lung Characteristics

Segment	Diameter in^2	Flow Area in^2	Transfer Area in^2
Trachea	0.75	0.40	
Bronchus	0.50	0.36	
Lobarbronchi	0.30	0.33	
Bronchioles	0.04	4.50	
Alveoli	0.016	1830	10^5

As can be seen in Table III the diameter of the airways decrease as branching occurs with an increasing cross-sectional area for flow. Gas exchange takes place in the alveoli which have a very large gas transfer area - estimated as approximately the area of a tennis court! This gas transfer area is much in excess of resting needs but is used during heavy exertion. Gas exchange takes place across the thin membrane of the alveoli. Thickening or scarring of this membrane or restriction or blockage of air passages due to age, disease, pollutants, irritants, etc., interfere with and decrease the rate of gas exchange.

The trachea and larger bronchi have cells on their surfaces with numerous projections called cilia. Coordinated wavelike motion of these cilia pump fluid and particulate matter to a point in the respiratory tract where it can be coughed up and expectorated or swallowed.

Speaking of particulate matter there is a selective deposition of particulate matter in the respiratory tract according to size as shown in Figure 4. Note that there is a minimum particle deposition size in the respiratory system around one-half micron median diameter.

Let us look at the medical effects of coming from power plants as shown in Table II. Table IV shows selected medical data for sulfur dioxide as well as primary air quality standards.

TABLE IV

Health Effect Data for SO_2

Air Quality Standard PPM	Time	Urban Concentrations PPM
0.03	Annual Mean	.01-.14
0.14	25 hr.	.1-.6

MEDICAL EFFECTS

PPM	Time	Symptoms
.8-1.0	Short	Odor perception
1-2	Short	Sensitive persons
5	Short	General response
10-15	1 hr.	Slow mucus remov- and increased airway resistance

It appears that SO_2 (and H_2SO_4 even more so) has three effects on the respiratory system. First of all, it serves as an irritant which over long periods of time, has potential for affecting the membrane and thus the gas exchange process. Secondly, it produces bronchoconstriction which increases airway resistance. This probably tends to confine SO_2 to the upper sections of the lungs particularly at higher SO_2 concentrations. Thirdly, it seems to decrease the pumping action of the cilia and thus slow down the removal of fluids and particulate material from the lungs.

Table V shows similar data for particulates.

TABLE V

Health Effect Data For Particulates

Air Quality Standard Micrograms/cubic meter	Time	Urban Concentration $\mu g/m^3$
75	annual mean	60-200
260	24 hr.	200-1400

MEDICAL EFFECTS

1. Intrinsically toxic
2. Obstructs respiratory clearance
3. Carries absorbed toxic substances

$\mu g/m^3$ (Plus SO_2) Annual Mean	Symptom
100-200	Respiratory illness
300-600	Bronchitic patients
2000	London episode (1962)

The chemical composition of particulates is ill-defined
(if at all) since they include a wide variety of substances. These substances may be toxic in their own
right including carcinogenic effects. In addition, their
place of deposition in the respiratory system depends upon
their size (Fig. 4). Furthermore, as particulates are
cleared from the respiratory system, they may be introduced into the gastro-intestinal tract via removal from
the respiratory system and subsequent swallowing; they
may go through the tissue into the blood stream; or
they may be retained in the respiratory system.
   Particulates may be toxic in their own right (asbes-
tos, lead, etc). Note that these toxic effects are not
necessarily limited to the respiratory system because of
the removal processes. In addition, particulate matter
may interfere with one or more of the clearance mechanisms
in the respiratory tract. There may also be a synergistic
effect between particulates and other pollutants since an
absorbed pollutant gas may be carried by the particulate
to regions of the respiratory tract not normally reached

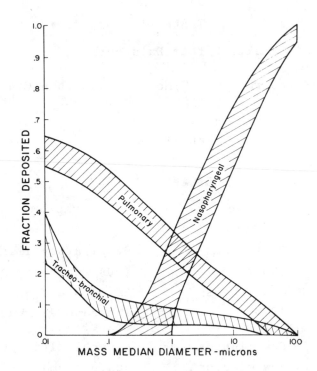

Fig. 4. Particulate deposition in lungs
according to particle size.

by the pollution. A good example of this synergistic
effect is SO_2. Normally, via absorption of liquids
and bronchio construction, SO_2 tends to be confined to
the upper portion of the lungs . However, SO_2 may be
carried to the alveolar regions when absorbed on very
small size particulate matter . The combination of SO_2,
particulate material and fog has proved to be quite
lethal and is commonly called smog .
    Photochemical smog (as opposed to smog) results
from a combination of sunlight, NO and hydrocarbons and
produces NO_2 oxidants. NO is formed in high temperature
combustion but, via atmospheric processes, (particularly
during photochemical smog) is converted to NO_2. No case
of human poisoning due to NO has been reported and it is
not a strong irritant for the respiratory surfaces although there is some evidence to indicate it affects the
oxygen carrying ability of the blood. However, NO_2 is
very reactive biologically with the effects shown in
Table VI.

TABLE VI

Health Effect Data For NO_2

Air Quality Standard PPM	Time	Urban Concentration PPM
.005	Annual Mean	.02-0.5

MEDICAL EFFECTS

PPM	Time	Symptom
.06-.1	Continuous	Respiratory illness
.4	Short	100% recognize odor
.8	Continuous	Affects alveoli of rats
5	10 min.	Increase air way resistance
150-200	Few minutes	Bronchiolitis obliterano
500	Few minutes	Acute edema

In general, exposure to low concentration of NO_2 seems to increase susceptibility to respiratory diseases. However, the Chattanooga experiments on susceptibility to respiratory disc1ses on which the NO_2 air quality standard were based are extremely suspect. Consequently, the whole question of low level affects of NO_2 is currently "up in the air". High concentrations of NO_2 produce muscular spasms of the respiratory tract, severe coughing, accumulation of fluid in tissue and delayed preliminary edema (accumulation of fluid in the tissue).

Hydrocarbons are not directly toxic except at very high concentrations or from a carcinogenic standpoint. They do however combine with sunlight and NO_2 to form photochemical smog. During photochemical smog ozone, oxidants (mixture of substances produced by photochemical reactions) and peroxyacl nitrates are produced with the predominant compound being ozone. All these compounds seem to be biologically active and therefore are of interest. However Table VII is listed as hydrocarbons (Ozone) since ozone is the predominant oxidant and hydrocarbons are controlled to control oxidants. The effects shown are for ozone.

TABLE VII

Health Effect Data For Hydrocarbons (Ozone)

Air Quality Standard PPM	Time	Urban Concentration
.08	1 hr.	.1-.6

MEDICAL EFFECTS

PPM	Time	Symptom
.01-.02	Short	Odor perception
.2	Long	No apparent effect
.30	Moderate	Threshold nasal, throat irritation
.50	Intermittent	Changed forced expiratory volume
.5-1.0	1-2 hours	Increased airway resistance
1-3	? hours	Fatigue, coordination loss
9	Short	Severe edema

At low levels ozone seems to irritate the respiration tissues which has long term implications. Higher concentrations seem to increasingly cause fluid accumulation in tissue (edema), increased airway resistance and other changes in pulmonary functions. Even higher concentrations produce severe fatigue, lack of coordination, and severe pulmonary edema.

Carbon monoxide effects the ability of the blood to deliver oxygen to the tissue rather than serving as an irritant to the respiratory system. After oxygen passes through the alveolar membrane it is picked up by the hemoglobin in the blood with each hemoglobin molecule having 4 oxygen-carrying sites. The hemoglobin molecules have a lifetime of around 120 days and form CO as they break down. This CO is carried back to the lungs by the hemoglobin on the same sites as the oxygen and expelled along with the CO_2 by partial pressure differences. Thus expelled air during resting will have a CO concentration of 6-8 PPM and this concentration will increase during exercise. Presumably because of the low concentration of CO in the blood the affinity of the CO for the hemoglobin is some 200-250 times that of O_2. Thus, when air containing CO is taken into the lungs, some of the CO is attached to the oxygen carrying sites on the hemoglobin thus decreasing the oxygen carrying ability of the blood. Also it is found that if one of the 4 oxygen-carrying sites is occupied by a CO molecule it is harder for the remaining oxygen molecules to get off the hemoglobin molecules when they reach the tissue (Haldane effect). The percentage of total available sites occupied by CO rather than O_2 is called percent carboxy hemoglobin (COHb).

Table VIII shows the effects of different concentrations of CO in the air supplied to the respiratory system.

TABLE VIII

Health Effect Data For Carbon Monoxide

Air Quality Standard PPM	Time	Urban Concentration
9*	8	15-40
35*	1	20-50

MEDICAL EFFECTS

PPM	COHb**	Symptom
15	3	None (controversial)
30	5	None (controversial)
10-35	2-6	Level of smoker
100	14	Headache
500	45	Collapse
1000	67	Death

*Exceeded only once per year.

**Equilibrium values.

It must be recognized that the body has a long time constant for CO, i.e., when a person is put in an atmosphere
containing CO at low to moderate levels it takes hours for
COHb to come to equilibrium values with the exact time dependent upon CO and exercise levels. The process seems to
be reversible. The symptoms at low concentrations are
listed as none in Table VII but as noted this is controversial. It should also be noted that lowered oxygen availability can be offset by increased blood flow if the cir-
culatory system has the capability of providing increased
flow. Thus the effect of CO is increasingly important if
the cardiac system is marginal.

In summary, at the concentrations normally found in
the atmosphere there is no immediate effect of pollutants
on healthy people but there is an effect on people in
marginal health. There is evidence and feeling among medical people that there is a long term effect, i.e., scarring
and thickening of the alveolar membrane and blockage of
passages which decreases the capacity of the respiratory
system so that over a period of years its capacity may become marginal. If the respiratory system is already marginal due to defects or disease there are immediate effects
even at low levels. There have been well publicized, short
term, high atmospheric concentration incidents such as the
1956 London smog which in a few days killed some 4,000
people in marginal health. These incidents cause concern
and demand action.

It seems that the capability of an industrial society
for better sanitation, sophisticated health care, etc.,
must be weighed against the negative effects of the pollutants resulting from this industrialization. The first
step in this "weighing" process is to consider methods of
reducing pollutants. In so doing we will need to consider
costs of all kind but we must proceed intelligently and
thoughtfully. For example, Fig. 5 shows that transportation (primarily the automobile) on a mass basis produced
some 40% of the man made air pollutants in the United
States in 1968.[5] However it does not seem logical to
equate a pound of sugar and a pound of strychnine. Accordingly if one weighs the pollutants by air quality
criteria a different picture of the importance of transportation (primarily automobile) pollution is obtained.
Fig. 6 shows that on an "effect" basis transportation
accounts for only 10-15% of air pollution rather than the
40% indicated in Fig. 5. However in many cases (Southern
California for example) the automobile is a major contributor even on an effect basis so let us examine the way in
which automobiles produce pollution and what can be done
about it.

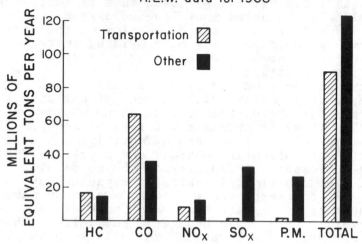

Fig. 5.   U. S. Air pollution on a
weight basis.

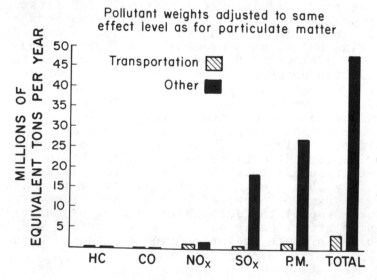

Fig. 6.   U. S. air pollution on relative effect basis.

THE AUTOMOBILE AND AIR POLLUTION

We have seen that on a mass basis the automobile contributes significantly to air pollution. The sources of
this pollution in the automobile are shown in Fig. 7.

Fig. 7. Sources of pollution from an
automobile.

The first step (1961 in California, 1964 nationwide)
in minimizing pollution from the automobile was to install
a positive crankcase ventilation (PCV) system as shown
schematically in Fig. 8 to treat the crankcase gases (predominately hydrocarbons) resulting from gases "blowing by"
the piston. As shown in Fig. 8 the vacuum in the intake
manifold is used to route the crankcase gases to the engine
combustion chamber where they are burned. Because the

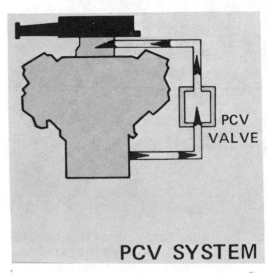

Fig. 8. Schematic drawing of a
PCV system.

intake manifold vacuum varies from nearly zero to as much
as 25" HG a variable area valve (called PCV valve) is used
to control the flow. If the PCV valve operates properly
the system is quite effective. Unfortunately, most people
do not clean or service the PCV valve in their car (inci-
dentally a very simple cheap operation). Thus a high percentage of PCV valves on older cars do not function properly.

Vaporizing fuel (again predominantly hydrocarbons) was
controlled nationally starting in 1972. One system used is
shown schematically in Fig. 9. While the car is stationary
fuel vapors from the carburetor and fuel tank are absorbed
by charcoal in a charcoal canister. When the engine is
started air is automatically drawn over the charcoal with
the purged vapors taken into the engine combustion chamber
to be burned. Thus the charcoal is ready for use again
when the engine is stopped. Such a system seems to be
roughly 75% effective and is being slowly improved with
additional development.

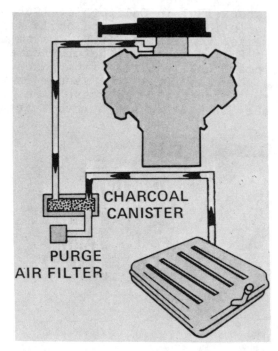

Fig. 9.   Schematic drawing of fuel
          vapor collecting system.

The pollutants appearing in the exhaust gas are formed in the combustion chamber of the engine as shown schematically in Fig. 10. Combustion is initiated at the spark plug with a flame spreading approximately radially outwards.

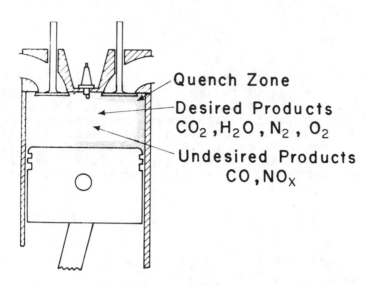

Quench Zone

Desired Products
CO_2, H_2O, N_2, O_2

Undesired Products
CO, NO_x

Fig. 10. Formation of combustion chamber pollutants.

However, because of water-cooled walls, the flame lacks a few thousandths of an inch of burning all the way to the wall. The net result is a "quench" layer of unburned or partially-burned fuel which is scraped off the wall during the exhaust stroke and pushed out with the exhaust gases by the piston. Some of the hydrocarbons are burnt up in the exhaust system. It should be clear that the quantities we are talking about are insignificant from an energy or engine efficiency standpoint but not from a pollution standpoint when large numbers of cars are present.

Carbon monoxide and oxides of nitrogen (primarily NO but designated as NO_x) are formed in the extremely hot (2700-2800 K) bulk gases. Carbon monoxide concentrations are primarily related to oxygen deficiency although there is some effect due to slow reaction rates. NO is not thermodynamically stable at room temperatures and should decompose during expansion. Practically speaking, its concentration in the exhaust gas is very heavily determined by formation and destruction reaction rates. It is generally accepted that the most important reactions are the so-called extended Zeldovich reactions given in Table IX.

TABLE IX

Extended Zeldovich Reactions

$$O + N_2 \quad \rightleftarrows \quad NO + N \qquad (1)$$

$$N + O_2 \quad \rightleftarrows \quad NO + O \qquad (2)$$

$$N + OH \quad \rightleftarrows \quad NO + H \qquad (3)$$

Fig. 11 shows schematically NO concentrations versus time in an engine. If equilibrium existed, very high concentrations (due to high combustion temperatures) would be reached during combustion with lower concentrations occurring as temperatures dropped during expansion. In real life, due to slow chemical reaction rates, high equilibrium concentrations are not reached but conversely essentially little or no destruction takes place during expansion.

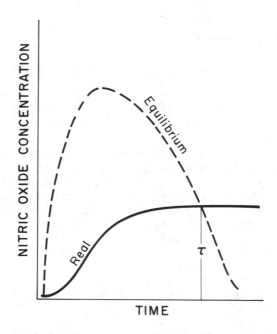

Fig. 11. Nitric oxide formation in Hypothetical combustion process.

It should also be mentioned that evidence shows that if the thought experiment of a stationary flame front (Fig. 12) were run and NO concentrations measured at

different distances from the flame front and extrapolated
to the flame front the extrapolated value at the temperature
discontinuity would not be zero. This only partially understood phenomena is known as instant NO.

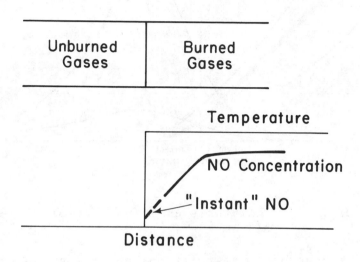

Fig. 12.   Illustration of formation
of instant NO.

Engineers working with engines tend to correlate
pollutant concentrations in the exhaust terms of air-fuel
ratios. Fig. 13 shows schematically the concentration of
the different exhaust pollutants as a function of air-fuel
ratio. Both hydrocarbons and NO_x are a function of other
engine parameters such as spark timing, load, etc., and
consequently are shown as bands. CO is affected only
slightly by these other engine parameters and consequently
is shown as a line. Note that the automotive engineer is
in a dilemna. If the mixture supplied to the engine is
rich (excess fuel) hydrocarbons and CO are produced. If
the mixture is moderately lean NO_x may be maximized. He
is further constrained by the fact that if lean mixtures
are used there may be a tendency for the engine to surge
(speed up and slow down without throttle change) as well
as to falter and stumble, i.e., driveability is not good.
Since surge is not normally so much of a problem on the
rich side most cars are operated slightly on the rich
side of a stoichiometric mixture.

88

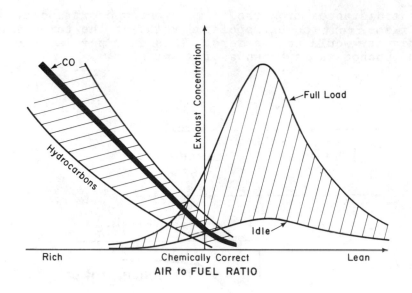

Fig. 13. Pollutants as a function of
A/F ratio.

Initial government restrictions on exhaust pollutants
required reduction of hydrocarbons and CO but not of NO_x.
Consequently, two control techniques were initially used.
The first was to use leaner mixtures which of course resulted in more NO_x. The second technique was to burn up
CO and hydrocarbons in the exhaust as much as practical by
conserving heat in the exhaust but primarily by making the
engine less efficient. By the First Law the energy in the
fuel must appear as work or in the cooling water or exhaust.
Thus less work out means a hotter exhaust and better opportunities for oxidation of fuel in the exhaust but of course
poorer gasoline mileage.

Reductions in NO_x were required in 1973. Exhaust gas
recirculation (EGR) was used as shown schematically in Fig.
14 as well as late spark when the transmission was in low
gears. The basic idea of EGR is to add inert exhaust gases
(no oxygen) to the air-fuel mixture to lower the combustion
temperature and thus decrease NO_x. Later spark timing
gives lower combustion temperatures which gives lower NO_x.
However there is an adverse effect on fuel economy.

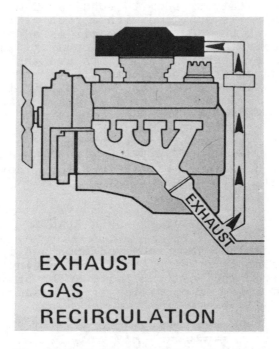

Fig 14. Schematic illustration
of EGR.

It appears that the original very severe 1975 regulations for hydrocarbons and CO and the 1976 regulations
for NO_x (both regulations have been postponed for one year
and interim regulations established) can only be met using
a dual catalytic muffler as shown in Fig. 15. Because

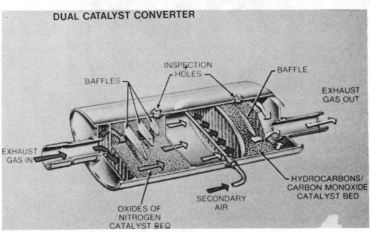

Fig. 15. Dual catalytic muffler.

NO$_x$ reducing catalysts seem to require CO or H$_2$ for operation the engine is supplied with a rich mixture. After passing through the NO$_x$ catalyst air is added and the CO and hydrocarbons destroyed by a second oxidizing catalyst .

In summary, exhaust pollutants are the most difficult pollutant source from the automobile to treat and, of the exhaust pollutants, NO$_x$ is the most difficult to treat . The original 1975 and 1976 regulations required approximately a 98% reduction in hydrocarbons and CO and approximately a 96% reduction in NO$_x^!$. Meeting these emission standards with reasonable performance, life and fuel economy is proving extremely difficult for the conventional spark ignition engine . Let us therefore look at alternative power plants .

ALTERNATIVE POWER PLANTS

Rotary engines have received much publicity in the press in recent years with the Wankel engine (Fig .16) having received the most publicity . The basic thermodynamic principles of operation are the same for the

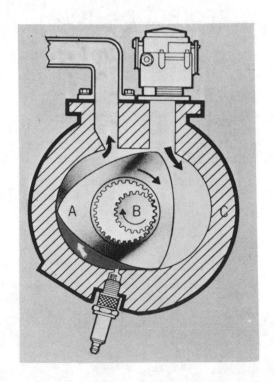

Fig. 16 . Schematic drawing of
Wankel engine .

rotary as for the reciprocating engine, i.e., take in an air-fuel mixture, compress it, burn it, expand it and push the resulting products out the exhaust. Its disadvantages are first a high surface-to-volume ratio in the combustion zone. This means a large quench volume as well as relatively high heat losses with correspondingly low fuel economy. This situation becomes even more pronounced as compression ratio is increased. Leakage is a problem particularly at the apex (corner of the triangular rotor) seal where the seal makes varying angles with respect to the sealing surface. Leakage at the apex seal can send unburned fuel to the gases being exhausted and unburned gases and products of combustion to the mixture being compressed. The products are beneficial in the sense that it is a form of internal exhaust gas recirculation. On the plus side the Wankel has a volume and weight about one-half that of the conventional engine and all of the exhaust gases from a single rotor come out the same short port with correspondingly low heat losses. As a result the exhaust gases are hotter and easier to treat. With increasing emphasis on fuel economy the future of the Wankel is not quite as bright as it looked a year ago since basically its exhaust pollutants are comparable to those from the conventional engine.

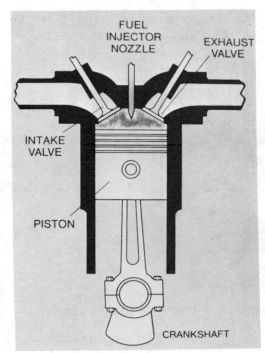

Fig. 17. Schematic drawing of diesel engine.

The diesel (compression ignition) engine (Fig. 17) is the proven work horse for commercial vehicles because of its outstanding fuel economy and long life. From an emissions standpoint it is low on hydrocarbons and CO - in fact, a diesel can probably meet the original 1975 standards for these two compounds without exhaust gas treatment. It is better than the spark-ignition engine from an NO_x standpoint but not good enough to meet the original 1976 standards. It does suffer from smoke, halitosis, overweight and high initial cost. There are some experimental diesel combustion chambers which produce minimum odor; smoke is being improved; and, as the output of spark ignition engines decreases

and fuel consumption increases, the outstanding diesel
engine fuel economy may more than offset its weight problems particularly if NO_x emission standards are relaxed
as EPA has recommended.

In a stratified charge engine the gases at the time
of combustion are not homogeneous - usually there is an
air-fuel zone and an air zone in the combustion chamber.
The diesel engine is a good example of a stratified charge
engine with the stratified charge in the diesel being
formed via the injected fuel. If, in addition to using
fuel injection, ignition is initiated by spark the engine
is often referred to as an hybrid engine. Two examples
of hybrid engines are shown in Figs. 18 and 19. These
particular engines are of more than casual interest because of the large amount of development work done on
them and because experimental versions of these engines
in Jeeps have met the 1975-76 emission standards when
using EGR and a catalytic muffler. In Fig. 18

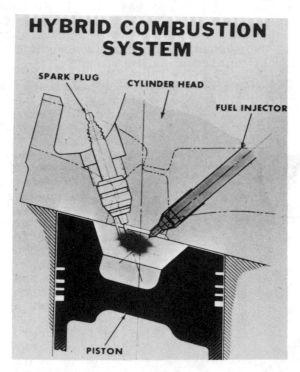

Fig. 18. Schematic drawing of Ford Proco Engine.

the air-fuel mixture is formed by the injector during compression with the volume of the air-fuel mixture increasing as load increases and ignition occurring by spark. In
Fig. 19 the air in the cylinder is given a swirl (rotary)

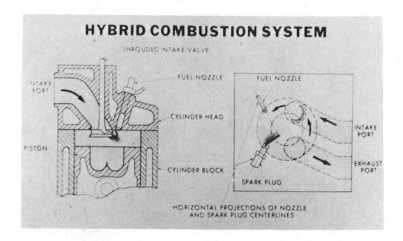

Fig. 19. Schematic drawing of TCP engine.

motion by the intake port; a stream of fuel is injected
parallel to this air motion to form a semi-circular ribbon of air-fuel mixture; ignition occurs via a spark; and
the length of the ribbon is increased as the load increases.
This arrangement is capable of burning a wide variety of
fuels with good fuel economy and it has the potential for
higher power output via supercharging. On the other hand,
particularly on a mass production basis, it is very difficult to keep the swirl rate and fuel injection matched
for all cylinders at all speeds and loads. Exhaust emissions are roughly comparable to those from a diesel.
     The simple Stirling cycle is illustrated in Fig. 20.
The practical Stirling cycle includes internal heat exchange for improved fuel economy but this is unimportant
for present purposes. As can be seen in Fig. 20 the engine uses the same working fluid (hydrogen or helium)
over and over again while heat is supplied to the cylinder via a steady flow external combustor. In a steady
flow combustor the quench zone is not scrapped off by a
piston so unburned hydrocarbons can be made low. CO is
low because some excess air can be supplied. NO_x, however, is comparable with a diesel engine. The Stirling
engine is competitive with the diesel engine from a fuel
economy standpoint but has some load control problems
under highly variable conditions since load control is
achieved by increasing or decreasing the average pressure
in the cylinder. Since the Stirling engine rejects only
a small amount of heat via the exhaust it requires a
radiator about twice the size of a conventional car.

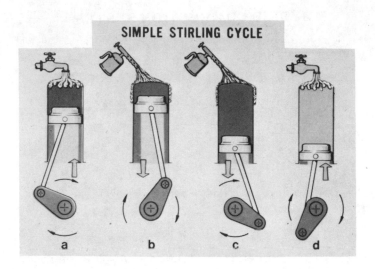

SIMPLE STIRLING CYCLE

a   b   c   d

Fig. 20.   Schematic of simple Stirling engine.

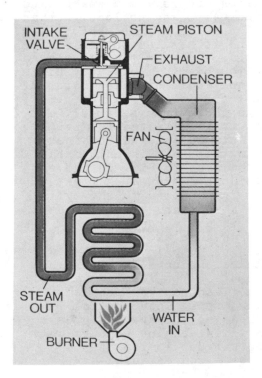

INTAKE VALVE
STEAM PISTON
EXHAUST
CONDENSER
FAN
STEAM OUT
WATER IN
BURNER

Fig. 21.   Schematic Diagram of steam engine .

The steam (vapor) engine (Fig. 21) like the Stirling engine uses the same working fluid over and over and has a continuous flow combustor. Water as a working fluid has a tendency to freeze in winter and needs high pressures and turbine speeds to even approach acceptable fuel economy. Higher molecular weight fluids (Freon, for example) have been suggested but they limit efficiency because they typically tend to decompose at temperatures above 300°C or so. Because of lower thermal efficiency and because of low heat rejection via exhaust gases the radiator must be even larger for the vapor engine than that required for the Stirling engine. Exhaust emissions are comparable or better than those from the Stirling engine.

The gas turbine (Fig. 22) is well known because of its use in jet engines. For vehicular uses an internal regenerator (shown in Fig. 22) is necessary from a fuel

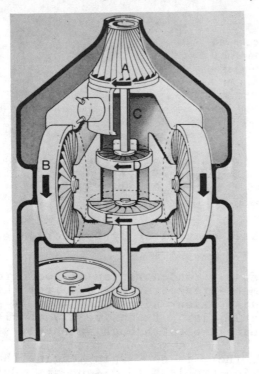

Fig. 22.   Schematic diagram of regenerative gas turbine.

economy standpoint but fuel economy is not really competitive even with the regenerator. The regenerator has been the source of a host of problems including leakage. Because the gas turbine uses steady flow combustion, CO and unburned hydrocarbons are low but NO_x is on a par with the diesel. First cost is judged to be high.

Although given much publicity several years ago, electric cars (Fig. 23) are now evaluated more realistically recognizing that a large battery volume is required to store the necessary energy. One must face the fact that hydrocarbons are a very efficient way to store chemical energy. In an electric car emissions come from and depend on the type of powerplant used, i.e., radioactive wastes if atomic energy is used and NO_x, particulates and SO_2 if coal is used. If coal is used there is not much difference in mass of pollutants emitted per mile traveled but their is a difference in type of pollutants (Table II) and the source of pollutants is concentrated

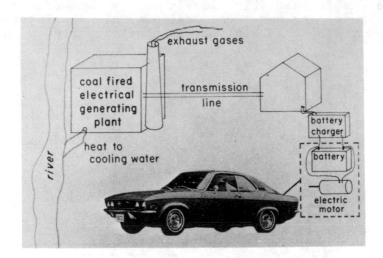

Fig. 23.   Schematic diagram of system for an electric car.

rather than diffuse. Concentration is good from an exhaust gas treatment standpoint but bad from the standpoint
of atmospheric dispersion and use of nature's pollutant
destruction capabilities. Extensive use of electric vehicles will require a breakthrough in energy storage batteries.

In summary, there is no "magic" engine that will
cause our pollution problems to disappear. Our only pollution solution lies in steady continued research and engineering development aimed at low emissions with good
fuel economy and performance. This solution will have to
be assisted by increasing use of less energy consumptive
modes of transportation including lighter cars, mass transportation, fewer trips and other energy conserving measures .

FUEL ECONOMY AND EMISSIONS

During the lifetime of the automobile energy has
been cheap and abundant . Recent political developments
plus ecological concerns have hastened the advent of increasing cost and shortage of energy. Since fuel economy must be among the compromises to be made and is much
in the news now let us take a short look at factors affecting the fuel economy of the automobile.

The two largest single factors affecting fuel economy are driving cycle and car size - predominantly weight.
Fig. 24 shows steady-speed fuel economy for three differ-

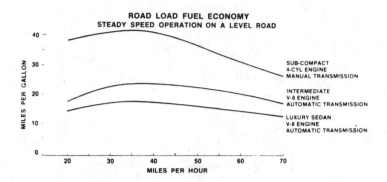

Fig. 24
Road load fuel economy for three different size cars.

ent size cars. Note the effect of speed on fuel economy
particularly on the smaller size car. Fig. 25 illustrates additional aspects of the relationship between
driving cycle and fuel economy. The 40 and 70 mph fuel
consumption points from Fig. 24 are shown in Fig. 25 as
well as data from an urban driving cycle both when the
car was hot and cold. This already wide range of fuel

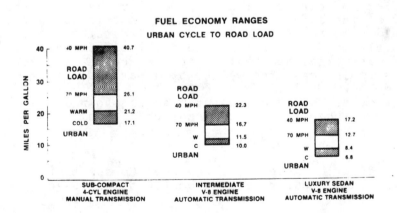

Fig. 25. Effect of driving conditions on fuel economy.

economy can be further increased by individual driving ha-
bits, i.e., rapid acceleration and deceleration.
    The effect of car weight is shown in Fig. 26 where
data taken by EPA during the driving cycle used for emis-

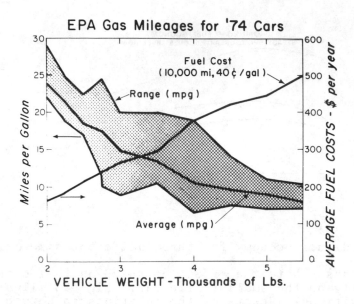

Fig. 26.   Effect of car weight on fuel economy using
          emission test cycle

sion testing are presented. Note that the emission driving cycle is a highly urban cycle as well as the considerable effect of car weight and the wide variation in fuel
economy at the same vehicle weight. In this connection
it should be noted that recent so-called safety legislation involving 5 mph crash without vehicle damage has increased car weight and that there has been a tendency for
stylists to increase the weight of a particular car model
with time.
        One other aspect of the problem is shown in Table X
which shows the effect of accessories on fuel economy.

TABLE X

Effect of Accessories on Fuel Consumption
(Intermediate car with automatic transmission)

	Urban Cycle	70 mph
Air Conditioning	1.5 mpg	1.0 mpg
Alternator	0.9 mpg	0.5 mpg
Fan	0.1 mpg	0.5 mpg
Power Steering	0.1 mpg	0.4 mpg

As indicated previously many of the engines changes
made to reduce exhaust emissions were intended to increase
exhaust temperature with the undesirable effect of increasing fuel consumption. Because engine changes, weight
changes, etc., were all being made over the same period of
time it is difficult to sort out individual affects on fuel
economy. Also, as we have previously seen, the cycle used
for testing affects fuel economy markedly. Two different
estimates made by two different persons of fuel economy as
a function of time are shown in Fig. 27. Ford has estimated
the decrease in fuel consumption when driving a city-subur-
ban cycle by using a regression analysis to give the effect
of emissions only with the results as shown in Table XI

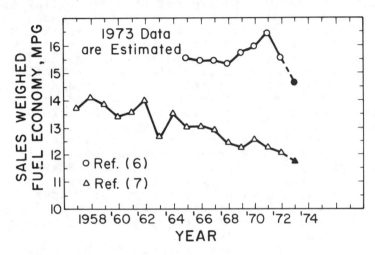

Fig. 27. Two different estimates of
fuel economy.

TABLE XI

Effect of Emission Controls on Fuel Economy

Year	Fuel Economy Change
1967	0
1968	0
1969	+2.2
1970	-1.3
1971	-4.5
1972	-7.6
1973	13.6

General Motors has talked about a 13% improvement in fuel economy in 1974 presumably due to use of catalytic mufflers but a portion of this gain may be due to a higher proportion of smaller cars . The best guess is that emission controls have directly caused a 10-15% increase in fuel consumption .

SUMMARY AND CONCLUSIONS

In summary, it seems clear that our best and most intelligent efforts are going to be needed to reduce energy consumption and air pollution while at the same time maintaining an acceptable standard of living . On the short term basis air pollutants are not harming healthy people although they can clearly be shown to have caused discomfort and death to people having marginal health . On a long term basis there are enough health warning signs that we must have concern with controlling air pollution .

With regard to the automobile, major reductions have been made in reducing carburetor, gas tank and crankcase pollutants; good progress in reducing CO and hydrocarbons from the exhaust; and slow progress in the more difficult problem of reducing NO_x from the exhaust . There is no easily distinguished "winner" among the alternatives to the spark ignition engine to power your automobile . The stratified charge engine seems to have the most advantages and least disadvantages except for the development required. Emissions are a significant factor in fuel economy but only one of many. Lighter and smaller cars seem desirable and inevitable in terms of energy, pollution reduced family size and traffic problems .

Reasonable compromises are needed . We cannot return to 1492 air and water purity unless we return to 1492 population, industrialization and life expectancy levels. However, equally clearly, we are locally overloading our "free" waste disposal system and must take steps accordingly. It will take steady continuing effort on all our parts to retain the best of industrialization and ecology. While some rhetorical excesses on all sides may be useful in emphasizing problems and solutions they can, if excessive, polarize people and harden positions so that the inevitable and necessary compromises are more difficult to achieve .

REFERENCES

1. P. S. Myers, SAE Paper 700182.

2. Lamont Eltinge, SAE Paper 730709.

3. Medical Effects of Air Pollution, SAE Papers 710297, 710298, 710299, 710300, 710301 and 710302.

4. "Conference on Health Effects of Air Pollution: Summary of Proceedings", National Academy of Sciences, Washington, D.C.

5. H. F. Barr, SAE Paper 700248.

6. Richard A. Place, Statement of Ford Motor Company, Submitted to the House Select Committee on Small Business on Motor Fuel Economy, July 17, 1973.

7. T. C. Austin and K. H. Hellman, SAE Paper 730790.

COMBUSTION MODELLING IN AUTOMOTIVE ENGINES

John B. Heywood
Massachusetts Institute of Technology, Cambridge, Ma. 02139

ABSTRACT

Before the emissions and energy crises were upon us, it was enough that automotive engines worked. We could develop in a leisurely way our basic understanding of engine combustion problems. It is now obvious that the current automobile engine's problems of high emissions and deteriorating fuel economy must be rapidly solved. It is likely that new engine concepts must be developed to satisfy the demand for improved air quality and better fuel utilization. To solve these problems effectively, we need a better analytic framework for our applied research and development efforts. Combustion modelling can provide this framework by improving our understanding of the engine combustion process, and then by developing predictive techniques which relate engine design and operating parameters to emissions and performance characteristics. This paper describes two engine combustion problems in some detail to demonstrate both the potential value of such modelling efforts, as well as the substantial gaps that exist in our current understanding of basic engine phenomena. The first problem is nitric oxide formation in conventional spark-ignition engines. The kinetic mechanism responsible for nitric oxide formation, and the coupling of this kinetic mechanism with the thermodynamics of the burnt gases during the combustion process are reasonably understood. The trade-off between engine performance and emissions can now be quantified sufficiently accurately for it to be useful to the engine designer. The second problem discussed is the combustion of a fuel spray injected into the engine cylinder as occurs in stratified charge or diesel engines. A complex sequence of processes—atomization, fuel droplet vaporization, fuel air mixing, ignition, flame propagation through a nonuniform mixture—must all occur before the fuel is fully burned. While we can currently quantify some of these processes in simpler geometries, our understanding of real engine processes is poor. Yet these types of engines probably offer the best fuel economy at given emission levels for automotive applications. Combustion modelling which leads to a better understanding of the details of these combustion processes will greatly assist the engine developer.

INTRODUCTION

What is meant by combustion modelling in this context? I mean the prediction of performance and emissions characteristics of automotive engines using computational procedures developed from a fundamental analysis of the combustion process. As physicists, long accustomed to basic research, you may well ask: what is so special about that? The answer is that only recently has this type of combustion modelling become a significant contributor to the practical engine development process. There are several reasons for the limited historical impact of basic combustion research. First, the combustion process in engines is unsteady, turbulent and three dimensional. Often the real physical and chemical processes which are important can only be modelled in much simpler geometries, and extrapolation of the results of such studies to the real situation has proved difficult. A second factor which has limited the application of combustion models is the wide range of conditions over which an automobile engine must operate. Normal driving covers the range from engine idle at low engine speed to wide open throttle at high engine speed. The relative importance of different combustion processes changes over this operating range; many different cases must be examined.

Perhaps the most important reason, however, is that until emissions and fuel limitations imposed significant additional constraints on the engine designer the need to understand the finer details of the engine combustion process was not that great. Automobile engine technology had evolved over many years. A vast background of experience on the effect of changes in engine design and operating parameters on performance had been accumulated. The final successful integration of the engine into a vehicle has been and will still inevitably be the result of an empirical optimization process.

The competing requirements of effective emission control, and improved fuel economy, have now changed this balance. Engine performance characteristics are not greatly dependent on the finer details of the combustion process. The emission characteristics are, however, as is obvious from the fact that significant emission control in the conventional spark-ignition engine has already been achieved through relatively minor adjustments in engine design and operation. Additionally, several new spark-ignition engine concepts appear to offer better emission control and fuel economy than the current engine, for the 1980's (1). The combustion processes in these new engines are even more complicated than in today's engine, and a better analytic framework will greatly assist orderly engine development programs. There is thus a need for a much better understanding of the automotive engine combustion process, the first important goal of combustion modelling. Such an understanding can then be used to relate engine performance and emissions characteristics to design and operating variables, the second important goal.

Let me show you the potential of combustion modelling by

describing two problems. One problem, the oxides of nitrogen (NO_x) emissions - fuel economy trade-off in a conventional spark-ignition engine is quite well understood, and an analysis of the combustion process inside the engine can yield useful quantitative results. The other problem, the combustion of a fuel spray inside the engine cylinder (as would occur in stratified charge engines or in diesel engines) is more complex and our understanding of the basic physical and chemical processes is limited. I will use this example to explain which processes are important, and indicate why our understanding of these processes needs to be greatly extended.

THE NO_x EMISSIONS - FUEL ECONOMY TRADE-OFF

Figure 1 shows a schematic of a conventional reciprocating 4-stroke cycle spark-ignition engine cylinder during the combustion, expansion and exhaust processes. The pollutant formation mechanisms are summarized. At the time of the spark, the fuel and air which were essentially premixed in the carburetor and intake, have been compressed almost adiabatically through the volumetric compression ratio. After the spark a turbulent flame front propagates through the unburnt mixture, further compressing the unburnt gas ahead and the burnt gas behind the flame. As the piston moves down during the power stroke, the gases expand and cool.

Nitric oxide, NO, the only oxide of nitrogen produced in significant quantities inside the engine, forms in the burning and burnt gases. The chemistry of NO formation in gas phase mixtures of O, N, C and H has been extensively studied in shock tubes, stirred reactors and flames, and rate constants for the important reactions have been determined (see Baulch et al. (2) for a critical review). Based on this work, kinetic models of NO formation in automotive engines have been proposed (3) - (5). The most important reactions to include are:

$$N_2 + O \rightleftarrows NO + N \qquad (1)$$

$$N + O_2 \rightleftarrows NO + O \qquad (2)$$

$$N + OH \rightleftarrows NO + H \qquad (3)$$

The first reaction is endothermic from left to right, and is slow relative to the hydrocarbon oxidation reactions. The second and third reactions are exothermic left to right, and fast.

NO forms in both the flame front and the post flame gases. However, the formation in the post flame gases is generally much more important. It is thus a reasonable approximation to assume the species O, O_2, OH, H and N_2 are in local equilibrium in the burnt gases, and that N is in steady state. It then follows that an equation for the rate of change of NO concentration can be derived (6). The formation rate depends on the burnt gas temperature, the equilibrium composition of the O-H system in the burnt gases (and thus the local pressure and fuel:air ratio), and the local NO concentration. This kinetic model must, therefore, be

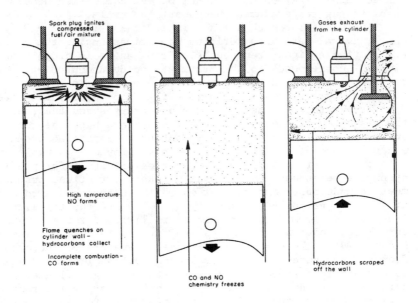

Fig. 1. Schematic of engine cylinder showing how pollutants are formed. The cylinder on the left shows the combustion process; the cylinder in the center shows the expansion process; the cylinder on the right the exhaust process.

linked with a thermodynamic model of the combustion process which predicts gas temperatures throughout the cycle.

The basis for such a model is indicated in Figure 2 (7), where several variables defining conditions inside the cylinder are plotted against crank angle. Crank angle is proportional to time; 0° corresponds to the crank at top dead center--the minimum cylinder volume. In this example, the mixture is sparked at -15°. As the mixtures burns, x (the mass fraction burned) increases from 0 to 1, and the cylinder pressure p increases and then decreases as the expansion of the gases dominates. x and p can be related by an energy balance for the cylinder contents. Sometimes p can be measured, and $x(\theta)$ calculated from $p(\theta)$ provides valuable information on the turbulent flame propagation process. For predictions of engine performance and emissions characteristics, assumptions must be made about the rate at which the charge burns to obtain $x(\theta)$ (8). Currently our ability to calculate the turbulent flame front motion from first principles is limited, though a simple turbulence model has recently been applied to engines with some success (9). The engine performance parameters can be obtained from the pressure-time curve. The conventional performance parameters are mean effective pressure (the average cylinder pressure which if applied to the piston throughout the power stroke would give the same torque; MEP is proportional to the shaft work delivered per cycle) and specific fuel consumption (lb fuel/horse power hour; SFC is inversely proportional to the thermodynamic efficiency).

The temperature distribution in the burnt and unburnt gases has a significant effect on the emissions characteristics of the engine. Figure 2 shows the gas temperatures in two elements which burn at different times, one early and one late in the combustion process. Because the state-time histories of parts of the fuel-air mixture which burn at different times are different, a temperature gradient is set up across the burnt gases. It has long been known from early spectroscopic studies (10), that temperature gradients of this magnitude exist inside the engine cylinder.

The NO kinetic model can now be coupled with this thermodynamic model of the combustion process, and NO concentrations in different parts of the mixture can be calculated. Figure 2 gives profiles for an early and late burning element. The results show that the formation of NO is rate controlled, and freezing of the NO chemistry occurs early in the expansion process as the gases cool. By integrating these frozen NO concentrations across the entire charge, the average exhaust NO concentration can be computed.

How do these predictions compare with measurements? Figure 3 shows one comparison (7); others have been equally successful (8). The results in Figure 3 also show the strong dependence of NO emissions on fuel-air ratio (ϕ in the figure is the fuel-air ratio normalized by the stoichiometric fuel-air ratio), and the effect of exhaust gas recycle (EGR). EGR is currently used to reduce NO emissions; it acts as a diluent and reduces the burnt gas temperatures, and hence the NO formation rates.

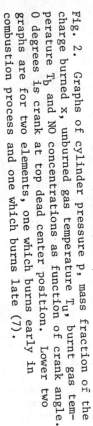

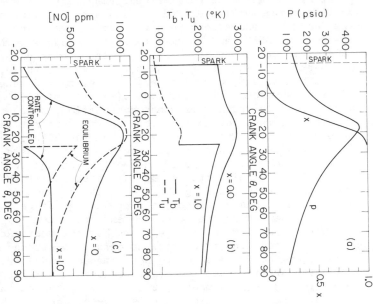

Fig. 2. Graphs of cylinder pressure p, mass fraction of the charge burned x, unburned gas temperature T_u, burnt gas temperature T_b and NO concentrations as function of crank angle. 0 degrees is crank at top dead center position. Lower two graphs are for two elements, one which burns early in combustion process and one which burns late (7).

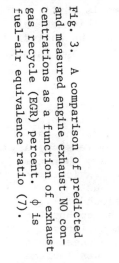

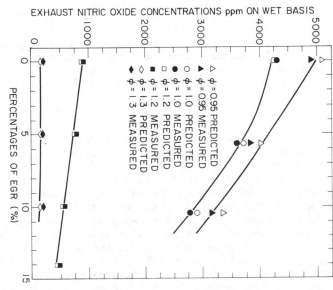

Fig. 3. A comparison of predicted and measured engine exhaust NO concentrations as a function of exhaust gas recycle (EGR) percent. ϕ is fuel-air equivalence ratio (7).

This type of calculation is now being done by many groups to examine the emission control - performance trade-offs. These combustion models allow numerical experiments to be carried out, and the effects of changes in a large number of engine design and operating variables to be examined. The value of such calculations is illustrated in Figure 4 (8). From a combustion model of the type described above, engine performance (brake specific fuel consumption, BSFC) and NO emissions (brake specific NO emissions, BSNO) have been calculated for different engine compression ratios and spark timings. Other parameters (fuel-air ratio, intake mixture conditions, engine speed, connecting rod length L/crank radius R, and combustion interval $\theta(x=0)$ to $\theta(x=1)$) have been held constant. Both spark-timing and compression ratio affect burnt gas temperatures; the model allows one to sort out the interactions between the two. The results show that the higher compression ratios with the spark retarded to closer to top dead center (TDC = 0°) give lower specific fuel consumption for the same mass NO emissions. Extensive studies of this type assist the engine developer in understanding the trends in the mass of experimental engine data he has collected, and provide guidance as to the more promising changes in engine design and operation.

With this example, I have shown how the engine combustion process can be modelled in sufficient detail to produce useful predictions of engine NO emissions - performance trade-off. A few words of caution are in order, however. Because many simplifying assumptions have been made, the accuracy of the calculations is limited; often calibration of this type of model with experimental data is necessary. The particular example I have chosen is quite well understood. Many people have contributed to the development of this understanding; I have chosen to describe results from my own and my colleagues' work primarily for convenience. Other engine emissions problems are not as well understood, and considerable basic research on the fundamental mechanisms is still required. An important element in the success of this type of modelling is the matching of the sophistication of the analysis and its assumptions, to the precision desired in the end result.

COMBUSTION OF FUEL SPRAYS

In diesel engines, and in stratified charge spark-ignition engines the fuel and air are not premixed; the fuel is sprayed directly into the air in the engine cylinder towards the end of the compression stroke. As a result of higher compression ratio, faster combustion process and smaller pumping losses than the conventional spark-ignition engine, these engines exhibit good fuel economy. Figure 5 shows a cross-section of one type of stratified-charge engine which is now being developed (11).

The important new element introduced into the combustion process in these engines is the fuel-air ratio nonuniformity. A complex and imperfectly understood sequence of processes must occur before combustion is complete. These processes include: atomization of the liquid fuel jet, air entrainment into the jet,

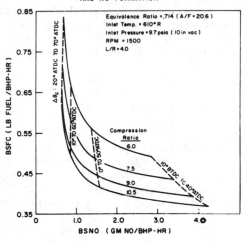

Fig. 4. Calculations of the effect of engine compression ratio and spark timing on the brake specific fuel consumption (BSFC) - brake specific NO emissions (BSNO) trade-off (8).

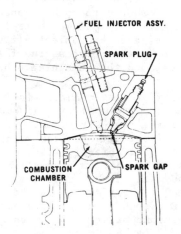

Fig. 5. Schematic of cylinder and cylinder head of 350 CID Ford PROCO fuel injected stratified charge engine (11).

fuel droplet vaporization, turbulent mixing of fuel vapor and air, ignition, flame propagation, the mixing of burning and burnt gases with excess air. The fact that the fuel-air ratio is not uniform throughout the engine cycle has an important effect on both emissions and performance. We do not yet understand how to put all these processes together to form a coherent model for this type of combustion process. We do not understand many of the individual processes which form a part of this whole.

Let me illustrate this complexity with a sequence of photographs of the diesel combustion process taken in a rapid compression machine. This machine simulates a single engine compression stroke and combustion process in a cylinder - piston combination driven by compressed air. Thus engine conditions can be correctly matched, but the cylinder geometry can be simplified. The photographs were taken through a plexiglass cylinder cap, looking down perpendicular to the piston face; a single diesel fuel spray was injected radially outward from the axis of the cylinder towards the wall. Figure 6 shows a sequence of four pictures taken during a single combustion process by Rife (12); the duration of the entire combustion process is about 15 msec. Figure 6A shows the developing fuel-air jet before combustion starts. The photograph shows light scattered from fuel droplets within the plume. The plume bends to the right because there is air swirl. The fuel evaporates close to the cylinder wall, and after an ignition delay combustion commences. Figure 6B shows the flame structure when about one quarter of the fuel has been burnt. The radiation from the flame recorded on the film is blue; soot is not yet being formed in the flame. The jet as it interacts with the wall can be thought of as an "umbrella;" the flame is now around the fringes and combustible fuel-air mixture continues to flow along the "handle." Figure 6C shows a later stage after yellow luminosity characteristic of soot burn-up has appeared. Now the rate of mixing of the fuel and air in the plume controls combustion, and the flame can be described as a "turbulent diffusion flame." Finally, near the end of the combustion process the flame flashes back to the base of the fuel yet, and the remaining unburnt fuel and the soot is consumed as shown in 6D.

How can we understand such a complex process quantitatively? In the absence of swirl, and with a number of simplifying assumptions we can model the average development of the fuel-air plume. With high pressure injection systems, the fuel rapidly atomizes to drops of order 20 μm diameter which entrain air sufficiently fast for relative motion between the two phases in the jet to be small. If it is further assumed that this unsteady jet can be treated as quasi-steady, then classical two-phase jet theories can be used to predict average jet characteristics. One such solution is shown in Figure 7 (12). Normalized velocity and concentration profiles across the jet at any axial position are assumed to be similar, and functions of ξ only. $\xi = 1$ is the outer boundary of the jet. $\xi = 0$ is the centerline, and the velocity along the centerline can be used to predict jet penetration. Concentration profiles (ϕ is the fuel-air equivalence ratio) can be computed, and the rate at which fuel

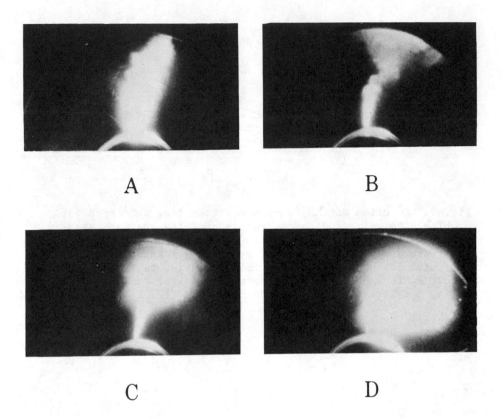

A

B

C

D

Fig. 6. Four photographs from a high speed movie of simulated diesel
combustion process in a rapid compression machine. View is along the
axis of the cylinder, looking onto the piston face. A single fuel
jet is injected radially outward from the axis. The duration of the
combustion process is 15 msec. 6A shows the fuel jet before combustion starts. A pair of flash bulbs provide the illumination and
light is reflected from the approximately 10 μm drops of fuel in the
developing fuel air plume. The jet bends to the right due to air
swirl. 6B shows the early stages of combustion. The flame is out
against the cylinder wall and the luminosity is blue, not yellow or
red. The jet is still illuminated by the flash bulbs. 6C shows the
middle stages of the combustion process. The luminosity is yellow-
white and comes from soot burn-up, the flame has moved back to half-
way along the fuel jet. Finally, in 6D, as combustion goes to
completion, the flame moves back to the start of the fuel jet as
the fuel flow rate decreases to zero. The luminosity is yellow-red
(12).

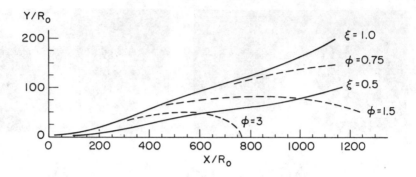

Fig. 7. Solution of the two-phase steady flow submerged jet problem. Velocity and concentration profiles are each assumed similar at different positions in the jet. The coordinates x and y are along and perpendicular to the jet axis. $\xi = 1.0$ is the outer jet boundary, $\xi = 0$, the x/R_0 axis, is the jet center-line. R_0 is the fuel nozzle radius. ϕ is the fuel-air equivalence ratio, $\phi < 1$ is leaner than stoichiometric (12).

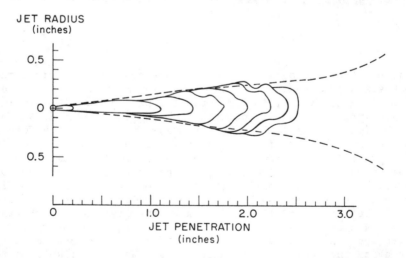

Fig. 8. Measured fuel jet profiles from high speed movies and calculated jet boundaries from the steady two-phase jet solution. Agreement is good (12).

and air mix across these concentration profiles into the combustible
range can be calculated. Figure 8 shows a comparison of the predicted jet profile with profiles obtained from photographs, and
agreement is good (12).

By careful analysis of the details of each component process we
can thus start to describe and quantify several important engine
operating parameters. Of course there is a long way to go before
the events occurring in a real engine of this type are adequately
understood.

RESEARCH AREAS FOR APPLIED SCIENTISTS

How can applied scientists contribute to our basic understanding of these types of problems? I use applied scientist rather
than physicist because the different disciplines which are important
in combustion modelling lie in the domains of physics, chemistry and
engineering science. Combustion theory has been under active development for several decades. Yet many gaps remain in our understanding of fundamental combustion phenomena, and the application of
what we do understand to practical devices such as automobile engines
has been limited by the enormous complexity of these problems.

What are the important areas for future basic research? Let me
introduce some of them by briefly reviewing the combustion process
in an engine. In engines where fuel is injected into the engine
cylinder, the details of the fuel air mixing process cannot yet be
predicted. A more accurate knowledge of the fuel distribution is an
essential preliminary to improving our understanding of the flame
structure and pollutant formation mechanisms in engines of this
type--the stratified charge and diesel. The details of the flame
structure itself need much more extensive analysis. Only the latter
stages of the hydrocarbon oxidation process for realistic fuels are
understood in a quantitative way. The initiation of the flame with
an electrical discharge has been treated empirically. Yet a better
understanding of the coupling between the energy in the discharge
and the combustion process would help our efforts to extend the lean
operating limits of today's engine--a direction which promises better
emission control and fuel economy. The effect of the turbulent flow
field inside the engine on flame initiation and flame propagation
cannot now be explained in a convincing quantitative way. These
processes obviously determine engine performance characteristics.
Only some of the pollutant formation mechanisms inside automobile
engines are well understood. While the NO and CO chemistry has been
examined in detail, the formation of soot, in a diesel engine for
example, cannot yet be predicted. Flame quenching, which is the
origin of unburnt hydrocarbons in spark-ignition engines, is still
treated empirically, and the subsequent mixing and oxidation of the
hydrocarbon rich quench regions in the bulk gases has not been
examined in detail. These are some of the important engine
combustion problems which need attention.

In 1827, Alexander Jamieson in his Dictionary of Mechanical
Science defined a flame as "an instance of combustion, whose colour
will be determined by the degree of decomposition which takes place.

If it be very imperfect, the most refrangible rays only will appear; if very perfect, all the rays will appear, and the flame will be brilliant in proportion to this perfection." While applied science has substituted precision for much of his poetry in the ensuing 150 years, we still grope for words when we try to explain precisely what happens in a real combustion process.

REFERENCES

1. Automotive Spark Ignition Engine Emission Control Systems to Meet the Requirements of the 1970 Clean Air Amendments (National Academy of Sciences, Washington, D.C., 1973).
2. D. L. Baulch, D. D. Drysdale, D. G. Horne and A. C. Lloyd, Evaluated Kinetic Data for High Temperature Reactions, Vol. 2 (Butterworths, 1973).
3. H. K. Newhall and E. S. Starkman, SAE Trans. $\underline{76}$, paper 670112 (1967).
4. P. Eyzat and J. C. Guibet, SAE Trans. $\underline{77}$, paper 680124 (1968).
5. G. A. Lavoie, J. B. Heywood and J. C. Keck, Comb. Sci. & Tech. $\underline{1}$, 313-26 (1970).
6. J. B. Heywood and J. C. Keck, Env. Sci. & Tech. $\underline{7}$, 216-223 (1973).
7. K. Komiyama and J. B. Heywood, SAE paper 730475, Automobile Engineering Meeting, Detroit (May 14-18, 1973).
8. P. Blumberg and J. T. Kummer, Comb. Sci. & Tech. $\underline{4}$, 73-96 (1971).
9. N. Blizard and J. C. Keck, SAE paper 740191, Automotive Engineering Congress, Detroit (Feb. 25-March 1, 1974).
10. G. M. Rassweiler and L. Withrow, SAE Trans., 125-33 (1935).
11. A. Simko, M. A. Choma and L. L. Repko, SAE paper 720052, Automotive Engineering Congress, Detroit (Jan. 10-14, 1972).
12. J. M. Rife, "Photographic and Performance Studies of Diesel Combustion in a Rapid Compression Machine," Ph.D. Thesis, M.I.T. (March 1974).

HIGH TEMPERATURE MATERIALS FOR AUTOMOTIVE POWER PLANTS

J. J. Harwood
Ford Motor Co., Dearborn, Michigan

ABSTRACT

Alternative power plants, emission control systems and control systems for optimizing performance of internal combustion engines have focused increased attention on automotive applications of high temperature materials. Ceramics and related materials are receiving particular attention. This paper presents an overview of some of the high temperature materials developments associated with catalytic devices for emission control systems, sensors for engine feedback control systems, small high-temperature gas turbines and solid electrolytes and ion transport ceramics for batteries and energy conversion systems. Opportunities for research in physics and solid state science, arising from the technological problems involved in these high temperature materials developments, are presented.

INTRODUCTION

The interactive issues of energy, environment and resources have introduced new dimensions in automotive materials. Classes of materials and fabrication techniques more commonly associated with "science based industries" are now undergoing intensive development and consideration for automotive application. In particular, exhaust emission control systems and the recent emphasis on fuel economy and energy conservation have focused on the potential of ceramics and related non-metallic inorganic materials for automotive powerplant usage. A major requirement for such applications is the ability to survive in harsh, high temperature environments.

It is the purpose of this paper to present an overview of some of the high temperature materials developments associated with emission control systems and exhaust gas treatments, control systems for optimizing performance and efficiency of internal combustion engines and advanced high temperature gas turbines for automotive vehicles. A special field of interest relates to solid electrolytes and ion transport ceramics for new high performance batteries and energy conversion systems. In keeping with the theme of this symposium the main objective will be to elucidate opportunities for research in physics and solid state science arising from technical problems involved in these high temperature materials developments.

Catalysts, Surface Science and Emission Control Systems

As you are well aware Federal legislation required that, beginning with the 1975 model year, automotive vehicles must exhibit a 90% reduction in hydrocarbon (HC) and carbon monoxide (CO) emissions from allowable emissions for 1970 model year vehicles. The legislation also required, beginning with 1976 model vehicles, a 90% reduction in oxides of nitrogen from 1971 vehicles. These datelines have been extended by recent governmental action.

The nature of these pollutants, as they are generated in an internal combustion engine, as a function of the air to fuel (A/F) ratio is shown in Fig. 1. The vertical dashed line is the

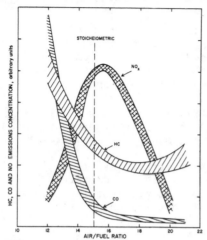

Fig. 1. Composition of exhaust gas as a function of A/F ratio.

stoichiometric ratio - just under 15 A/F - for which complete combustion is possible. Much of the effort in the automotive industry to meet these standards has revolved around systems for treating the exhaust gases after they leave the combustion chamber. The most promising system appears to be an oxidation catalyst for the HC and CO and the use of exhaust gas recirculation (EGR) for the control of oxides of nitrogen. The EGR system serves to dilute the combustion products with inert gas taken from the exhaust, thereby reducing the combustion temperature and decreasing the formation of the oxides of nitrogen. The more stringent NO_x standards for 1976 may require a reducing catalyst to treat the oxides of nitrogen.

A rather straightforward representation of the means of removing all of the constituents from the exhaust stream consists of the dual catalyst system of the general type shown in Fig. 2. If the engine is operated on the fuel-rich side of stoichiometry so that the atmosphere at the first catalyst is reducing, that is with a low oxygen concentration and an excess of CO, certain catalysts will effect the removal of the oxides of nitrogen. With the addition of air between the two beds, the excess carbon monoxide and hydrocarbons are then catalytically oxidized to give an exhaust of N_2, CO_2 and H_2O.

DUAL BED AXIAL-FLOW CONVERTER

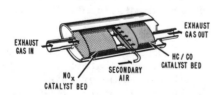

Fig. 2. Representation of dual catalyst system.

A catalyst system essentially consists of an alumina ceramic substrate, either pellets or monolithic cellular structures, on top of which is deposited a high surface area alumina wash coat, followed by the active catalytic material. The most promising choice at this time for the oxidizing catalyst is platinum or platinum-palladium alloys. Ruthenium-based catalysts, other platinum group metal catalysts and certain types of base metal oxides appear promising as NO_x reduction catalysts.

The most severe problem encountered in such catalytic systems is insuring that the catalyst will possess reasonable activity after long mileage accumulation. This requires stability of both the substrate and the active catalytic surface. The stability of the catalyst is limited in two major ways. First, the fuels and engine lubricants contain elements that poison the catalyst, e.g., lead, phosphorus and sulfur. Second, the active catalytic particles tend to coagulate or coalesce at the high operating temperatures, thereby reducing the effective surface area and the overall activity.

NO_x reducing catalysts, such as ruthenium or base metal oxides, are quite sensitive to sulfur poisoning indicating the necessity for the availability of low sulfur fuels. The platinum type oxidizing catalysts, although largely insensitive to sulfur in the fuel, are strongly affected by other impurities such as lead. It thus becomes important to be able to quantitatively assess the effects of various impurities in the fuels upon the activity of the catalysts and to develop laboratory test methods that reasonably simulate actual vehicle exhaust conditions.

Fig. 3 shows the activity for oxidation of hydrocarbons and carbon monoxide for a platinum catalyst after 24,000 simulated miles of operation. The fuel contained three different levels of lead - sterile, 0.03 g/gal and 0.05 g/gal. A monotonic decrease in the activity of the catalyst with increasing lead content is evident. These laboratory data have been quantitatively confirmed by engine and vehicle tests. For comparison, present gasoline stock contains roughly 2 g/gal of lead. A catalyst subjected to fuel with such a high lead level would become ineffective after a few hundred miles. Clearly the availability of "lead-free" gasoline will be a necessity for the utilization of catalytic devices for emission control systems.

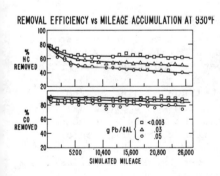

Fig. 3. Influence of lead content upon activity of platinum catalyst.

What does this technological area mean to physics and surface science? Until recently surface physics and catalysis, as related scientific fields, have had little interaction. The field of catalysis has been largely dominated by surface chemistry and we are not embarrassed to note that much of the automotive advances rely upon a strongly empirical basis. If we are to develop cheaper and less exotic catalytic materials to substitute for platinum and palladium, if we are to understand catalytic reactivity and poisoning mechanisms and if we are to design more efficient and more stable catalysts, then the techniques and the methodologies of physics in probing the electronic, structural and chemical nature of catalytic surfaces must play significant contributions.

We invite the contributions of theoretical physics to provide new insights into the understanding of chemisorption and catalysis. From an experimental point of view we are convinced that such ultrasensitive tools as field ion microscopy, field emission microscopy, low energy electron diffraction, Auger electron spectroscopy, ultraviolet photoelectron spectroscopy, ESCA, thermal desorption and mass spectroscopy are vital to attack a number of fundamental problems in catalysis.

Perhaps I can best illustrate this point with a few examples of studies underway in our own laboratory.

a. <u>Mechanism of Surface Reactions</u> - The mechanisms of surface reactions can be deduced from kinetics of reactions and to some extent from accompanying structural and surface composition changes. N. Gjostein and R. Ku are studying the reduction of NO_x on ruthenium surfaces. Auger electron spectroscopy is being used to monitor the kinetics of nitrogen and oxygen adsorption as a function of temperature and time. Using the same crystal, various surface species then can be thermally desorbed or removed by chemical reactions. The reactions are followed by mass spectrometry and the residual species on the surface are monitored by AES.

These experiments have shown that nitric oxide rapidly dissociates on Ru surfaces at $T > 90^{\circ}C$.

$$NO \text{ (gas)} \rightarrow NO \text{ (ads)} \xrightarrow{\text{rapid}} N \text{ (ads)} + O \text{ (ads)}$$

Associated LEED studies indicate that the adsorbed oxygen forms ordered patches, but the nitrogen is highly surface mobile. The nitrogen atoms can pair up to form an activated complex which readily desorbs at $T > 150^{\circ}C$.

Therefore, if it were not for the buildup of dissociated oxygen on the surface, the reduction reaction would proceed indefinitely and Ru would be an excellent catalyst without the need for a reducing agent in the feed gas, i.e. without the presence of CO as a reducing agent. CO apparently acts to remove the dissociated oxygen adsorbed on Ru, which cleans or frees the surface of oxygen and allows the dissociation of NO to proceed.

This simple mechanism has obvious implications with respect to the selectivity of Ru towards N_2 production rather than NH_3 formation, when hydrogen is present in the gas stream, as compared to Pt which favors the formation of NH_3 under similar conditions. The relative kinetics of hydrogen with adsorbed O or N atoms on ruthenium crystal specimens, to form respectively H_2O or NH_3, can be followed with these same techniques.

b. <u>Reactivity on Alloy Surfaces</u> - Metallic impurities are also important in determining the efficiency of a catalyst.

In order to relate catalytic activity to alloy composition, it is not sufficient merely to establish the average bulk composition, but more specifically the surface composition. P. Wynblatt and V. Sundaram have been examining the surface composition dependence of the catalytic activity of a series of Pt-X alloys. Some of their results are shown in Fig. 4 for the oxidative reactivity of these alloys for CO. Note that the reactivity of platinum is reduced, under the experimental conditions, for all of the alloying elements investigated.

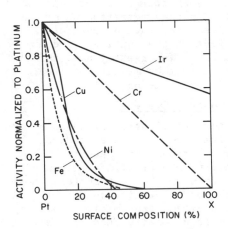

Fig. 4. Surface composition dependence of catalytic activity of Pt-X alloys.

The case of Pt-Cr alloys is particularly interesting. Under the experimental conditions used, pure Cr does not display any measurable activity and activity of Pt-Cr alloys increases linearly from pure Cr to pure Pt, i.e. each Pt atom in the surface of a Pt-Cr alloy is just as active as a Pt atom in a pure Pt surface. On the other hand, the activity of alloys of Pt with Cu, Ni and Fe all fall below the Pt-Cr curve, indicating that the electronic and bonding interactions between these atomic species and Pt impair the ability of Pt to catalyze the oxidation of CO. Auger electron spectroscopy is a potent tool for studying such effects. We have good evidence that adsorbed oxygen greatly influences the degree of Cu segregation in Pt-Cu alloys. Such findings may help clarify some of the puzzling features associated with multicomponent catalysts.

c. Stability of Supported Metal Crystallites - P. Wynblatt and N. Gjostein also are studying the coalescence behavior of the active catalytic particles on supported ceramic substrates as one of the features of catalytic deactivation. The highly dispersed metal particles, with radii in the range of 50 Å - 500 Å, are inherently unstable because of the strong tendency for reduction of total surface free energy. Under high temperatures, coalescence occurs with an increase in average particle size.

Studies of the detailed mechanisms of particle growth and active area loss are difficult on the complex technological catalyst systems. They have developed a model catalyst system consisting of Pt particles deposited on thin alumina films, in which the particle growth process can be followed by transmission and scanning electron microscopy and Auger electron spectroscopy. Through such studies of the kinetics of particle growth and particle growth laws they aim to identify the determining interparticle mass transport mechanisms. Such information may provide a fundamental basis for improving catalyst stability.

d. Surface Bonds of Chemisorbed Species - It has long been known that only a handful of metals such as Pt, Pd, Ru, Ir, Rh, Os, Ni and Fe exhibit high catalytic reactivity. But within this group of metals, reaction rates can vary by 8 orders of magnitude. As yet, there is no central theory of catalysis that enables relating electronic properties to this huge variation of reaction rates.

E. Sickafus is using Auger spectroscopy line shape analysis to

120

identify the molecular orbitals characterizing chemisorbed species,
e.g. oxygen and sulfur on nickel. Related LEED studies reveal how
the chemisorbed species are geometrically arranged on the surface.
The combination of such techniques are a start at providing the
types of information necessary to formulate a basic theory of
chemisorption, fundamental to a better theory of catalysis.

The techniques and theories of surface physics and surface
science are powerful tools to develop a more thorough understanding
of what goes on at the surface of a catalyst. This is essential if
we are to be able to reliably predict such things as the effects of
impurities. But the task of translating the understanding of the
interactions that take place in an idealized environment into an
understanding of systems that must live in the non-idealized environment of a vehicle exhaust is formidable. We urgently need this,
however, for only then will we be able to make meaningful predictions
of behavior under the variable conditions that occur in a vehicle.

Sensors and Engine Control Systems

Control of engine performance for optimization of emission and
fuel economy features has focused attention on engine feedback
control strategy and devices. What are needed are sensors to be
able to convert basic engine parameters into electrical signals
which can be used as inputs to actuators which can translate these
electrical inputs into actions which control engine operations.
Ignition timing, fuel metering, EGR flow and transmission shifting are
some of the parameters being adapted to automatic feedback control
techniques. Improved sensor materials with stability, reproducibility
and reliability in combustion product environments are key elements.

Again let me illustrate opportunities for physics research with
a brief discussion on the potential of oxide type semiconductors for
A/F ratio feedback control sensors. The preceding discussion on
catalytic converters indicated the importance of stringent control
of A/F ratio for effective performance. The objective of feedback
control is to hold a particular chemical composition of the exhaust.

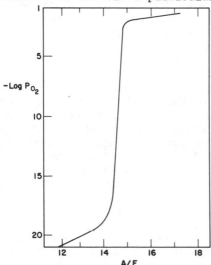

This changes abruptly near the
stoichiometric point. Fig. 5 shows
the calculated equilibrium oxygen
partial pressure P_{O_2}, of exhaust
gas at 700°C as a function of A/F
for a representative fuel composition. The reason that catalyst
performance is so extremely sensitive
to changes in A/F is that the catalyst is sensitive to P_{O_2}, which
varies almost discontinuously with
A/F near the stoichiometric ratio.
This also indicates that if the fuel

Fig. 5. Oxygen partial pressure of
exhaust gas as function of A/F ratio.

metering system could use P_{O_2} as its input information instead of A/F, it could ensure optimal chemical conditions even if P_{O_2} was not measured very precisely. Direct measurement of P_{O_2} is advantageous in that it changes in a parallel manner as the catalyst to changes in fuel composition and ambient temperature.

Two types of oxygen sensors are under development. One type, being carried out by Bendix in the United States and Bosch and Phillips in Europe, uses a closed hollow cylinder of stabilized zirconia (ZrO_2). This type of sensor functions as an electrolytic cell generating a voltage that depends primarily upon the ratio of the oxygen pressures at the two surfaces of the cylinder. One surface is exposed to the exhaust and the other is exposed to the atmosphere.

Another type, which we have pursued in our laboratory, uses titania (TiO_2) ceramics as the active material. This type of sensor responds to changes in the partial pressure of oxygen in the exhaust gas by changes in its electrical resistance. Other transition metal oxides and some rare-earth oxides can also be used.

At high temperatures ($\sim 700^\circ C$) TiO_2 becomes a non-stoichiometric semiconductor with the number of charge carriers dependent upon the density of oxygen vacancies in the material. These, in turn, depend directly upon the partial pressure of oxygen in the surrounding gas. The characteristics of a TiO_2 sensor as a function of A/F are shown in Fig. 6. The right hand scale shows the variation in resistance (indicated by the experimental points); the left hand scale shows the variation in oxygen partial pressure (indicated by the solid curve). The TiO_2 sensor is particularly useful in the region near and slightly rich of stoichiometric values of A/F, since the steep slope of the curve indicates that the TiO_2 resistance will be a very sensitive, specific indicator of A/F in that range. It might be noted that TiO_2 sensor

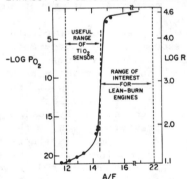

Fig. 6. Resistivity of TiO_2 sensor as function of A/F ratio

devices have been exposed to dynamometer engine exhaust at $700^\circ C$ for almost 1000 hours without damage or irreversible changes in the resistance vs. A/F curve.

A variety of effects and phenomena within the domain of physics and surface sciences obviously are involved with sensor materials and devices and all too often are not adequately enough understood to achieve devices with the efficiency and durability and stability required. Impurity and compositional effects in these transition metal oxide semiconductors and the effects of

122

combustion environments (in the range of 400-1000°C) on the electrical and surface properties call for more detailed research attention. The latter may be particularly important, since there is some intriguing evidence that the catalytic properties of the sensor material itself plays an intrinsic and not well understood role in its behavior.

Another area of opportunity for technological or engineering physics is the important need for concepts and devices for over-temperature and early failure warning systems for catalytic converters. The catalytic action is exothermic and if the input gas composition is too rich (due to misfires or other engine malfunctions) catastrophic temperatures can be reached very quickly with catalyst burnout. Even mild conditions of overheating degrade catalyst efficiency. What are required are fast response indicators of catalytic converter failure which give a warning signal when the degradation of the converter reaches an unacceptable level, and overtemperature warning devices which produce warning signals when the converter temperature exceeds a certain value.

High Temperature Gas Turbine

The small gas turbine provides a challenging opportunity as an alternative power plant to compete in the high volume automotive market. To do so, a number of major problems associated with conventional gas turbines must be solved. These are excessive oxides of nitrogen emissions, poor fuel economy at part power and excessive initial engine cost and size.

Gas turbine efficiency increases with peak cycle temperature. One way, therefore, of improving fuel economy is to increase turbine inlet temperatures. Fig. 7 shows a plot of specific fuel consumption vs. pressure ratio for a typical regenerative vehicular gas turbine. At typical pressure ratios in the range 4 to 6, an increase in turbine inlet temperature from 1800°F to 2500°F would improve specific fuel consumption by over 20%. Increasing temperature also reduces air consumption per horsepower; raising the turbine inlet from 1800°F to 2500°F would improve specific air consumption by 50%. Another way of saying this is that for the same

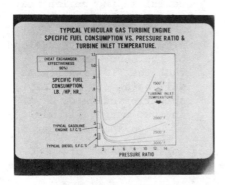

Fig. 7. Gas turbine specific fuel consumption as function of temperature.

engine size, that is the same air flow, the maximum power would be almost double. Thus, there is much incentive to increase turbine engine operating temperatures.

At Ford Motor Company, for example, a major program has been underway for several years to develop a high temperature (2500°F inlet gas temperature) small gas turbine. The key feature of this program is the focus on the design and application of ceramic materials and components for the hot end of the turbine. Fig. 8 shows a schematic of a vehicular gas turbine engine and high- lights the major components. The compressor, combustion chamber, regenerator, stator, nose cone and turbine rotor are major ceramic components under development. Fig. 9 illustrates some of these ceramic parts. Since 1971 this program has been accelerated through

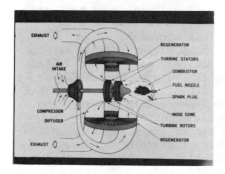

Fig. 8. Schematic of a
Gas Turbine engine.

Fig. 9. Gas Turbine
Ceramic Components

the support of the Advanced Research Projects Agency which has parallel interests in the design and utilization of brittle, ceramic materials for structural applications.

A principal advantage of the high temperature turbine is the potential for significantly increased fuel economy with multi-fuel capability. In addition new designs and innovations employed in the continuous combustion of the turbine indicate a reasonable likelihood of meeting low emission levels, including NO_x require- ments.

Hundreds of ceramic materials have been screened and the most promising of these materials are now being evaluated as actual com- ponents under both simulated and actual turbine engine conditions. Low coefficient of expansion ceramics, such as lithium aluminum silicate, are being used for regenerator components and silicon nitride and silicon carbide have been selected as the candidate materials for the high temperature and high stress conditions associated with turbine stators and rotors.

It is beyond the scope of this paper to present a detailed review of the materials requirements and developments and the problems remaining to be solved. But perhaps two brief examples of the high temperature materials effort can feature the contributions which physics research and techniques may provide.

A key determinant of the success of this program is the capability to develop fabrication techniques for ceramic materials which are amenable to low cost, high volume production output of complex shapes, and which will result in components with desired structural integrity, reproducibility and reliability of properties and durability in performance. A wide variety of fabrication methods are being evaluated including slip casting, pressing and sintering, hot (isostatic) pressing, reaction sintering, extrusion, injection molding and chemical vapor deposition. Exciting potential has been demonstrated in the use of polymeric materials and polymeric fabrication techniques to produce shapes which are later converted to ceramic forms by appropriate conversion techniques, e.g. carbonization, followed by siliciding. REFEL (silicon carbide material) developed in England is an example of this approach.

Major problems, which can use the help of physicists, lie in the detailed characterization of these silicon nitride and silicon carbide materials and in understanding the bonding mechanisms which underlie their fabrication processing. In order for the aggregate of small ceramic particles to stick together and to densify, the transfer of materials through surface diffusion is required. Materials like Si_3N_4, however, do not sinter under normal conditions. Therefore, how and in what form diffusion takes place in these materials and ways to enhance such diffusion to promote sintering needs to be understood in a more systematic way. Diffusion and sintering were popular topics for physics research in the post World War II period; understanding of these phenomena in these newer types of materials provides new challenges for similar efforts.

If such materials cannot be properly densified by conventional sintering approaches, an obvious resort is to employ "gluing" substances. A physical understanding of the nature of bonding via composition doping agents and of the ensuing grain boundary and interphase boundary phenomena which occur under high temperature/ stress service conditions, e.g. grain boundary creep, is required.

Reaction sintering is a process whereby fabrication of the ceramic part is accomplished by simultaneous composition reaction synthesis and sintering. This is a useful technique for preparation of both Si_3N_4 and SiC shapes. There is the need to understand the mechanism of growth of individual crystals as well as their sintering behavior. SiC particles, for example, do not bond properly unless the orientation of the crystals is the same. This problem clearly is related to the features of epitaxial growth.

H. Sato and his colleagues have done extensive research on the polytype structures of SiC, characterization of long period structures, stacking faults and twinning in these structures, and the epitaxial growth aspects of SiC during reaction sintering. Electron transmission microscopy has been a dominant tool for these studies. Reaction sintered SiC materials are generally produced from a

compacted mixture of silicon carbide particles and carbon by introducing molten silicon to convert the carbon into SiC. They have shown
clearly that the orientation of the newly formed β SiC is an important
condition involved in the bonding of the original α SiC particles,
Fig. 10. The β SiC grows epitaxially on the surface of α SiC, as well
as independently, but bonding of the new β SiC grains to each other
also only occurs under favorable orientation conditions.

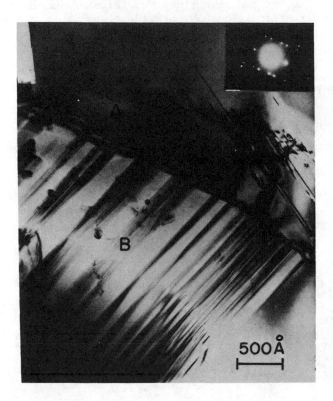

Fig. 10. An electronmicrograph of grain boundary between newly
formed β-SiC(3C) grain (B) and original α-SiC(4H) grain (A).
Lines in the α-SiC grain indicate stacking faults while dark bands
in the β-SiC grain indicate twin boundaries. The inset shows
superposed diffraction patterns of these two structures obtained at
the boundary. The orientation relation of these two grains indicate that the β-SiC grain has grown epitaxially on the α-SiC grain.

Interestingly enough, Sato and his colleagues have also
observed and identified, by electron transmission microscopy, many
long period structures in SiC. An example is shown in Fig. 11.

126

These are always observed in hot pressed material, frequently in extruded-reaction sintered REFEL type SiC, but never in material reaction-sintered by a process developed in our own laboratory unless subjected to a very high temperature post anneal. The origin of such long period structures is still highly speculative and the role of processing techniques and grain size is puzzling. It is essential to understand how the order is maintained over such a long distance. But, at the same time, knowledge of such structural alterations and transformations may provide clues as to the subtle effects of variations in processing methods.

Fig. 11. Electronmicrograph revealing stripe patterns which indicate the direct resolution of the structural lattice-period of 303R long-period polytype. 303R means a structure with a rhombohedral symmetry of 101 close packed double-layers of Si and C in a primitive rhombohedral unit cell or a structure with three hundred and three double-layer periods in the direction of the hexagonal C-axis. The regular spacing between the fine lines is about 255 Å indicating the 101 layer period. Heavy lines indicate stacking faults. The inset shows a series of diffraction spots obtained from the same area of the crystal. The number of spots within the unit reciprocal layer spacing is 101.

In a more general way, the effects of processing upon composition and microstructural details and the structure dependence of properties and behavior of these ceramic materials is still a wide open field. The correlation of physical properties with mechanical properties may result in ways to better control structure and properties.

W. A. Fate has been measuring elastic property data of these materials by pulsed ultrasonic techniques. Some of his early data indicate that shear modulus measurements may be highly valuable in understanding some of the structural and mechanical property behavior which occur at high temperatures. Fig. 12 shows the shear modulus and attenuation data for hot pressed silicon nitride for an ultrasonic frequency of 10 MHZ. As the temperature is raised above 1000°C a large increase in attenuation is observed until at least about

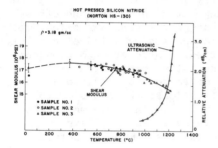

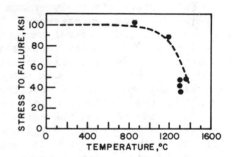

Fig. 12. Shear Modulus and Ultrasonic Attenuation of Hot Pressed Si₃N₄ as Function of Temperature.

Fig. 13 Temperature Dependence of Flexural Strength of Hot Pressed Si₃N₄.

1250°C the sound absorption is so large that higher temperature data cannot be obtained. Note in Fig. 13 that the flexural strength properties of similar material decrease markedly in the same temperature region as the attenuation increases. It is believed that the high temperature attenuation is due either to melting of an impurity promoted amorphous phase present at the grain boundaries of the material or to an elastic grain boundary effect similar to the creep related internal friction well known in metals. In the former case the attenuation should increase without limit as the temperature is raised, while for the latter the attenuation should go through a maximum and decrease at sufficiently high temperatures. Further experiments are underway to clarify this point and to correlate these effects more specifically with strength properties over the same temperature range.

These ceramic materials are intrinsically brittle and obviously non-destructive testing methods will play an important screening out defects and inherent flaws which would result in premature failure in service. Ultrasonic pulse-echo, X-ray xeroradiography and acoustic emission techniques are being explored for their NDE capability. This area of engineering physics is most significant since predictable reliability may well be the ultimate key to the successful structural utilization of brittle, ceramic materials.

Solid Electrolytes and Energy Conversion Systems

Solid electrolytes have many important applications as materials for solid state batteries. Following the Ford announcement in 1966 of the development of a sodium-sulfur battery with significantly higher energy and power characteristics than conventional lead acid batteries, numerous programs have been initiated in many countries along similar lines. This battery utilizes one of the well known solid electrolytes, β and β''-alumina. The method of fabrication of β and β''-alumina has since been constantly improved.

Solid state electrolytes are nothing but high ionic conductors at relatively low temperatures. There has been a constant effort to explore new solid electrolytes and several promising solid electrolytes in addition to β-alumina have been reported. There are two areas of research on solid electrolytes which especially require the help of solid state physicists. One is to find or to develop additional types of solid electrolytes for practical applications; the other is to understand the origin of high ionic conductivity in these materials.

The study to find new solid electrolytes is more or less a crystallographic study. Solid electrolytes have common structural characteristics which provide ample numbers of vacancies through which ions can migrate. Depending on the type of crystal, the distribution of vacancies is sometimes very directional. For example, there are solids in which diffusion occurs in one direction only. This is sometimes called "tunnel diffusion" and is now being studied in many laboratories. In the case of β and β''-alumina, the diffusion is two-dimensional. A three-dimensional system would have better characteristics if appropriate materials can be found. Therefore, in order to explore these areas, a proper characterization of crystal structures of appropriately chosen potential candidate materials is required. This type of research definitely needs the help of physicists.

On the other hand, for the understanding of the second problem, a theoretical background of many-body problems, which is now the essence of solid state theory, is required. If the main characteristic of solid electrolytes is the migration of ions through available vacancies, the interaction among migrating ions and, hence, the cooperative nature of diffusion is the central problem. The cooperative feature of ionic migration can enhance the diffusion rate substantially compared to that expected from the conventional random walk theory. According to a theory developed by H. Sato a large number of available vacancies for ionic migration and the repulsive interaction among migrating ions are largely responsible for the high ionic conductivity. In fact, the conventional theory of diffusion, based on the random walk approach, is not capable of treating such complicated features of diffusion and the origin of the high ionic conductivity. The understanding of such complicated features of diffusion is thus essential to explore new solid electrolytes.

CONCLUSION

In this abbreviated overview of some of the high temperature materials developments associated with emission, fuel economy and energy features of automotive power plants, I have attempted to indicate some of the problem areas now occupying the attention of physicists and solid-state scientists. We believe that there are challenging opportunities, of both a scientific and technological nature for work on these and related topics. I can only conclude with the obvious: we need all the input we can obtain from the solid state community to improve the understanding of these phenomena and to accelerate their technological exploitation.

ACKNOWLEDGMENT

The author is indebted to his many colleagues at the Ford Scientific Research Staff and High Temperature Turbine Department for their inputs and assistance in the preparation of this paper.

ENERGY REQUIREMENTS FOR HIGH SPEED GROUND
TRANSPORT SYSTEMS

John T. Harding

Federal Railroad Administration, Washington, D.C. 20590

ABSTRACT

The Department of Transportation is investigating a
number of technical alternatives for future high speed
ground transportation systems. In several of the nation's
most heavily travelled corridors it has become clear that
continued reliance on air transportation will result in
saturation of corridor air facilities within the present
decade. Additional airports can be tolerated only at such
great distances from population centers that their usefulness for short haul corridor traffic is seriously impaired.
Additionally, devoting scarce and expensive suburban land
to such use is not cost effective relative to ground transportation. It is hard to believe, but a simple calculation
shows that the 18 square kilometers of land occupied by
New York's JFK airport alone equals the area traversed by
a 30 meter wide right of way extending from Boston to
Washington via New York and Philadelphia. To carry the
peak hour traffic expected on this route by 1985 requires
a capacity of 4000 seats/hour. This is equivalent to 10
Boeing 747 departures in one hour for corridor service
alone! By comparison Japan's double-track, high-speed rail
line can provide up to 34,000 seats per hour, and has in
fact carried 553,000 passengers in a single day.

In addition to conventional rail systems which can
provide speeds up to perhaps 300 km/h, still higher speeds
are possible with vehicles suspended and guided by non-
contacting methods designated "tracked levitated vehicle"
(TLV) systems. These include tracked air cushion vehicles,
and magnetically levitated vehicles employing either servo-
controlled electromagnets or superconducting magnets. The
energy requirement for such high speed systems is very substantial and for this reason it is particularly important
to see that they are as efficient as possible. Aerodynamic
drag is the predominant source of power consumption.

Continued growth in intercity travel resulting from increased population and disposable income is expected to result in saturation of existing transportation modes in several of the nation's corridors by 1985. Total transportation demand projections through 1995 are shown in Table I[1] for five representative travel corridors. The highway mode, auto, bus etc. accounts for about 89% of the intercity passenger trips in the United States, while the air mode is responsible for about 8% [2]. Because any significant increases in capacity for either of these modes requires vast amounts of land, they cannot be relied on to provide the total solution for increased capacity needed in the 1980's.

Most people are well aware of the enormous land requirements for highways and parking; the fact that air transportation consumes significant quantities of land may appear ironic. Yet consider the area occupied by the following airports; New York JFK-18 km^2 (7 square miles), Washington Dulles-41 km^2 (16 square miles), and Dallas-Fort Worth, 75 km^2 (29 square miles). It should be obvious that such parcels of land are no longer available within a convenient distance of any major metropolitan area, yet existing airports are rapidly reaching ultimate capacity. Despite maximum improvements to enhance capacity at existing Northeast corridor airports, Table II[3] shows that the average amount of time that a passenger must delay his departure on account of congestion is expected to approach one hour at the New York airports by 1985. A similar situation is avoided at Washington National by imposing a quota on departures. The alternative is Dulles airport, located 42 km from central Washington.

High speed guided ground transportation can provide the capacity needed in the 1980's within modest land requirements. The construction of Japan's high speed railroad on the 515 km Tokaido corridor between Tokyo and Osaka expropriated only 10 km^2 of land. 85 million passengers were carried 27 billion passenger-km in the year ending March 31, 1972[4]. On peak days traffic has exceeded 500,000 passengers. The line carries no freight.

The characteristics of high speed guided ground transportation which result in such high capacity are the ability to entrain vehicles, to operate at very short headways, and to achieve very low operating costs. The high capital costs associated with such systems become less significant with increased volume. To illustrate, Japan's New Tokaido Line was completed in 1964 at a cost of $380X10^9$ yen ($1056 million). Financial results for the fiscal year ending March 31, 1972[4] are as follows:

	10^9yen	(10^6 dollars)
Revenue	199	(553)
Expenses		
Operating costs including maintenance	51	(142)
Interest and Depreciation	40	(111)
Profit	108	(300)

TABLE   I

TOTAL TRANSPORTATION DEMAND PROJECTIONS

(millions of annual passenger trips by all modes)

Corridor	Year			Percent Growth	
	1975	1985	1995	1975-1985	1985-1995
Chicago-Detroit	6.2	8.0	10.2	29%	27%
Portland-Seattle	3.1	4.4	5.8	42	32
San Diego-Los Angeles	33.7	49.7	68.0	47	37
San Diego-Sacramento	65.3	94.8	128.3	45	35
Washington-Boston	203.0	300.0	444.0	48	48

TABLE II
PASSENGER DELAYS AT
NORTHEAST CORRIDOR AIRPORTS

	Year	Passenger Operations/Hour		Average Delay Minutes/ Passenger	% Passengers in Peak Periods
		Airport Capacity	Average Demand		
Kennedy	1970	5,200	2,370	8.7	25
	1975	8,260	3,767	9.0	25
	1985	16,220	10,160	46.8	60
Laguardia	1970	2,680	1,232	7.7	22
	1975	3,880	2,046	20.0	41
	1985	6,760	5,060	Saturated	—
Wash. Nat'l	1970	2,330	1,142	12.6	33
	1975	3,280	1,567	11.0	29
	1985	5,880	1,940	2.1	4
Wash. Dulles	1970	2,220	252	< 1.5	0
	1975	4,620	567	< 1.5	0
	1985	9,090	2,100	< 1.5	0
Boston	1970	3,600	1,120	1.9	2.8
	1975	4,950	1,800	2.9	8
	1985	10,480	4,500	6.8	20

The short term answer to our nation's high speed ground transportation needs is typified by the upgrading of passenger rail service in the Northeast corridor, first with Metroliner service and more recently by provisions of the Rail Services Reorganization Act of 1973 which, by means of improvements in right of way and passenger equipment, will reduce the trip time from Washington to New York to 2 hours and from Boston to New York to 2 1/2 at a cost of several hundred million dollars. The response to the introduction of Metroliner service in 1969 was a prompt reversal in the long term decline in through rail passengers [5] between New York and Washington even as air traffic encountered a hesitation in its growth (figure 1).

In less densely travelled corridors, major investments in improved trackage and electrification may not be economically warranted. An attractive concept in such cases is an "improved passenger train" (IPT). The improvements sought are those that would allow a 240 km/h cruise speed on existing railroad rights-of-way. To be sure the generally dilapidated conditions of trackage and wayside facilities must be upgraded for high speed operation, as was done for the Metroliner between Washington and New York, however new alignments are to be avoided. What this requires is an advanced suspension system which allows the train to negotiate the sharp curves of the existing routes at high speed with safety and passenger comfort, and also a high power-to-weight propulsion system which permits spending maximum time at full speed. Such improvements are not yet state of the art. The turbine powered, improved suspension trains which Amtrak operates between Chicago and St Louis, and New-York and Boston cannot perform above 150 km/h without major realignments of those routes.

Improved rail can provide a superior alternative to highway with respect to speed and cost, but what about those corridors where the predominent flow between major cities is via air rather than highway? Figure 1 presents the trend of origin-destination air traffic between the metropolitan areas of the California [6] and Northeast [7] corridors where the airplane has outpaced the automobile. The demonstrated need for short trip times between these cities is incompatible with the speeds allowed by steel wheel on steel rail. Consequently, for the long term, an alternative which offers the short trip time of air with the huge capacity of rail will be needed.

FRA's Office of Research, Development and Demonstrations is evaluating several technological alternatives to meet this need. All involve vehicle suspensions which avoid mechanical contact with the guideway. Accordingly they are designated "tracked levitated vehicles" or TLV for short. Typical TLV concepts are illustrated in figure 2.

ANNUAL ORIGIN-DESTINATION TRAFFIC VOLUME
FOR THREE CITY PAIRS

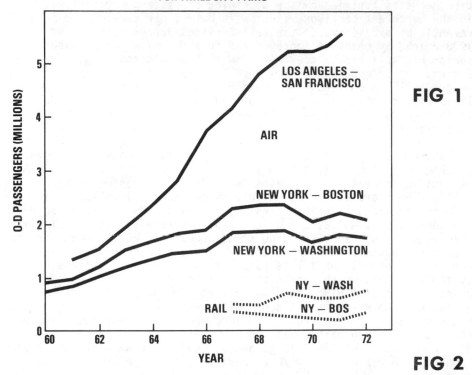

FIG 1

FIG 2

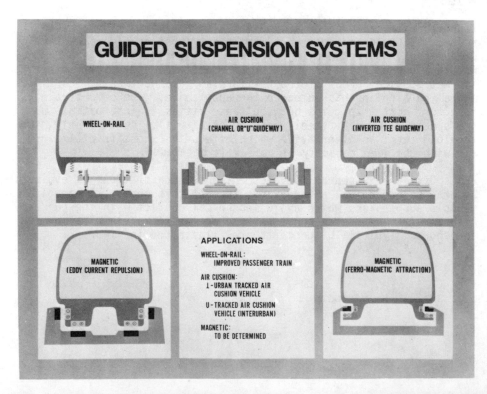

Lift and guidance forces are provided by cushions of pressurized air
in the case of the tracked air cushion vehicle (TACV), or by electromagnets in the case of the tracked magnetically levitated (Maglev)
vehicle. The Maglev suspensions are of two kinds. Attraction Maglev
utilizes electromagnets on the vehicle attracted upward toward elevated steel rails on the guideway. The instability of the attractive
force is overcome by servo-controlling the magnets in response to
signals from gap sensors. Repulsion Maglev employs high current density superconducting magnets on the vehicle which floats over an
aluminum guideway. Eddy currents are generated in the aluminum by
the forward motion of the magnets and these produce a stable repulsive force on the magnets. In common with the airplane, a critical
speed, typically 70 km/h, is required for lift-off.

Overall power versus speed for the TLV systems just described are
given in figure 3.[8] Aerodynamic drag accounts for the majority of
the cruise power at high speed. It has been assumed that the
aerodynamics are identical for all TLV's. In reality there may be
important differences. The repulsive Maglev system enjoys a much
larger air gap than the other systems, while attractive Maglev can
use a less enclosed guideway, either of which is bound to have an
effect on air drag. The balance of the cruise power is that consumed by the support and guidance systems. For the TACV system,
support and guidance power includes the cushion compressor power and
associated momentum (or captation) drag power (to bring captured air
up to vehicle speed). For repulsive Maglev, eddy currents dissipated
in the guideway give rise to a magnetic drag which varies as the lift
divided by speed and aluminum thickness. By comparison cryogenic
power for the superconducting magnets is trivial. Attraction Maglev
requires large reactive power to stabilize the suspension but the
dissipated power is only a few kilowatts per tonne. More important
is the eddy current dissipation in the steel support rails. This is
a complex function of speed, rail conductivity and magnet geometry,
and results in reduction in lift as well as drag. Considerable
uncertainty exists in the attraction Maglev suspension power requirements at speeds above 300 km/h.

Energy requirements for alternative transportation systems are tab-
ulated in Table III and shown graphically in figure 4.[9] On a seat-
mile basis, long trains use the least energy and jet planes the
most. This is an obvious consequence of the fact that aerodynamic
drag predominates. The beneficial effect of train length on energy
consumption is shown in figure 5,[10] while the severe penalty of
speed is shown in figure 6.[11]

It must be firmly borne in mind that these alternative travel modes
are not necessarily equivalent substitutes for each other. Different
travelers have different needs. For some, speed is of primary concern, for others, door to door convenience or cost. Auto, conventional rail, IPT, TLV and airplane each appear to have a distinct clien-
tele and a viable role in a balanced national transportation network.

MOTIVE POWER REQUIREMENTS
FOR TLV SYSTEMS

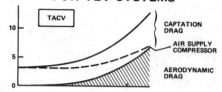

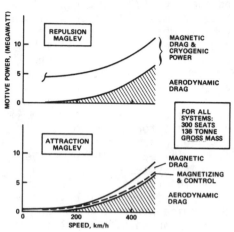

FIG 3

FOR ALL
SYSTEMS:
300 SEATS
136 TONNE
GROSS MASS

TABLE III

GROUND TRANSPORTATION MODES
ENERGY AND POWER CONSIDERATIONS

Mode	Number of Seats	Vehicle(s) Gross Mass Tonne	Cruise Speed km/h	Number of Vehicles	Output Cruise Power MW	Aero Drag Power MW	Efficiency (Fuel to Motive Pwr)	Specific Energy MJ/ Seat–km
Auto (U.S. Compact)	4	1.4	97	1	.016	.010	.17[6]	.86
Bus	53	12.7	97	1	.13	.084	.32[5]	.29
Metroliner	382	477	200	6	2.21	1.53	.25[2]	.41
NTL	1407	865	210	16	–	–		.28
Turbotrain	144	116	240	3	1.07	.85	.16[3]	.68
	144	116	275	3	1.49	1.23	.16	.85
	326	200	200	7	1.19	.91	.16	.40
IPT	300	200	200	6	.88	.63	.28[4]	.20
	150	100	240	3	.94	.76	.28	.33
	300	200	240	6	1.58	1.23	.28	.28
	600	370	240	11	2.66	2.02	.28	.23
	900	540	240	16	3.75	2.83	.28	.22
	300	200	275	6	2.21	1.78	.28	.34
	300	200	240	6	1.58	1.23	.32[5]	.25
TLV: TACV	100	45	485	1	5.65	3.66	.25[2]	1.66
	300	136	485	3	12.6	6.64	.25	1.25
	600	273	485	6	24.6	12.6	.25	1.14
	800	364	485	8	30.1	14.1	.25	1.11
Repulsion MAGLEV	300	136	485	3	11.2	6.64	.25	1.11
Attraction MAGLEV	300	136	485	3	8.8	6.64	.25	.86
DC–9–30[7] (for Ref.)	115	45	900		–	–	–	1.66

[2] All–Electric Propulsion;

[3] Aircraft Gas Turbine: Efficiency n = .18, Drive Train Efficiency n_{DT} = .87;

[4] Regenerative Gas Turbine: n = .32, n_{DT} = .87

[5] Diesel: n = .37, n_{DT} = .87;

[6] Otto Cycle: n = .20, n_{DT} = .85;

[7] 300 Mile Trip, Including LTO Cycle

SPECIFIC ENERGY REQUIREMENTS FOR VARIOUS GROUND TRANSPORTATION MODES

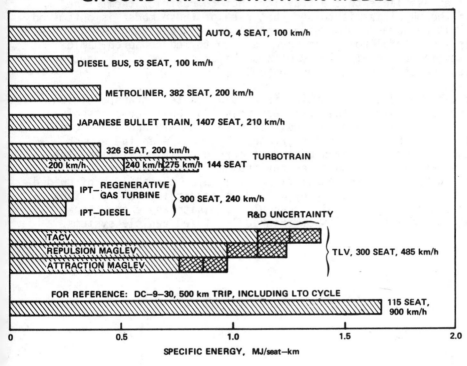

FIG 4

ƎFECT OF TRAIN LENGTH ON ƆIFIC ENERGY CONSUMPTION TLV AND IPT MODES

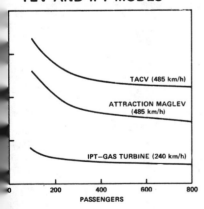

FIG 5

MOTIVE POWER REQUIREMENTS FOR VARIOUS HIGH SPEED GROUND TRANSPORTATION SYSTEMS

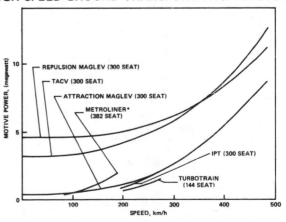

FIG 6

The proper function of the government is to encourage the expansion of choices available to the traveler, not to restrict them. The task of the physicist is to see that each of these modes is optimized so that it consumes a minimum amount of energy consistent with serving its most useful role. Improved design to reduce aerodynamic drag, and increased efficiency of propulsion systems can have the greatest impact on energy consumption.

High speed ground transportation is a solution to a transportation and land use problem. The near term solution, which consists of improving rail passenger service, can have a timely and beneficial effect on the current energy shortage. By 1985, when the demand for much shorter trip times outstrips corridor air facilities, the nation's capacity to produce energy from its vast resources should be back in balance with consumption. This plus the ability of guided ground transportation to employ wayside electrical power, and thus to be independent of petroleum, should eliminate any energy constraint on the introduction of tracked levitated vehicle systems.

References

1. U.S. Department of Transportation, OST Office of Systems Analysis and Information "High Speed Ground Transportation Alternatives Study" p. 5-7 (1973) NTIS Accession # PB 220079

2. 1967 Census of Transportation, Vol I, <u>National Travel Survey</u>, Bureau of the Census, U.S. Dept. Commerce, Washington, D.C. (1970).

3. U.S. Department of Transportation "Recommendations for Northeast Corridor Transportation" Vol II, p. 5E-4. (1971).

4. Data on the New Tokaido Line furnished to DOT by Japanese National Railways, Tokyo, November 1972.

5. U.S. Department of Transportation, FRA Office of Economics "Rail Passenger Statistics in the Northeast Corridor" (issued annually)

6. California Public Utilities Commission "Intrastate Passengers of Scheduled Air Carriers Between Major Metropolitan Areas" Form 1511, 1960 through mid 1972.

7. Civil Aeronautics Board "Domestic Origin - Destination Survey of Airline Passenger Traffic, Domestic" (1960 through 1972). Corrigenda to same dated October 9, 1973 (The latter corrects gross under-reporting of Eastern Air Lines shuttle data for 1971 and 1972)

8. Reference 1, p. 3-22

9. Reference 1, pp. 3-16 and 17 (revised)

10. Reference 1, p. 3-24

11. Reference 1, p 3-23

OPPORTUNITIES FOR PHYSICISTS IN THE AREA OF
ELECTRIC POWER DELIVERY

Allan Greenwood
Rensselaer Polytechnic Institute, Troy, N. Y. 12181

ABSTRACT

A physicist, wishing to make innovative contributions
to the field of electric power delivery can approach the
subject in at least two ways. He may familiarize himself
with what his experienced engineering colleagues believe
to be the barriers impeding progress and attempt to apply
his talents to those areas closest to his personal expertise. Alternatively, he can acquaint himself with the
overall functions and goals of transmission systems and
attempt to devise quite new and different solutions to
fulfill these functions and attain these goals, unimpeded
by preconceived notions. Both approaches are considered.

A power delivery system is described in a quite
fundamental way, in terms of function rather than hardware so as to point up the limitations of present methods
of energy transmission and hopefully stimulate thought on
new methods.

A spectrum of physical problems are cited which ob-
struct progress towards such goals as higher transmission
efficiency, lessening of environmental intrusion and improving system reliability. Potential solutions ranging
from solid state physics to plasma physics are discussed.
The important area of materials, a traditional frontier
in almost any circle of engineering, is examined.

The point is made that a strong dialogue must be established between physicists and engineers if the essential relevance of their work is to be achieved.

INTRODUCTION

The function of an electric power delivery system is to convey
electrical energy from the points of its generation to the points of
its ultimate use. Electrical energy is most readily generated and,
for the most part, most conveniently used, as alternating current.
For these reasons a-c is the preferred mode of operation in most instances. It also happens that, for a variety of reasons, it is more
desirable to perform the function of bulk power delivery (transmission) at much higher voltages than those used for both its generation
and its delivery to the consumer (distribution). The ability to
transform up and down in voltage with alternating current is therefore a further advantage for its use.

Whilst the power transmission and distribution lines and the
transformers that connect these with each other, with the system
generators, and with the consumers' equipment, are the most obvious

components of a power delivery system, there are in fact many others.
For example, there is a vast quantity of switching equipment used to
direct the power down the desired routes and isolate faulty sections
of the system when short circuits develop. There is equipment to
monitor and control the system and equipment such as surge diverters
and lightning arresters to protect terminal apparatus against damaging overvoltages.

It is the purpose of this paper to introduce the physicist to
electric power delivery systems so that he can see them as they are
and how it appears they are going in the years ahead. In particular,
it will acquaint him with some of the constraints that impede progress in the development of such systems. These are economic and
sociological as well as technical. Hopefully, as a consequence, some
physicists will be motivated to apply their science to the business
of power delivery in such a way as to remove or lessen the sum of
these constraints and thereby help to develop better systems for the
future.

A physicist, wishing to make innovative contributions to the
field of electric power delivery can approach the subject in at least
two ways. He may familiarize himself with what his experienced engineering colleagues believe to be the barriers impeding progress and
attempt to apply his talents to those areas closest to his personal
expertise. Alternatively, he can acquaint himself with the overall
functions and goals of transmission and distribution systems and
attempt to devise quite new and different solutions to fulfill these
functions and attain these goals, unimpeded by preconceived notions.
Both approaches are considered.

CALIBRATION

As part of the familiarization process just described, it is
useful to become calibrated to the dimensions of power delivery systems. At the present time the electric power systems of the United
States deliver approximately 1.9 trillion kilowatt hours per year
(it is perhaps worth reminding the reader that the unit of electrical
energy, the kilowatt hour, for which we typically pay a few pennies,
is the equivalent of a 180-lb. man climbing a ladder three miles high
with an efficiency of 100%). The load has been growing for some time
at an annual rate of between 7 and 8%, which means that the energy to
be delivered doubles about every ten years. This rate is twice that
of the gross national product, two and one-half times the rate of
growth of energy consumption as a whole and six times the rate of
population growth. It is to be expected that the current energy
dilemma will affect this growth to some extent. On the other hand,
it is likely that there will be shifts to electrical energy from
other less environmentally acceptable forms; this will have an off-
setting effect.

The power handling capability of a power transmission line
varies approximately as the square of the operating voltage (1).
Thus a 230 kV transmission circuit can carry approximately three
times the power of a 138 kV circuit and a 500 kV can carry more than
four times the power of a 230 kV circuit. The highest voltage in

commercial use in this country is 765 kV. The surge impedance loading of such a line (a convenient way of expressing the power handling capability of such circuits) is in excess of 2000 megawatts.
This would roughly supply the peak load requirements of a city of
one million people. In practice, of course, such a population center
would not entrust its lifeline to one circuit.

Typically, the current capacity of a power transmission line is
of the order of 1000 to 3000 amperes. Under severe fault conditions
when a short circuit occurs close to a source of generation, the
current may momentarily rise to many times this value, 40,000 A is
not unusual and systems are being planned where the fault current
level is in the range of 60,000 to 80,000 A. Recognizing that these
are root mean square values, it will be appreciated that the peak
asymmetrical fault current can approach more than twice these numbers. The electromagnetic forces brought into play on closely spaced
conductors when carrying such currents are simply enormous. These
figures are also helpful in giving calibration on the interrupting
capability required of circuit breakers that must remove such faults
from the system.

SOME IMPORTANT CONSIDERATIONS

Traditionally, the electric utility industry has seen its task
as that of supplying electric power needs of an energy-hungry society
at minimum cost. The requirements for this, stated briefly, have
been:
1) Meeting the load.
2) Minimizing the cost.
Meeting the load means satisfying the need for energy when and
where it arises and doing so with only modest fluctuations of voltage
at the customers' terminals as the load varies. In speaking of cost,
one means the total cost, that is, the investment required to install
the system and the cost of operating it, most importantly the cost
of transmission and distribution losses.

Implicit in the first requirement is "availability" which depends upon the reliability of the system. Reliability has become
increasingly important and should now be stated as a third requirement.

In the relatively recent past another factor has assumed very
considerable importance. This can be stated succinctly as:
4) Minimizing the intrusion on the environment.
The forest of utility towers that dominate certain parts of our
landscape are no longer acceptable, overhead crossings of rivers are
being restricted, and there is a growing sentiment against pole-
mounted distribution equipment in residential areas.

It will be recognized that quite often these requirements, 1
through 4, will be in conflict with each other. To provide redun-
dancy in order to increase overall reliability, or to install an
underwater cable instead of an overhead crossing for a river, set
requirements 3 and 4 in conflict with requirement 2. Moreover, these
requirements are not static, as situations change so will the requirements change. For example, at a period of acute energy shortage

and rising costs for energy, greater attention must be paid to energy conservation which means reducing losses in a power delivery system.

FUTURE TRENDS IN ELECTRIC POWER DELIVERY SYSTEMS

It seems almost inevitable that as the demand for electrical energy increases, the voltage at which it is delivered will have to be increased. The historical pattern as far as transmission (2) is concerned clearly indicates this, indeed, the interval between the introduction of new voltage levels has been diminishing. If these trends were to continue one would suppose that the next higher transmission voltages would be somewhere between 1100 and 1200 kV. The development of such a system will be an immense task.

A great deal of work to elucidate the problems of operating at ultra-high voltage (UHV) is already underway (3,4). A notable example is the project being conducted by the General Electric Company under the auspices of the Electric Power Research Institute (5). Some of the results of this effort and some of the problems remaining to be solved will be described later.

Most UHV research and development to date has been focused on transmission lines. Ultimately a matching effort will have to be made with respect to terminal equipment, most notably power transformers and circuit breakers.

The raising of voltage will not be limited to bulk energy transmission but will extend through the range of operating voltages down to the distribution level. Until a relatively short time ago, 15 kV was considered an upper limit for electrical energy distribution. In recent years more and more distribution circuits have been put into operation at 34.5 kV and this trend is expected to continue. However, for the most part, the needs here are for more innovative engineering design work rather than for basic physical research.

Continuing in the area of transmission, it is likely that greater use will be made of direct current. High voltage direct current transmission (HVDC) has a number of distinct advantages (6). For instance, with a-c it is necessary to insulate against the peak of the sinusoidal voltage and the transients associated therewith, although of course, this is considerably higher than the effective rms value of the voltage, for which an equivalent d-c line would be insulated. It is not necessary with d-c to cyclically charge and discharge the capacitance of the system at the power frequency; this is particularly important in cables where the capacitance is relatively high and the charging current is correspondingly great. This factor virtually eliminates the use of cable in lengths greater than 20 or 30 miles in a-c circuits.

There are a number of operating advantages associated with d-c; for one thing it is conducive to limiting short-circuit currents. For another, it is possible to provide asynchronous ties, that is, interconnections between a-c systems that are not running in synchronism. During periods of disturbance in an a-c system when machines are swinging relative to each other and there is a danger of loss of synchronism, the presence of d-c lines with their associated control can be most helpful in stabilizing the situation.

The outstanding disadvantage of d-c is the very considerable cost of the conversion equipment. As will be described later, great strides have been made in recent years, but some major breakthroughs are required if the cost of the equipment is to be greatly reduced.

Environmental considerations mandate greater use of underground as against overhead for power transmission and distribution. This is a situation where environmental needs and cost meet head on. At the distribution voltage level, undergrounding may be two to three times more expensive than overhead construction, but at transmission voltage it may be ten to twenty times more costly than overhead. Inspite of this, there will be a trend to more underground. A considerable amount of innovative work is currently underway and in the cable area the Electric Research Council (now the Electric Power Research Institute) has established an experimental test site at Waltz Mill near Pittsburgh, which is operated by the Westinghouse Corporation.

Another important trend is the movement towards compaction. The scarcity and cost of real estate in many areas make it absolutely necessary to make the most out of existing power corridors and new routes that may yet be found. Compaction is expected to be applied to overhead construction per se as well as the natural compaction that occurs when a circuit is put underground. Where overly conservative designs have been used for overhead transmission and distribution, they are expected to be replaced by more compact designs and some entirely new concepts are being considered. One such proposal (7) that would give a low profile and therefore less obtrusive UHV transmission lines would make use of a restricted access right of way (it is possible to walk freely under most existing high voltage power transmission lines). The practicality of this is still open to question, however, it may find additional interest in combination with other utility services such as high speed and limited access highways, where the inconvenience of restricted access can be shared with other functions.

Compaction will not be restricted to transmission circuits alone, but will encompass terminal equipment as well. Existing high voltage transformer and switching stations often cover a considerable area. In metropolitan areas such sites are extremely expensive if they can be found at all. Moreover, some people find such structures aesthetically offensive. Compaction here means compressing all of the equipment into a relatively small space and containing it in the metal cubicles which can be approached with impunity from the outside. Some so-called "minisubstations" of this kind are already in existence overseas (8).

FUNCTIONAL DESCRIPTION OF SOME IMPORTANT POWER DELIVERY COMPONENTS

Before attempting to identify those areas where physical and technological barriers are impeding the progress of development in power delivery systems, a few of the more important components of such systems will be described functionally. This should provide a better idea of the problems involved when they are later described and perhaps offer some incentive for developing entirely new concepts that could serve the same functions.

A transmission line can be viewed as an energy guide comprising a number of parallel metallic conductors immersed in a dielectric medium. Most frequently the dielectric is air, though in cables it may be oil-impregnated paper, new polymers or other materials. An overhead line must be supported in the dielectric medium with adequate space between the conductors and between the individual phases and ground, which includes all grounded structures. This is usually accomplished by suspending the phase conductors from towers by strings of porcelain or glass insulators. These insulators, together with the ambient air, form an insulation system for the line. The geometry of the system dictates that the electrical gradient is greatest at the surface of the conductors. This limits the design in as much as an excessive gradient will result in an unacceptable corona loss and electromagnetic "noise."

If losses are neglected, the power measured by overall terminal quantities in any means of power transmission must be equal to the surface integral of density in some form of stress over a cross section of the transmitting medium. In an electric transmission line this can be stated in terms of the Poynting vector:

$$S = E \times H \tag{1}$$

In Eq. (1), S is equal to the power density in watts per square meter, E is the electric field strength in volts per meter, and H is the magnetic field intensity in amperes per meter. Since Maxwell's equations relate E and H by the intrinsic impedance of the transmission medium Z_i, and magnitude of the Poynting vector can be expressed in terms of E and Z_i as follows:

$$S = \frac{|E^2|}{Z_i} \tag{2}$$

The integral of the Poynting vector over the entire cross-section through which the line passes must be equal to the power as measured at the line terminals; i.e.,

$$|U||I|^* = \int_x \int_y S_z \, dx \, dy \tag{3}$$

It is instructive to examine the density limits implied by Eq. (2), knowing that Z_i for air is 377 ohms and that the field intensity, E, is limited by the electrical breakdown of air, approximately 30 kV per centimeter. A calculation of this limit shows (9) the limiting power density to be approximately 12,200 MW per square meter. Densities of this magnitude are approached only very close to the surface of the conductors in traditional power transmission line construction. It is estimated that on typical 3-phase lines over 95% of the energy stored in the field surrounding the conductors is contained within a radius equal to 5% of the phase-to-phase spacing.

This is, of course, a consequence of the line geometry, but it indicates how relatively inefficiently the transmitting medium is being used.

The second principal component of interest is the transformer. In principle, it is the simplest of devices, being no more and no less than a steel-cored mutual inductor, whose function would appear to be completely described by Faraday's Law. In fact, a modern power transformer is an extremely sophisticated piece of equipment. It is an aggregate of copper (or aluminum), grain-oriented steel and a complex multicomponent insulation system which sometimes must simultaneously sustain extremes of thermal, electrical and mechanical stress. The design of such equipment requires a most judicious compromise between the needs of heat transfer, dielectric integrity (under transient conditions the electrical stresses can be very non-uniform), and mechanical strength to withstand the crunching mechanical forces that attend short circuits. It is extremely doubtful that a transformer will ever be replaced by anything but another transformer, on the other hand, as will be discussed later, there are ample opportunities for innovation, especially in the area of materials.

One of the most challenging components in a power system from a physicist's point of view is the power circuit breaker. The function of this device is to close and open power circuits, but most especially, to interrupt fault currents when short circuits occur.

When the circuit breaker is opened, current is not interrupted immediately, it continues to flow through an arc which is at once established between the separating contacts of the breaker (10). The arc comprises a body of extremely luminous, very hot gas or high pressure plasma which derives its electrical conductivity from the presence of free charge carriers. Compared with metallic conductors, conductivity of the circuit breaker arc is quite low, but it is sufficient to support the current with an arc voltage which, in most instances, is small compared with the system voltage. Typical voltage gradients in circuit breaker arcs are in the range of tens to hundreds of volts per centimeter length of the arc. The core temperature is likely to be in the range of $10-15,000°K$.

Typically, the core current density is in the range of 5000-10,000 amperes per square centimeter. Taken with the voltage gradient this implies a power density of the megawatts per cubic centimeter. This energy, of course, comes from the power system and in "steady-state" conditions is dissipated in conductive, convective, and radiative losses from the arc. To interrupt the current the arc must be controlled, indeed, it must be quenched. Put another way, the energy balance just described must be destroyed. This is the function of the elaborate arc-control devices of every circuit breaker be they gas-blast or what have you.

Fortunately, at least in a-c circuits, the current passes through zero twice each cycle. At this instant the input to the arc is momentarily zero. The attack on the arc may be launched almost as soon as the arc is established, but it is only during this critical time around current zero that interruption can be effected. Basically, the objective is to so reduce the conductivity of the arc that, inspite of a considerable voltage being impressed across the contacts after current zero (11), the current is not reestablished. The principal way in which this is done is by cooling the arc since the conductivity is extremely temperature sensitive. In the critical range of temperature associated with thermal ionization, conductivity changes perhaps 10^8 or 10^9 times faster than the temperature (10). It is instructive to determine the cooling rate required. Experiment has established that to reduce the electrical conductivity to neglibible proportions the arc-plasma must be cooled to about 2500°K (12). This must be accomplished during a very brief interval around current zero, for under certain circumstances the recovery voltage impressed across the residual plasma by the power system can be applied at the rate of several kV per microsecond. In such situations the cooling time may be as short as 10 microseconds or even less. These figures indicate that the arc must be cooled at the rate of about 10^9°K/sec. From this point of view we see the circuit breaker functionally as an extremely efficient gas cooler.

OPPORTUNITIES FOR PHYSICISTS

Areas in need of basic research run the spectrum from high vacuum, through the gas phase, into liquids for dielectrics and finally into many different aspects of the solid state. On the temperature scale these extend from the temperature of liquid helium to that of a fully ionized plasma. So far the material has been presented technologically or by device. What follows will cut across these lines and be arranged more by discipline.

Opportunities for Gas Discharge Physicists and Aerodynamicists

Great strides have been made in recent years in obtaining a better understanding of circuit breaker arcs. Many of the contributions made have come from investigators with fundamental training in physics and have been both theoretical and experimental in nature. In most types of circuit breaker the arc exists in a high-speed, usually supersonic, gas flow with regions of both turbulent mixing and laminar flow. Further effort is required to better understand this complex system, particularly in electronegative gases such as sulfurhexafluoride which are presently being used as switching ambients. In some situations conditions are further complicated by the introduction into the arc of other materials from ablating, nonmetallic nozzles. There is a general need for better diagnostic techniques to study conditions during the brief interval of a few microseconds around current zero. A more refined mathematical model to describe conditions at that time is also desirable, as are better flow visualization techniques.

Another area of interest to the gas discharge man is corona. This represents a loss from overhead transmission lines and other high voltage equipment as well as being a source of radio and television interference. Recently, increasingly severe problems have been experienced with audible noice from overhead conductors. On some 765 kV circuits this can sound like the igniting of fire crackers. More work on this phenomenon is required.

Whereas in the past the criteria for the design for overhead line insulation have been set by lightning or switching surge considerations, indications are that the 60 Hz operating voltage of the line will be more important at UHV levels. This reflects the importance of insulator contamination. When such contamination is present, and it is a common occurrence, extremely complicated conditions can be set up on the surface of insulators. Under foul weather conditions some contaminants will combine with atmospheric water to form conducting films which partially bridge sections of the insulation producing highly nonuniform voltage distributions in the insulation system. These, in turn, lead to scintillations or small surface discharges which ultimately develop into insulator flashovers (13).

The subject of electrical breakdown will be addressed in a companion paper. However, it is perhaps worth pointing out a phenomenon quite often experienced by those experimenting with UHV transmission lines. It is observed that when breakdown occurs through the air as a consequence of switching surges, the path taken by the discharge is not necessarily the shortest one available to it (14). Incidents have occurred where the nearest ground point was 26 feet away, but the breakdown took place to a grounded object between 80 and 150 feet away.

Generally speaking, progress in UHV overhead power transmission line design is being accomplished by quite elaborate, and sometimes quite costly experimentation. Of necessity, the design criteria being developed are of an empirical nature. There is a continuing need for basic studies to put this information on a more secure, physical foundation.

Opportunities in Low Temperature Physics

The trend towards compaction and the need to conserve energy which were referred to earlier in this paper have led to work being undertaken in the area of cryogenic power transmission. It is common knowledge that the resistivity of metallic conductors decreases with temperature and that with so-called super conductors it disappears altogether at a sufficiently low temperature. The purpose of the present work is to exploit these facts in new underground cable technologies. In the natural course of events cables cannot conduct as much power as overhead lines of comparable conductor cross section because of the problem of extracting the heat generated by ohmic losses from a system which is relatively well insulated, thermally, by the dielectric and the surrounding earth. Reducing the losses helps in this regard as well as being a virtue in itself.

Two approaches are being pursued, the first is the resistive cryogenic transmission system (15,16). Here the cable would be operated at the temperature of liquid nitrogen (approximately 80°K), at which temperature the resistivities of aluminum and copper are substantially lower than the temperature values. A short length of prototype cable of this type has already been constructed and successfully tested. It is cooled by liquid nitrogen and insulated by Tybek tapes. A three-phase cable in an 18" pipe has been designed to carry 3500 MVA at a system voltage of 500 kV.

The second method is the superconductor approach (17,18). Research has shown that niobium has sufficiently low a-c losses to suggest its use for an a-c transmission system and work is underway with the object of demonstrating the feasibility of building an economic system for the transfer of power in the order of 1000 to 10,000 MVA at voltages in the range of 138 to 345 kV. This system envisages the use of niobium plated copper conductors supported on dielectric spacers. Liquid helium at about 4°K would provide both the electrical insulation and the cooling medium. A thermal insulating system combining the use of high vacuum and insulation will be necessary. Refrigeration plants would be installed at intervals of approximately 5 miles.

There are of course many problems to be solved before such a system can become operational. Of special interest will be any advances that physicists can make in higher temperature superconductors.

Opportunities in Solid State Physics

Solid state physics is probably today's fastest growing branch of physics. The work on superconductors just mentioned would probably come in this category. Another area is semiconductors.

Only a few years ago electric power conversion equipment used mercury arc devices for rectification and inversion. The biggest HVDC system presently installed in this country uses equipment of this type. Since its installation this particular technology has been almost completely superceded for HVDC transmission by equipment based on the solid state thyristor. Embryonic devices of a little more than a decade ago, silicon thyristors now function in power converters that handle hundreds of megawatts (19,20). This is an extraordinary record of achievement.

Remarkable as this advance has been, there remain many opportunities to improve device performance. The voltage capabilities of present devices is relatively low (a few kV), so that for high system operating voltages many devices must be corrected in series. This presents technical and economic problems that could be greatly reduced were it possible to operate individual modules at a much higher voltage.

The forward volt drop of thyristors in their conducting state represents a significant energy loss when the device is carrying its rated current. The cause of energy conservation and efficiency would be served if semiconducting materials with a lower forward volt drop could be developed.

The technology of lightning arrestors and surge diverters is a also dependent upon semiconducting materials, the one most commonly used being silicon carbide. The particular feature that is sought in this application is a highly nonlinear volt/current characteristic. New, stable materials of this kind that would increase the flexibility and the design of surge suppressing devices would be warmly welcomed by power engineers.

Opportunities in the Materials Area

In almost any technology it is true to say that the frontier is
only pushed forward when new material problems are solved. This is
assuredly true in many areas of the power delivery business. This
subject is really all-embracing, it includes conductors, insulators
and the semiconductors referred to in the last section. Indeed, it
covers all aspects of solid state, liquid phase and gas phase also.

In the materials area physicists would probably operate as a
member of a team, working with chemists or metallurgists. His contribution might be theoretical, or experimental as in the area of
evaluation and measurement.

As an alternate to overhead transmission and to avoid some of
the problems associated with the high capacitance of conventional
underground cables, so called compressed gas insulated bus systems
are being developed. The conductors are usually concentric cylinders
with the outer cylinder grounded. The principal insulation between
the cylinders is compressed gas, but of course, ingenious insulating
supports must be provided at quite frequent intervals. Sulfurhexafluoride is the principal gas being used, but other gases and mixtures of gases are under study. The behavior of such gases as
coolants and as dielectrics in extremely nonuniform fields is of
considerable interest. Such gases will be integrated into the compact terminals mentioned earlier. As operating voltages are progressively increased and the trend to compaction continues, a great
effort must be made to wring the most out of insulation systems of
this type.

Oil has been the principal dielectric in electric power components for many years. It is used in transformers, circuit breakers
and pipe-type cables, in all of which it serves the dual purposes of
insulant and coolant. New liquids for general or special duties are
being sought. The knowledge of electrical breakdown in liquid
dielectrics is considerable but far from exhaustive. This is a very
fertile field for the physicist.

The requirements of future electric power delivery equipment
will place ever more stringent requirements on solid insulation.
Improved polymers and ceramics will be necessary which, under the
trends of increasing voltage and energy density, will have to perform
near their ultimate potentials. To determine these limits and to
prepare components which will perform at these limits, will require
not only expanded knowledge of the responses of the materials to
electrical, thermal and mechanical stresses, but studies of interaction of these stresses and response of the materials to their combined effects.

Magnetic steels continue to play a very important part in transformer technology. The improvement in these materials as a consequence of better basic understanding has had a significant effect on
transformer design and performance over the last two decades. This
experience encourages the belief that more can yet be done. The matter of voltage control on long transmission lines at high voltage
through magnetically saturating devices is currently being given much
attention. These may well require special magnetic characteristics.

Needs of this kind are often the mother of invention and therefore
generate more fundamental work in materials development.

Opportunities in the Area of Heat Transfer

There is scarcely a piece of power equipment whose design is not
thermally limited. Whilst it is important to minimize losses, one
must also recognize that losses will inevitably exist. The extraction of these losses, which appear as heat, determines the size and
therefore the cost of such components as transformers, reactors,
power circuit breakers, etc. More use will have to be made of
devices such as heat pumps. It is to be expected that much of the
progress that will be made will derive from innovative engineering
design. A recent example of this is the electromagnetic pumping of
liquid metal through the heat sink and across the surfaces of silicon
wafers in large solid state conversion equipments. However, better
designs cannot be fully exploited without a fundamental knowledge of
processes going on. This is where the physicist can help. Familiarity with the needs sporns new ideas from fresh minds.

The opportunity areas outlined above necessarily overlap; heat
transfer amd materials would be a case in point. Most good electrical insulators are also reasonable thermal insulators. Bearing
in mind the physical phenomena involved, this is perhaps not surprising, but it is a distinct disadvantage in most pieces of electric
power equipment. Perhaps a harder, fundamental look at this area to
see if one cannot "have one's cake and eat it" would be worthwhile.
The consequences of developing a practical engineering material with
good dielectric strength and good thermal conductivity would be considerable.

SUMMARY AND ASSESSMENT

It may appear at first sight that this paper simply calls for
"more of the same", and that, like a trolley car, the industry does
not have much space in which to move. To some extent this is true.
The electric utility industry is the ultimate in capital intensive-
ness, it must therefore be relatively conservative. The life expectancy of power equipments is typically thirty years; radical
change can be expensive; serious error can be disastrous. However,
if one looks back over the last twenty or thirty years, one observes
that a very considerable evolutionary change has been taking place.
Moreover, new forces are now at work that mandate a continuation of
this process. It is probable that events will still proceed in an
evolutionary rather than a revolutionary manner.

Unquestionably there will be opportunities for contributions
from people in the physical sciences in the many areas that have been
outlined. It is most likely that these will be made by people
working alongside experienced power engineers and colleagues from
other disciplines. Such interdisciplinary groups now exist within
manufacturers' organizations and some utilities; their efforts have
already borne fruit.

Finally, there is still a place for the entrepreneur, who, from a different vantage point and with new ideas, can introduce from time to time some innovation which may have far-reaching effects.

REFERENCES

1. O. J. Elgerd, Electric Energy Systems Theory: An Introduction (McGraw Hill, New York 1971) p. 49.
2. D. D. MacCarthy et al, E.H.V. Transmission Line Reference Book (Edison Electric Institute, New York 1968) p. 3.
3. H. C. Barnes and B. Thoren, CIGRE (1970) Report No. 31-01.
4. J. Clade et al, CIGRE (1970) Report No. 31-05.
5. J. G. Anderson and J. M. Schamberger, CIGRE (1970) Report No. 31-07.
6. E. W. Kimbark, Direct Current Transmission, Vol. I (Wiley, New York 1971).
7. F. Reggiani, CIGRE (1972) Report No. 31-00.
8. G. Mauthe, R. Ottischnig and K. D. Schmidt, Brown Boveri Rev. $\underline{57}$ (1970) 562.
9. L. O. Barthold and H. C. Barnes, Electra $\underline{24}$, 139 (1972).
10. W. Rieder, IEEE Spectrum, $\underline{7}$, No. 7 (1970) 35.
11. A. N. Greenwood, Electrical Transients in Power Systems (Wiley, New York 1971).
12. T. H. Lee et al, CIGRE (1968) Report No. 13-06.
13. A. J. McElroy et al, Trans IEEE. PAS-89, 1848 (1970).
14. K. W. Priest et al, Transmission (General Electric Company, Schenectady) Dec. 1969, p. 6.
15. S. H. Minnich and G. R. Fox, Cryogenics (June 1969).
16. M. J. Jeffries and K. N. Mathes, Trans IEEE, PAS-89, 2006 (1970).
17. R. W. Meyerhoff, ASME Publication No. 72-PWR-7 (1972).
18. H. M. Long and J. Notaro, Jour. App. Phys. $\underline{42}$, 155 (1971).
19. F. H. Ryder et al, Proc. Am. Power Conf. $\underline{32}$, 866 (1970).
20. D. M. Demarest et al, Proc. Am. Power Conf. $\underline{35}$, 1170 (1973).

DIELECTRIC BREAKDOWN PROBLEMS
IN ELECTRIC ENERGY TRANSMISSION AND STORAGE

Thomas W. Dakin
Westinghouse Research Laboratories
Pittsburgh, Pennsylvania 15235

INTRODUCTION

The present electrical transmission and distribution systems in the USA and in many other nations has been developed gradually over about 75 years. Its purpose, of course, is to deliver electric power from generating stations to distant domestic and industrial customers. The vast extent to which the transmission system has grown is probably little noticed or appreciated by the average layman or even scientist, until some outage briefly alerts his attention with the sudden interruption of that electric power he has become accustomed to accept as routinely there.

Electric transmission voltages have been steadily climbing to accommodate the increasing amount of power handled, and the distances to which it must be transmitted. There is an unavoidable loss of power in the I^2R losses in the conductors. If the percentage loss due to this resistance is held constant, the power transmission capability of a line increases with the square of the line voltage. Thus, the desirability of increasing transmission voltages is obvious. This increase in voltage, however, does not come free, since the amount of electrical insulation and space required to contain the voltage has to be increased usually more than in proportion to the voltage. The dielectric breakdown problems associated with this insulation and voltage containment is the subject of this paper.

Electric generator voltages are commonly in the range of 13.8 to 34 kV, (phase to phase), with a predominance toward the lower end of this range. Due to the peculiar design problems of generators, generator voltages have been increasing less rapidly than transmission voltages. From the generator the voltage is transformed up, often in one step, to the main transmission voltage, if the generator is remote from the area of power use. AC transmission line voltage levels are now 69, 138, 245, 362, 550, and 800 kV, phase to phase, respectively. To indicate how rapidly transmission voltages have been increasing, the first commercially used 550-kV lines were energized in the mid 1960's, and the first 765-kV lines about 1970. There are 1100-kV and 1500-kV lines being considered. The phase, or line to ground voltage is $1/\sqrt{3}$ of this value for the usual 3 phase systems. At the consumer end of the high voltage line, the voltage is transformed down, usually in several steps, eventually to industrial use voltages, which may be as high as 13.8 kV for large motors and to the familiar 230 and 115 volts of domestic use.

The highest voltage electrical transmission lines have been particularly valuable for economically transporting power from remote energy sources, such as water power generating stations, like the Columbia River to California and Arizona, northern Canada to Montreal,

Toronto, New England and New York, and coal mine mouth stations in the Appalachians to the East Coast, etc. Such applications are likely to continue to increase as new remote energy sources are developed. Another very important application of the highest voltage lines is the intertie of generating systems from one part of the country to another to permit sharing of power during partial interruption of power on any one system or different peak load times from time zone differences.

It should be emphasized that all of the existing high voltage transmission systems, including not only the transmission line itself, but also the connected apparatus: the transformers, circuit breakers, lightning arresters, and capacitors have been engineered and designed to operate with reliability. The problems of dielectric breakdown with high voltage have been accommodated by the engineers using adequately large spacings. The spacings used have been usually based on empirical tests and trials on experimental lines and apparatus. Therefore, the need for scientific investigation into dielectric breakdown is prompted not by any apparently critical limitation of existing system vol ages, but by the need for still higher voltage lines and more compact, esthetically concealed, more efficient and reliable systems.

DIELECTRIC BREAKDOWN GENERAL CONSIDERATIONS

In scientific and theoretical investigations of dielectric breakdown, there has been a tendency to concentrate on uniform idealized electric field situations, where we now have a somewhat better understanding of the mechanism. Even here, however, there is inadequate understanding of the surface electron emission and internal space charge effects to permit reliable calculation and prediction of breakdown except in a semi-empirical sense.

The real world of high voltage apparatus and transmission lines usually involves divergent electric fields, where dielectric breakdown proceeds as a filamentary plasma streamer, involving copious photo ionization ahead of the streamer, as well as electron impact multiplication of electrons. The Meek and Raether theories of streamer breakdown in gases, which are now more than 35 years old, still furnish the clearest insight into this complex process, but these theories are still inadequate for making quantitative predictions. There is overwhelming evidence that, even in apparently uniform fields, particularly in gaps of more than 1 cm, the development of space charges from a primary electron avalanche or from emission points on the surface, that conditions for streamer breakdown quickly develop. Thus, it seems likely that breakdown in most practical situations is by a streamer-like mechanism.

In divergent electric fields, the threshold voltage for initiation of breakdown is determined by the highest electric stress location. Therefore, it has become very important for design purposes to be able to calculate or estimate by electric field plotting methods the electric field in complex geometries. Determination of the initial electric field distribution is only adequate for estimating the field after development of a conducting plasma streamer bridging part of the insulation gap and creating a new dynamic, rapidly changing, electric field distribution. There has been no adequate solution to this very complex problem.

Another aspect to the dielectric breakdown problem in transmission systems are the overvoltage requirements. 550-kV transformers, circuit breakers, etc., operating steadily at 290 kV to ground, must withstand a lightning impulse voltage of 1550 to 1800-kV 1 minute at test voltages of 600-700 kV to ground and switching overvoltages of about twice the line to ground voltage. The dielectric breakdown strength variation with time of insulating materials is unfortunately not usually in the same ratio as these expected voltages. Thus, the insulation must be designed, on the basis of empirical testing, to accommodate the worst conditions.

AIR INSULATION

Air is the primary dielectric of overhead high voltage lines. There are several problems with this situation: partial discharges or corona extending from the surface of the conductor, flashover under switching overvoltages, and flashover of insulator strings under wet polluted conditions. Partial discharges or corona causing radio interference is usually not a serious problem under dry conditions, since the diameter of the conductor is usually made large enough that the clean surface electric stress is not much above the critical stress for local air breakdown, but with accumulated dirt, and when wet with water droplets local points of higher stress may cause serious radio, and even audible, noise. The problem is more serious with higher voltages because, this author believes, the electric field around larger conductor diameters used at higher voltage decreases more slowly, going away from the conductor, thus leading to longer discharge streamers. A practical solution improving this situation is needed. Fair weather corona losses are usually only a few kilowatts or less per 3 phase mile of a 500 kV line. This is a small percent of the conductor resistance losses, but the corona loss may increase to levels of the order of 100 kilowatts per 3 phase mile in heavy rain.

It was discovered in the 1960's that switching overvoltage impulses of intermediate rise times and durations (about 250/2000 microseconds) between lightning (1.5/40 microseconds) and 60 Hz cycles led to lower breakdown voltages in very large gap divergent fields than with either lightning or 60 Hz. This phenomenon is related to the mechanism of streamer discharge, where it is believed qualitatively that the propagation time of streamers with short lightning type pulses is too short to complete breakdown for gap dimensions where longer switching impulses can. With the slower rise times of a 60 Hz cycle, there is sufficient lateral spreading of the discharges to develop a space charge reducing the field. A better analysis of this effect is needed.

Lightning breakdown mechanisms, as they relate to the direct stroke onto transmission lines or indirect induced voltage on the transmission line overvoltages due to this cause, continues to be a very important problem which would benefit from further basic investigation.

Flashover of wet, conducting contaminated insulators is a continuing problem for those areas such as the sea coast. A great deal of empirical work and testing has been done on this problem and some basic work. More basic research on wet contaminated surface flash-over is needed.

COMPRESSED GASES

In the past two decades, there has been a continuing growth in the use of enclosed compressed gas insulation, starting with high voltage current transformers and now more widely with circuit break-ers, using primarily SF_6. Compact high voltage switching and inter-connection stations are now growing in use to improve the esthetic appearance and save land area in or near cities. This gas has about 2.4 x the dielectric strength of air at atmospheric pressure, and it is commonly used at pressures of 3 to 4 atmospheres. There are several problems with compressed gases where the average electric stresses have been increased to where minor conductor surface imper-fections and particularly free conducting particles can cause local high stresses sufficient to cause abnormally low breakdowns. Avoid-ing these effects in large systems under factory and outdoor field installation conditions is a difficult practical problem. The mech-anism by which conducting particles initiate breakdown is still quite obscure. Free particles can be charged by contact with conduc-tors and oscillate wildly in open gaps. Even in clean uniform fields, compressed SF_6 (and other gases) show large departures from Paschen's law related to the conductor or electrode surface condition. The de-parture from Paschen's law, showing a declining rate of increase of breakdown voltage with pressure, appears to be closely related to the mean surface electric stress in the gas. The departure often starts at about 150-200 KV/cm and is attributed to surface field e-mission at local higher stress points, but the mechanism is still to be clearly elucidated.

LIQUID BREAKDOWN

Liquids, particularly mineral oil, and in a lesser amount, polychlorinated biphenyls (PCBs), are widely used in high voltage apparatus such as transformers, circuit breakers, and capacitors. These fluids have quite high dielectric strength when tested in small clean volumes, but show a serious decline in dielectric strength with increasing volumes under stress. This effect has been thoroughly analyzed using the minimum value (Gumbel and Weibull) statistics, but this statistical approach does nothing to explain the basis for the effect. It is assumed that there are defects of varied size lead-ing to breakdown due to the largest defect. These defects may be conducting particles floating and moving about in the oil, but this has not been clearly shown. There are many questions unresolved re-garding the mechanism by which these particles initiate breakdown. The dispersion of breakdown voltages and the area effect is much greater with 60 Hz than with impulse voltages, a feature which must be considered in the explanation. Typically, the statistics indi-cate a decline in mean breakdown voltage proportional to the stan-

dard deviation and the log of the area or volume ratio: (It is difficult to separate area and volume effects). This effect practically
prevents the use of higher stresses and more compact apparatus. The
breakdown of compressed gases and liquids have similar characteristics. Indeed, breakdown tests of CO_2 and SF_6 carried at pressures up
to the liquid range show no significant discontinuity on going from
compressed gas to liquid. Also, compressed gases show a declining
breakdown strength with area as do liquids.

The prebreakdown electric fields in insulating liquids may be
drastically altered due to stress induced fluid motion and conduction.
This motion is due to injected charges from electrodes, as well as
existing molecular ions and particles in the liquid. These effects
are greater with dc but also occur with ac and transient voltages.
Consideration of these effects, which have been demonstrated with Kerr
effect measurements, is also essential to understanding liquid breakdown.

SOLID BREAKDOWN

In recent years, extruded resins such as polyethylene, cross-
linked polyethylene and ethylene-propylene rubber have been used widely as conductor insulation in coaxial arrangement for voltages as
high as 115 kV in the USA. Cast and molded resins have been used to
completely insulate transformers and as bushings up to 115 kV. The
advantage is a more compact structure, elimination of impregnation
with oil (as with paper insulation) and a metal oil container. Extruded resin cables have been experiencing in some areas a so-called
"treeing" type breakdown which must be avoided. The treeing appears
to be of two types, one involving growth of tree-like electric discharge channels into the resin from the surface. Such breakdown
patterns resemble streamer growth discharges in gases, and can be
readily reproduced by applying high voltage to needle points cast or
molded into the solid resin. The second type of treeing appears to be
due to penetration of filamentary channels of water or other conducting chemicals into the resin. It has been suggested that the electrical discharge tree is started at a surface or included conducting
imperfection presenting very high local electric stress. The electrochemical or water tree may penetrate the resin by the high electrostatic forces starting at local sharp cracks or cavities. In neither
case is the mechanism of breakdown in this important problem proven.
It seems that the pre-detection of this potential problem in new
cables is not possible with present partial discharge detection techniques, and there is apprehension of delayed failures.

In all high voltage applications of resin or plastic insulation,
gas discharges in cavities or gaps must be avoided to prevent decomposition and breakdown by the discharges. Partial discharge detection
methods developed particularly for this purpose have usually been
successful in avoiding breakdown by this mechanism.

COMPOSITE SYSTEMS

In composite systems involving liquid filled or impregnated

insulation with solid barrier films or sheets, such as transformers and capacitors, dielectric breakdown is usually initiated in the medium having the lower dielectric constant and weaker dielectric strength; this is the liquid. Liquid breakdown leads to formation of gas bubbles and subsequent breakdown of the solid immediately, if the stress is high enough, or delayed, if partial discharges persist. In present transformers, dielectric breakdown, if it occurs at all, is more probably a result of gas bubbles subsequently developed from accidental faults such as metal inclusions, or overheated high resistance conductor connections or stray magnetic field heating.

Power capacitors are used to compensate series inductance on long transmission lines, or shunt inductance of loads, such as motors, on distribution lines. They are operated at continuous ac (rms) stresses of the order of 500 kV/cm average on polypropylene films. This is by far the highest stress used in any electric power industry application. They are expected, however, to withstand overvoltages well above this. DC capacitors are able to withstand much higher continuous stresses due to better voltage distribution on the film resulting from conduction in the fluid impregnation medium. Partial discharges at the foil edges, which is avoided by good processing and design, or imperfections in the capacitor film are the most likely causes of dielectric breakdown. However, power capacitors have an excellent record of reliability.

SUPERCONDUCTING TRANSMISSION LINES

Superconducting lines present similar dielectric breakdown problems to those of conventional cables and compressed gas lines, with an obvious limitation of materials. Liquid helium or possibly liquid hydrogen are the only fluids, and tests to date indicate that at least a 115 kV line is feasible. Operation of helium at supercritical pressures should avoid gas bubble formation which would reduce the strength. Particle and solid insulator surfaces can still be a problem. A cast epoxy bushing termination for a 115 kV line has been made and is about to be tested. Preliminary tests of the dielectric strength of epoxy at liquid nitrogen temperatures indicate its dielectric strength is undiminished and that axial temperature gradients from 4° K to 300° K can be tolerated.

SUMMARY

This brief review of dielectric breakdown problems in high voltage transmission has been able to cover only the more important problems which exist, but which are for the most part presently being met by suitable engineering design. From a physics standpoint, there are many interesting problems. They relate predominately to divergent electric fields and streamer plasma breakdown in all three media: gas, liquid and solid. In all these media, initiation of the streamer breakdown appears to occur from surface imperfections or particles in the media. A better understanding of the breakdown mechanisms would presumably suggest methods of making more efficient designs for future power transmission from new energy sources to the consumer.

The scientific and technical papers discussing the problems mentioned in this paper have been mostly contributed to the IEEE Transactions on Power Apparatus and Systems, the Transactions on Insulation, and the Annual Conference on Electrical Insulation and Dielectric Phenomena sponsored by the Division of Engineering of the National Research Council - National Academy of Sciences. The latter conference publishes annually a Digest of Literature on Dielectrics which can be consulted for references.

PHYSICS OPPORTUNITIES
IN
ELECTROCHEMICAL ENERGY CONVERSION

Elton J. Cairns
General Motors Research Laboratories

ABSTRACT

Fuel cells and batteries promise convenient, efficient power generation and energy storage if they can be developed as low-cost, long-lived systems. The efficiency of these devices can be higher than that for heat engines because they are not Carnot-cycle limited.

Fuel cells are under development for use as electrical substations, and as small-scale power sources. The problem areas for these systems include the development of porous electrodes with good electrochemical activity, transport characteristics and long life, the evolution of new cell designs for high performance, and the identification of stable, inexpensive materials of construction.

Rechargeable batteries offer an efficient means to store off-peak electrical energy in utility networks, and propel automobiles, if they can be developed to have a low cost ($\sim$\$4-12/MJ) and a long life ($\sim$3 yr). The major candidates are high-temperature (350-400°C) batteries with problems in the areas of materials of construction, insulators, seals and electrode design.

INTRODUCTION

Electrochemical energy conversion has been at various times a fashionable and exciting area of research and development activity. During the period of the late 1950's to mid-1960's, NASA and the Department of Defense placed strong emphasis on the use of fuel cells for missions of a few days' to a few weeks' duration, because fuel cells offered the lowest system weight for a given amount of energy produced over that period of time (usually at the rate of a few hundred watts to a few kilowatts). High-performance batteries (especially rechargeable batteries) have also been in great demand for a large number of applications, primarily space and military, where price was of secondary importance.

The long-range goal of electrochemical energy conversion workers has been the development of devices which would have widespread application in the non-military and consumer sectors of the economy. The two largest consumer demands for energy and power are those for the home (typically 1 kW average power and about 8 kW peak power for electrical energy) and the automobile (typically 20 kW average power and 80 kW peak power for urban-suburban driving).[1,2] The usual power levels for fuel cells and batteries have been lower than the above figures, the lifetimes have been relatively short (with some exceptions), and the costs have been prohibitively high. In the case of vehicle propulsion, the specific power (power per unit weight), and the specific energy (energy per unit weight), are the key performance parameters upon which judgments can be made concerning the technical attractiveness of electrochemical power sources. The specific energy of fuel cell systems readily can be made high enough to be attractive, but the specific power has been lacking. The converse situation has been true for rechargeable batteries.

The increasing concern over air pollution and more recently over the shortage of imported petroleum has focused attention on the use of fuel cells and batteries as clean, efficient devices for the generation and storage (respectively) of electrical energy in the national electrical power grid,[3] and for the propulsion of vehicles. Electrochemical energy conversion is attractive from a pollution point of view because fuel cells do not emit oxides of nitrogen or carbon monoxide, and batteries do not emit NO_x, CO, hydrocarbons, or any of the other common air pollutants. With some batteries, precautions must be taken to prevent the emission of small amounts of electrolyte. The efficiency of energy conversion and storage is becoming a central issue, providing an additional incentive to develop fuel cells and batteries. Electrochemical energy conversion devices are not limited by the Carnot-cycle efficiency because they are not heat engines. They are theoretically capable of converting all of the Gibbs free energy of reaction to electrical energy. In practice, it is not uncommon for fuel cells to convert 60% or more of the Gibbs free energy of combustion of a fuel such as hydrogen (or a hydrogen-rich gas mixture) into electrical energy, and for batteries to convert 90% of the Gibbs free energy of reaction into electrical energy. Rechargeable batteries are particularly attractive candidates for vehicle propulsion because they can be recharged from the existing electrical grid during off-peak hours (at lower cost) and can therefore make use of coal or nuclear energy as the primary energy source, rather than precious petroleum. In addition, there is a significant efficiency advantage in using coal or nuclear energy to charge batteries rather than to produce a fuel for a heat engine.

In order to assess the status of electrochemical energy conversion technology, it is important to have a set of performance and cost goals for each potential application. For the present discussion, applications in the electrical power grid and vehicle propulsion will be considered. Table I shows performance and cost goals for some large potential markets, and an indication of candidate electrochemical systems. In the case of fuel cells, two possible applications are a) electrical substations, especially smaller substations in areas such that it would be difficult or expensive to install transmission lines to central stations (e.g., mature urban areas or remote areas), and b) electric family automobiles, requiring a range of more than about 300 kilometers between refuelings. For the electrical substation, a life of at least 160 Ms (5 years) is important for economic reasons, and this life might be compatible with a capital cost of $150-200/kW.[8, 9] Such a fuel cell system

Table I. Performance and cost goals for fuel cells and batteries

Application	Candidate System	Average Specific Power W/kg	Maximum Specific Power W/kg	Specific Energy kJ/kg	Operating Time Ms(yr)	Life Ms(yr)	Cost $/kW	Cost $/MJ
Fuel Cells								
Electrical substation	Indirect HC/air	20	40	–	130 (~4)	160 (~5)	150–200†[8,9]	–
Family automobile	Indirect NH$_3$/air	50	200	1000+	8 (1/4)	100 (~3)	10–15**[1,2]	~3[1,2]
Batteries								
Off-peak energy storage	Na/S, Li/S	10–20	50	400–800	110 (~3.5)	130 (~4)	140*[3]	4[3]
Urban automobile	Pb/PbO$_2$, Zn/MnO$_2$	40	100	160–360	8 (1/4)	100 (~3)	20**[4-7]	5–13[4-7]
Family automobile	Li/S, Na/S	50	200	600–800	8 (1/4)	100 (~3)	10–15**[1,2]	3–4[1,2]

* Ten-hour rate assumed

** At maximum power

† At average power

would probably have an average specific power of at least 20 W/kg of
system weight. In the case of the family automobile, the specific
power of the fuel cell system is very important, since this determines the acceleration capability of the auto. A minimum peak specific power of 200 W/kg is necessary for safe mixing with existing
traffic. [1,2] Under normal driving conditions, an average of
50 W/kg would be required. In order to have a range of at least
300 km, the fuel tank should be sized to provide a system specific
energy of at least 1000 kJ/kg. Since most first owners retain
their automobiles for about 100 Ms (3 years), it is considered
reasonable to replace the battery about every 100 Ms, which might
justify a cost of $10-15/kW.

The possible applications for batteries considered here are
off-peak energy storage in the electrical power grid, [3] and electric
automobiles. [1,2] Off-peak energy storage permits the existing
base-load central power stations to be used at peak efficiency for a
greater fraction of the time, and decreases the need for the use of
lower-efficiency intermediate cycling plants and peak power units
(such as gas turbines and diesel engines). Furthermore, the use of
batteries for off-peak energy storage permits the storage of energy
near the demand centers, removing the need for peak-load capability
in the transmission and distribution lines, at a considerable cost
saving of up to $200/kW. [3] Such batteries should last at least
$130 Ms (4 years) and cost about $4/MJ, in order to compete with
such equipment as pumped hydroelectric or gas turbines. [3] These
batteries will probably have an average specific power of 10-20 W/kg,
and a specific energy of 400-800 kJ/kg. The overall efficiency of
electrical energy storage in these batteries when operated at the 4
to 14 hour rate will be above 70%, [10-12] which is higher than that
of pumped hydroelectric energy storage.

Battery-powered automobiles have existed for more than seventy
years, but do not comprise a significant fraction of the automobile
population because the only rechargeable batteries available at a
tolerable cost have been Pb/PbO_2 (lead-acid) batteries, which have
shown too short a lifetime under deep-discharge conditions (80% or
more of the energy withdrawn), and have been too heavy per unit of
energy stored (too low a specific energy - near 100 kJ/kg). [12] Now
that the petroleum shortage is widespread, and prices are increasing
dramatically, smaller, lower-performance automobiles are becoming
more popular. This situation makes it worthwhile to reexamine the
possible role of the limited-performance electric urban automobile
using a Pb/PbO_2 battery designed specifically for automobile propulsion. Higher-performance electric automobiles would be feasible
if advanced batteries such as rechargeable Zn/MnO_2, Na/S and
Li/S were successfully developed. Reasonable goals for urban
electric automobile batteries are a specific energy of 160 kJ/kg to
360 kJ/kg and a maximum specific energy of 100 W/kg, with a lifetime
of at least 100 Ms (3 years), and a cycle life of 1000 deep cycles.
An electric family automobile which would be used part of the time
at highway speeds, and would be required to have a range of

164

200-300 km,* would need a battery with a specific energy approaching 800 kJ/kg a specific power of 200 W/kg, and a lifetime of 100 Ms, costing about $3/MJ of energy storage capability. [1, 2]

In the remainder of this paper, the status of fuel cells (for electrical energy generation) and rechargeable batteries (for electrical energy storage) will be reviewed, and the problem areas for each will be discussed, with reference to the goals outlined above. Opportunities for contributions by physicists will be noted in the discussion.

FUEL CELLS

Since there are a number of definitions of fuel cells which differ from one another, it is appropriate to indicate that in this discussion, a fuel cell is taken to be an electrochemical cell in which the Gibbs free energy of reaction between a fuel and oxygen, preferably from air, is converted directly and usefully into (low-voltage) direct-current electrical energy. [13] The fuel and air (or oxygen) are stored separately, and fed to the fuel cell as needed. The fuel is a fossil fuel or a fuel derivable therefrom, such as hydrogen, ammonia, or methanol.

The fuel cells that have been developed for space and military applications are significantly smaller, orders of magnitude more expensive, and shorter lived than is appropriate for the applications being considered here. This means that it has been necessary to continue a significant, long-range research and development program to build upon the results of the space-and military-oriented fuel cell activities. Progress has been made in the areas of electrode performance, content of precious-metal electrocatalysts in the electrodes (and therefore cost), lifetime, and sizes of cells and batteries of cells. Typical values representing the current status of fuel cell performance are shown in Table II. The values in the table are intended to represent the state of development of fuel cells, and are typical, rather than specific values for any particular fuel-cell system.

The largest fuel cell activity is that of Pratt and Whitney Aircraft, funded by the gas industry, the electric utility industry, and United Aircraft. The gas industry portion has centered on the development of a 12.5 kW fuel cell system, using methane (natural gas) as the primary fuel to a steam reformer, which produces a hydrogen-rich gas mixture for use in a hydrogen/air fuel battery. [9] The overall system efficiency is about 40%, and the lifetime has been about 15 Ms. More recently, the electrical utilities have funded a program at P & W for the development of a 26 MW fuel-cell system (based on 16-kW building blocks), which will use a liquid hydrocarbon as the primary fuel to the steam reformers. [8, 9, 14] The fuel cells will operate on the hydrogen-rich reformer product and air. This program appears to be still in the prototype stage for the 16-kW building block unit.

* With about 25% of the curb weight of the vehicle as battery.

As can be seen from Table II, fuel cells typically contain a few milligrams of platinoid elements per square centimeter of electrode area, and yield a power density of about 0.07 W/cm^2, corresponding to 0.07 g/W, which gives an electrocatalyst cost (at $\sim$\$5/g) of \$350/kW, far too high for electrical substations or automobiles. Therefore, a key problem area is electrocatalyst cost. Since the world production of platinoid elements is near 30,000 kg/year, there would not be enough of these precious metals produced per year to build 10^6 kW of generating capability, or 25,000 full-size automobiles. Clearly, a means must be found to use much less platinoid electrocatalyst per unit of power, or a plentiful, inexpensive platinoid element substitute must be found before fuel cells can be used for the production of significant amounts of power. This has been a central focus of a large fraction of the continuing fuel-cell R&D activity. An important part of the effort to make better use of precious-metal electrocatalysts has been the study of the structure and mechanism of operation of porous gas-diffusion electrodes. Physicists have made significant contributions to the theory and improvement of porous electrodes, and their continuing attention is needed here.

The general problem of effective porous gas electrodes involves the provision for and promotion of intimate contact (on a molecular level) between the reactant (fuel or oxygen), the electrolyte, and the electrocatalyst, and the removal of the products of reaction, and electrical current. Mass, heat, and momentum transport all should be optimized, and the electrocatalyst should be placed only where it is most effective. This is a very complex problem which cannot be adequately described here, but there is a large volume of literature available. [13, 15]

In addition to electrode structure, cell designs which minimize inefficiencies and weight are especially important for those applications (such as automobile propulsion) which require high specific power. The weight of the electrolyte, and the hardware must be minimized. New ideas for cell design concepts might be an important contribution from physicists.

Because of the fact that fuel cell electrolytes are usually strongly acidic or strongly alkaline, materials of construction have been a continuing problem. Such expensive materials as tantalum, polytetrafluoroethylene (PTFE), and gold have been used for space and military applications. More recently, use has been made of carbon and other less expensive materials, but more work is needed here.

Physicists can make contributions to the development of fuel cells by helping to solve the problems of expensive electrocatalysts, electrode structure, cell designs, and inert materials of construction.

RECHARGEABLE BATTERIES

Rechargeable batteries can make significant contributions to the easing of the energy crisis and air pollution, provided that the goals set out in Table I can be met. Not only must the cost goals be

Table II. Status of fuel cell development

Characteristic	Acid [1] Electrolyte	Alkaline [2] Electrolyte
Fuel	Impure H_2	Impure H_2 [3]
Oxidant	Air	Air [3]
Electrode size cm x cm	20 x 20	20 x 20
Electrocatalyst		
Fuel electrode	Platinoid	Platinoid
Oxygen (air) electrode	Platinoid	Non-platinoid (Ag, others)
Electrocatalyst loading		
Fuel electrode mg/cm^2	∿2	∿1
Oxygen (air) electrode, mg/cm^2	∿2-4	1-2
Cell operating voltage, V	0.6-0.75	0.7-0.9
Average current density, A/cm^2	0.1	0.1
Maximum current density, A/cm^2	0.2	0.3
Average power density, W/cm^2	0.07	0.08
Maximum power density, W/cm^2	0.15	0.25
No. of cells in a stack	tens	tens
Total power, kW	10-20	5
Specific power, W/kg	20	20
Lifetime, Ms	∿15	∿18

[1] Typically, 85 w/o H_3PO_4 at 80-140°C or 3N H_2SO_4 at 25-80°C
[2] Typically, 30 w/o KOH at 50-80°C
[3] Alkaline-electrolyte cells cannot tolerate significant concentrations of CO or CO_2 in the fuel or air streams.

Table III. Status of rechargeable battery development

Characteristic	Pb/PbO_2	Zn/NiOOH	Na/S	Li/S
Operating temperature, °C	ambient	ambient	350	370-400
Open circuit voltage, Volts	2	1.7	2.1	2.4
Avg. current density, A/cm^2	0.01	0.01	0.1-0.2	0.1-0.5
Capacity density, kC/cm^2	0.13	0.1	1	1-1.5
Max. power density, W/cm^2	0.3	0.3	1	2
Cycle life, cycles[†]	300	300	1000+	∿1000
Lifetime, Ms	>100	100***	>10	∿10
Max. electrode size, cm^2	∿7500	∿300	∿100	∿100
Max. battery size, kW	many	a few	20	<1
Specific energy* kJ/kg	120	180	310[21]	440
Specific power** W/kg	100	200	150	30

[†] Deep discharge - 80%
* At the five-hour rate
** Peak power
***Separator hydrolysis limiting

met, but the question of adequate supplies of materials must be
faced. This combination of requirements causes a number of otherwise attractive electrochemical systems to be eliminated from consideration.* The familiar Pb/PbO_2 battery might be used for a limited number of electric urban automobiles or off-peak energy storage
stations, but very widespread application of Pb/PbO_2 batteries would
be unlikely in view of the performance and cost of these batteries,
and the supply of lead. [6]

The goals of Table I for rechargeable batteries are compatible
with the projected characteristics of such high-temperature batteries as sodium/sulfur and lithium/sulfur. [12, 15] Sodium, lithium,
and sulfur are all very plentiful and are inexpensive enough to be
compatible with the cost goals. The sodium/sulfur and lithium/
sulfur batteries operate at temperatures in the range 350-400°C, and
laboratory results indicate that they should be capable of storing
600-800 kJ per kg of battery weight.

Table III summarizes the status of development for some rechargeable batteries. The lead-acid battery represents a mature
technology which has shown rather gradual improvements over the last
several years. Very large cells and batteries have been built for
various types of applications, including electrical power in certain
areas of large cities such as Chicago. For application to urban
electric automobiles, improvements in the cycle life for deep-
discharge conditions, and minimization of the internal resistance
are important.

The zinc/nickel oxide battery is a relatively recent development which is not yet commercially available on a production basis.
This battery offers an improved specific energy (perhaps ultimately
as high as 230 kJ/kg). Because nickel is not in great supply, this
battery can be viewed as a step in the direction of a rechargeable
zinc/manganese dioxide battery which may have a specific energy as
high as 310 kJ/kg, if the problems with the rechargeability of the
MnO_2 electrode can be satisfactorily resolved.

The high-temperature cells (sodium/sulfur and lithium/sulfur)
offer promise of very high specific energy ($\sim$800 kJ/kg) and high
specific power (>200 W/kg), which could yield a high-performance
family automobile. The status of these cells is shown in Table III,
which indicates that these cells still have inadequate lifetime.
Sodium/sulfur cells have been assembled into a battery of 20 kW size,
which powered an electric van, but the specific energy was relatively
low and the lifetime was short. Nevertheless, this was an impressive
demonstration of a new battery which continues to show promise.

The most common type of sodium/sulfur cell[16 - 18] makes use of a
ceramic electrolyte which conducts sodium ions (called beta-alumina,
and having the formula $Na_2O \cdot xAl_2O_3$, where x can be in the range 5 to
11). The overall cell reaction is the formation of sodium-sulfur
compounds from liquid sodium and liquid sulfur. Sulfur is a very

---

* For example, cadmium, silver oxide, and other electrode materials
are too expensive and are in short supply.

poor electronic conductor and therefore requires the use of a good
conductor to distribute the electons to the sites of the electrochemical reaction in the sulfur compartment. The design of the
sulfur electrode for minimum weight and maximum utilization of
sulfur at high rates is a significant problem, requiring the optimization of electronic resistance, ionic resistance, and mass transport. Physicists can aid in this complex problem.

The beta-alumina electrolyte has been found to crack after
having been operated in sodium/sulfur cells for a number of charge-
discharge cycles (usually 100-300 cycles) at moderate current densities. The exact failure mechanism(s) is(are) not definitely established, but the formation of sodium deposits which penetrate the
beta alumina is one common cause of failure. Other problems with
beta alumina may be caused by the presence of various harmful impurities. Physicists can aid in gaining a more complete understanding
of the modes of failure of the beta alumina, and means for avoiding
failure, through improvements in the beta alumina or the identification of a superior alternative.

Because sodium reacts rapidly with oxygen and with moisture, it
is necessary to hermetically seal sodium/sulfur cells. The development of materials and methods for hermetic seals and electrical
feedthroughs (to remove the current) is another area requiring at-
tention--perhaps from physicists. Other problem areas include the
identification of inexpensive, corrosion-resistant, electronically
conductive (metallic?) materials for use as cell casings, which will
be in contact with sulfur and sodium-sulfur mixtures. Some stainless
steels show promise for this application, but high-resistance sulfide
layers tend to form, and the stainless steel tends to be corroded.

The lithium/sulfur cell, which uses a molten-salt electrolyte,
is at an earlier stage of development than the sodium/sulfur cell.
Although single cells having an electrode area of about 100 cm^2 have
been successfully operated, and have shown specific energies of about
440 kJ/kg, [19] the lifetime and cycle life remain below the goals, and
no batteries of significant size have been constructed. The problem
areas for this cell include the same general problems of the sulfur
electrode as are present in the sodium/sulfur cell, plus the problem
of the solubility of sulfur-bearing species in the electrolyte. The
solution to this latter problem is being sought through the use of
additives to the sulfur, [20] and the use of metal sulfides. [19] The
composition and design of the sulfur electrode must be optimized
from the standpoint of high sulfur utilization at high current densities, and high rates of mass transport with minimum electronic and
ionic resistance. This complex problem deserves the attention of
physicists and chemists alike. Because this cell must be sealed from
the atmosphere, high-temperature, corrosion-resistant seals and feedthroughs are needed. Advanced cell designs, minimizing weight, are
important to the ultimate success of these cells in propulsion applications. Physicists could make contributions to both of these areas.

Batteries for the applications being considered here must all be
less expensive than they are now, and must have longer lives under
the rigorous conditions of deep discharge. Those which are

candidates for propulsion applications must have a maximum specific
energy and specific power. To these general problem areas must be
added the specific problems indicated above for each system.

CONCLUSIONS

Fuel cells and batteries show promise for helping to alleviate
the energy crisis and air pollution, but a number of significant
problems remain to be solved before their promise can be realized.
High cost and short lifetimes can be traced to a number of problems.
Physicists have made and can continue to make contributions to the
solution of a large fraction of these problems. The possible impact
of successful electrochemical energy conversion devices on our
society is too great to allow these problems to be ignored.

REFERENCES

1. M. L. Kyle, H. Shimotake, R. K. Steunenberg, F. J. Martino,
   R. Rubischko, and E. J. Cairns in 1971 Intersociety Energy Conversion Eng'g Conf. Proceedings, Soc. Automotive Engrs, New
   York, p. 80, 1971.
2. E. J. Cairns, et al., Argonne National Laboratory Report,
   ANL 7888, Dec. 1971.
3. M. L. Kyle, E. J. Cairns, and D. S. Webster, Argonne National
   Laboratory Report, ANL 7958, Mar. 1973.
4. C. C. Christiansen, in Proc. 2nd Intl. Elec. Vehicle Symp.,
   Atlantic City, Nov. 1971.
5. D. L. Douglas, in Proc. Symp. on Power Systems for Elec.
   Vehicles, Columbia Univ., New York, Apr. 1967.
6. P. Ruetschi, J. Electrochem. Soc., 108, 297 (1961).
7. J. Bjerklie, E. J. Cairns, H. J. Korp, C. Tobias, D. G. Wilson,
   and C. Zener, Report of the Panel on Alternate Power Sources to
   the Committee on Motor Vehicle Emissions, National Academy of
   Sciences, Wash., D.C., Apr. 1973.
8. W. J. Lueckel, L. G. Eklund, and S. H. Law, IEEE Transactions
   for Power, Apparatus, and Systems, PAS 92, 230 (1973).
9. W. Podolny, Presentation to the Subcommittee on Energy R&D
   Policy of the House Space and Astronautics Committee, July 1972.
10. E. J. Cairns, E. C. Gay, R. K. Steunenberg, H. Shimotake,
    J. R. Selman, T. L. Wilson, and D. S. Webster, Argonne National
    Laboratory Report, ANL-7953, Sept. 1972.
11. C. A. Levine, W. E. Brown, and R. G. Heitz, Presented at The
    Electrochem. Soc. Meeting, Boston, Oct. 1973, Abstract No. 11;
    See also Extended Abstracts, 73-2, 33 (1973).
12. E. J. Cairns and R. K. Steunenberg, in Progress in High
    Temperature Physics and Chemistry, 5, C. A. Rouse, Editor
    (Pergamon Press, N. Y., 1973), p. 63 ff.
13. H. A. Liebhafsky and E. J. Cairns, Fuel Cells and Fuel
    Batteries (John Wiley & Sons, N. Y., 1968).
14. Wall Street Journal, Dec. 21, 1973.
15. G. Sandstede, Editor, From Electrocatalysis to Fuel Cells
    (Univ. of Wash. Press, Seattle, 1972).

16. N. Weber and J. T. Kummer, in Proc. 21st Ann. Power Sources
     Conf., (PSC Publications Committee, Red Bank, N. J., 1967).
17. J. L. Sudworth, M. D. Hames, M. A. Storey, and M. F. Azim,
     Presented at 8th Intl. Power Sources Symp., Brighton,
     Sept. 1972.
18. J. Fally, C. Lasne, Y. Lazennec, and P. Margotin, J.
     Electrochem. Soc., 120, 1292 (1973).
19. H. Shimotake, W. J. Walsh, N. P. Yao, J. D. Arntzen, and
     J. W. Allen, Presented at The Electrochemical Society Meeting,
     Boston, Oct. 1973, Abstract No. 13.
20. W. J. Walsh, E. J. Cairns, and J. D. Arntzen, Presented at The
     Electrochemical Society Meeting, Chicago, May 1973, Abstract
     No. 254.
21. R. W. Minck, in Proc. 7th Intersociety Energy Conversion Eng'g
     Conf., Amer. Chem. Soc., Wash., D.C., 1972, p. 42.

THE HYDROGEN ECONOMY CONCEPT

Dr. Derek Gregory
Institute of Gas Technology
3424 S. State St., Chicago, Ill. 60616

As the United States' supply of nonpolluting fossil fuels begins to dwindle, the gas industry has begun a search for alternative sources of clean-burning energy, such as the gasification of coal and importation of liquified natural gas (LNG). Coal is considered to be this country's major fuel resource in the near future, but many projections suggest that the rate at which the U.S. can produce coal, as well as the other fossil fuels such as oil and natural gas, will reach a peak in the next 20 to 40 years and will then decline.

Nuclear and solar energy are the only effectively abundant energy sources that can be considered to fill the country's long-term energy gap. But almost all work to harness these energy forms is directed toward using them to produce electric power. The U.S. is already increasing its electricity use at 2 1/2 times the rate of its overall energy growth. We are thus heading for an all-electric economy.

An all-electric economy is undesirable for a number of reasons: 1) Electricity transmission lines are unsightly and expensive --- prohibitively expensive for underground lines; 2) Bulk storage of electricity is not widely available, yet its use is essential to the harnessing of solar power; and 3) Because electric energy is not portable, it is not ideally suited for vehicle and aircraft propulsion.

For the past 10 years, IGT has been studying a synthetic chemical fuel that can easily be produced from nuclear or solar energy and other readily available sources: hydrogen.

Hydrogen can be produced from water by the addition of energy and can be oxidized back to water to give up this energy as heat or electric power directly. Since the only direct combustion product of hydrogen is water, which is already present in such abundance and is so mobile on the earth's crust, no permanent dislocation could be produced in the ecology by using the hydrogen-water cycle this way. It is the perfect ecology fuel, a virtually pollution-free recyclable fuel. So, while hydrogen is not a naturally occurring fuel or a primary source of energy, it is an attractive medium for storage and transmission of an energy source. It is a means of making such energy sources as nuclear, solar, and low-grade fossil fuels available to the consumer in a clean, convenient, and flexible way.

The concept of hydrogen as a fuel is not new. In 1933, English scientists proposed using off-peak electric generating capacity to generate electrolytic hydrogen which would be used for powering automobiles, thus relieving both air pollution and oil importation problems, two problems that also are particularly timely today.

Water can be converted to hydrogen and oxygen in three ways. In the electrolyte method, a direct electric current is passed through an electricity-conducting water solution, freeing hydrogen and oxygen separately at the electrodes. In the second method, water can be heated to a high temperature where it will spontaneously split into hydrogen and oxygen. In the third method, thermochemical splitting, water enters into a series of chemical reactions in which the intermediate chemicals are always recycled and the overall reaction produces hydrogen and oxygen in separate steps. A few such processes under study at present appear to be capable of operating at temperatures well within the reach of nuclear reactors or solar furnaces.

Today, industrial hydrogen is made from natural gas and other fossil fuels by reacting them with steam. The conversion of coal, oil shale, and low-grade fuels to hydrogen is technically possible today and could certainly become the source of hydrogen fuel in the near future. In this way, hydrogen could act as a common fuel to bridge the gap between the fossil fuel age and the nuclear or solar age.

Because of the concern over finding a supplemental energy source and an associated delivery system, an extensive nuclear breeder-reactor program has been established. Almost all the work directed toward harnessing nuclear (both breeder and fusion), solar, and geothermal energy is concerned with the generation of electricity. However, this will lead to severe problems in both siting generating stations and providing transmission lines from them to the ultimate users. As the amount of power to be transmitted increases, overhead power lines become more expensive and more unsightly. Long-distance underground transmission of electricity is prohibitively expensive.

TRANSMISSION

If we chose to make hydrogen rather than electricity from our supplemental energy sources, it can, unlike electricity, be stored near the load center and its lower transmission cost would allow greater freedom in siting generating stations. Gaseous fuels are relatively cheap to transmit in underground pipelines; hydrogen is no exception. Existing natural gas lines could carry the same energy content as natural gas for only a small penalty in pumping costs.

A 36-inch pipeline can carry 11 times the energy capacity of a single-circuit 500-kilovolt overhead line at a fraction of the cost. In fact, the savings in transmission cost over typical distances more than compensate for the extra cost of turning electricity into hydrogen at the power station. Thus with today's technology, one could deliver hydrogen energy more cheaply to the average customer than the present price of electricity and have the by-product oxygen free.

The storage capability of hydrogen is of extreme importance in being able to even out both the daily and seasonal peaks. Hydrogen can be stored in underground reservoirs or liquefied and stored in insulated tanks the same way that natural gas is stored today. The largest tank for liquid hydrogen, located at Cape Kennedy, has a 900,000-gallon capacity. This is only about one-twentieth of the sizes normally used for LNG, but it is about 75 percent of the energy capacity of the world's largest pumped-hydroelectric power system for the production and storage of electricity at Ludington, Michigan.

Storage costs for hydrogen are 2 or 3 times higher than for natural gas but could perhaps be reduced by large-scale engineering techniques. The storage costs for pumped hydroelectricity are many times higher than those for the storage of liquid hydrogen.

The use of hydrogen as a fuel presents both some problems and some distinct advantages. There seems to be no reason why pure hydrogen could not be used for all purposes served by natural gas today: it burns smoothly and easily when mixed with air, and it can be burned in domestic and industrial appliances similar to those used for natural gas. In addition, because hydrogen burns without noxious exhaust products, it can be used in an unvented appliance without hazard; thus, a home heating furnace could conceivably operate without a flue, thereby saving the cost of a chimney and adding as much as 30 percent to the efficiency of a gas-fired home-heating system.

Engines operate well on hydrogen. Already, several teams are converting automobile engines to run on hydrogen. The main modifications required are to carburetion and ignition. At IGT we have already run an unmodified Chevrolet engine on this fuel. In the laboratory it has been shown that the hydrogen engine produces far less nitrogen emissions than a gasoline engine.

These are just some of the reasons why IGT, under A.G.A. sponsorship, is studying the concept of what it calls the Hydrogen Economy, in which hydrogen becomes the master fuel, feeding all the applications met by natural gas today and more. In addition, hydrogen is a vital raw material for the petrochemical and metallurgical industries. For the perhaps 20 percent of our energy needs best met by electrical power, we can consider the use of fuel cells, ideally suited to hydrogen, or advanced steam turbine systems, located close to the load centers. But about 80 percent of our needs, which are for heat, will be met by burning the hydrogen directly.

SAFETY

One other aspect of hydrogen use that should be mentioned is safety. Ever since the Hindenburg airship disaster in 1937, hydrogen has had a questionable image in the public's eye. At the time, because any kind of air travel was still a novelty, the mishap seemed more spectacular than it actually was. Of the 97 people on board, 62 of them escaped, probably because the hydrogen fire, which only lasted just over a minute, produced very little radiation, most of what was produced going straight upward.

Admittedly, hydrogen can be a very hazardous and dangerous material, but it has been used so extensively in industry and aerospace that very clearly defined codes of practice have been developed. With odorization to make leaks easily detected and with proper handling techniques, pure hydrogen should be no more hazardous than the old town gas, or manufactured gas, which was composed of 50 percent hydrogen and which had the added hazard of toxity because of its carbon monoxide content.

Although the price of hydrogen is high today compared with that of fossil fuels, hydrogen from non-fossil sources should become relatively cheaper as the fossil fuels become more expensive. One of the major attractions of hydrogen is its flexibility because it can be produced from so many different energy sources. Such flexibility will be of very great importance to the energy industry in the next 50 years or so when very considerable changes in energy sources must be made. Government officials have warned of the twofold danger of importation of energy: a very unfavorable balance of payments in our economy and a possible national security problem. With hydrogen, we have the cleanest chemical fuel possible which can be produced from wholly domestic energy sources without the need for importation.

While interest in this Hydrogen Economy concept is rapidly growing within the energy industry, the problems in the implementation of such a system are immense and would have to be thoroughly planned well in advance of the day when the concept will be a reality, not just a blueprint for the future.

IGT is not alone in studying such a system. It is seriously being considered by the American Gas Association, the Atomic Energy Commission, several electric utilities, universities, and the Euratom organization in Italy. Many of the sceptics who have been persuaded to look into the scheme in detail have themselves become "hooked" and are enthusiastically looking at conversion to hydrogen within a fairly short time scale.

Although years of research and work are ahead of us, we feel that the great potentiality of such a system certainly justifies this time and effort on our part.

BREEDER REACTORS--THE PHYSICIST'S CONTRIBUTION

M. W. Dyos
Westinghouse Electric Corporation
Madison, PA 15663

INTRODUCTION

The existence of the current generation of fission-powered reactors is a result of considerable theoretical and experimental research in the basic processes by the physicist. This research has led to the generation of accurate nuclear data, realistic integral experimental programs, operational experience, and sophisticated calculational methods. The "reactor physicist" or "nuclear analyst" is an integral part of the overall team in the design and operation of fission-powered reactors.

The physicist who is involved in the development of the breeder reactor has realized the elusive dream of the early alchemist who had attempted to transform worthless materials into gold. Today's physicist has achieved this objective with the breeder reactor by generating plutonium from the waste product of the light water reactor enrichment process to creat a virtually unlimited supply of fuel to provide power for the future. The next step is the development of an economic power system which can be used in the electric utility environment.

What I would like to do in this presentation is to identify those areas where the physicist can make a considerable contribution in meeting the future energy requirements by a successful development of the Liquid Metal Fast Breeder Reactor. The term "physics" is, in terms of the lay community, interpreted in much too narrow a concept. The reactor physicist who is involved in design, experimental interpretation, or the development of calculational models is required to have a considerably broad spectrum of talents. The reactor physicist employed in the design of the LMFBR must be cognizant of areas of expertise ranging from economics through radiation damage to mechanical and thermal/hydraulic design. The experimental physicist involved in measuring nuclear data, or in the performance of integral experiments, must be familiar with electronics and computer science. The physicist working to develop calculational methods to predict the performance of reactors must be familiar with numerical analysis and computer science and have a considerable mathematical background. It can be seen from these examples that

176

the implementation of physics requires knowledge of
areas considerably outside of those normally referred
to as "physics".

What I will do in the first part of this presentation is cover the basic physics problems associated with
the design and development of the LMFBR Demonstration
Plant and outline the outstanding problems which exist
in going from today's knowledge to a successful commercial LMFBR. I will cover the existing information
regarding the nuclear characteristics of the reactor,
as well as the experimental programs which are being
performed to understand basic phenomena associated with
neutron and photon transport.

The physicist is required to establish a design
which provides a viable compromise between a high
plutonium production rate and an economic system. These
two goals are not necessarily compatible. Having solved
many of the basic questions relating to the achievement
of this modern alchemy, he must now become part of a
design team which can harvest the benefits of the early
research.

The production of plutonium in a breeder reactor
is often stated in terms of a quantity called the breeding ratio, which is the rate of production of fissle
material divided by the rate of destruction of fissle
material. This ratio is obtained as a result of complex
calculations, and a value of the order of 1.20 to 1.25
is required to achieve an economic and self-sustaining
growth to keep pace with the demand for electrical power
generation. A second quantity, which is used to compare
reactor designs, and is extremely important in matching
the rate of production of fissle plutonium to the rate
of demand of power generation, is the doubling time.
Basically, this quantity is the time required for a
breeder reactor to produce an amount of fuel equal to
its initial loading. We believe that it is possible to
design an LMFBR which will have a doubling time comparable to the doubling time of electrical power generation
in this country, which is of the order of 10 years.

REACTOR PHYSICS OF THE LMFBR DEMONSTRATION PLANT

In order to put the physics activities in prospective, I will give a brief description of the reactor for
the Clinch River Breeder Reactor Plant. This reactor
will produce 975 MW of thermal power and provide steam
at 900°F. The source of this heat is the reactor core
which is contained within the reactor vessel. Sodium
enters the reactor vessel through the lower nozzles
at 710°F flowing upward through the core and exiting

through the upper nozzles at 970°F. The core is composed
of fuel assemblies. The fuel, which consists of mixed
oxides of plutonium and natural uranium, is contained
within 0.23" diameter stainless steel tubes. The core
section, which produces the majority of the thermal
power, is 3 feet long and consists of 198 of these
assemblies, each containing 217 fuel rods. Above, below,
and around the core are blanket regions consisting of
depleted uranium oxide, whose purpose is to capture
neutrons that leak from the core so that the uranium-238
in the blankets is converted to valuable plutonium. This
configuration is surrounded by steel structures to protect the structural components against radiation damage.

The fission process, which occurs in the core, is
designed to be a controlled chain reaction. This control
is achieved through movable control assemblies which are
similar to the fuel assemblies, but contain boron carbide. The responsibilities of the reactor physicist
consists of predicting the fuel composition to achieve
and maintain criticality, to specify the neutron flux
and power distributions which result from competition
between neutron production, absorption and leakage, and
to specify the location and amount of the control
materials. He must also specify how many, and which,
fuel assemblies should be replaced at each refueling,
and determine the fuel production as a function of time.
Establishment of the dynamic performance of the reactor,
both in normal transient and potential accident situations that might arise and to establish the reactivity
effects associated with temperature changes which provide feedback to maintain the stability of the system,
are also his responsibilities.

The physicist must also specify and analyze systems
to monitor the performance of the reactor and to insure
that the reactor is securely shutdown and monitored
during refueling operations.

The responsibilities of the physicist do not end
with establishing the distribution of neutrons in the
core. In the LMFBR, peak neutron fluxes are of the
order of 7×10^{15} n/cm^2 sec and result in high neutron
fluxes at the structural components. It is thus essential to predict the distribution of neutrons at the
structural components and to establish how much non-
structural shielding must be provided around the core
to mitigate against radiation damage and also to
eliminate biological effects to the operating personnel.
The physicist must ensure that sufficient attenuation
is provided to reduce the neutron flux at the core edge,
which is the order of 10^{15} n/cm^2 sec, to that at the
operating stations, where the total flux cannot exceed
10 n/cm^2 sec. The calculational methods and accuracies

required of the nuclear data to accurately establish an attenuation of 14 orders of magnitude are prodigious.

EXPERIMENTAL AND DEVELOPMENT PROGRAMS

At the present time, considerable research is being performed on the reactor physics of fast breeder reactors. A coordinated program is being pursued in the U.S. The ultimate objective of the program is the development of analytical methods, nuclear data, and the performance of integral experiments required for the design, construction, and operation of practical commercial LMFBRs. Key questions addressed are those of safety, economic performance, and reliability. The U.S. approach in transforming this very general objective into a concrete program is based on a number of explicit assumptions. These can be stated as follows:

a. The neutronic design of fast reactors should ultimately be based on accurate nuclear data and rigorous mathematical methods. Near term efforts must use appropriate integral measurements to supplement and validate the design and safety work obtained using approximate data and methods.

b. Integral measurements are the prime means of confirming data and design methods. Approximate mockups can provide useful confirmation of data, methods and designs, and are essential where a detailed calculational capability is not yet demonstrated.

c. The basics of reactor theory are in general known. There are selected exceptions, such as the effects of neutron polarization effects and multilevel resonance treatments, in neutron interaction with matter which require significant theoretical advances. Substantial work remains to be done in developing effective and efficient means of applying known theory to practical design and analysis situations, and the range of validity of various approximations is frequently not well delineated.

NUCLEAR DATA

One of the most successful aspects of the U.S. Reactor Physics program has been in the area of evaluated cross sections. This is a broadly based and comprehensive program including measurements, evaluations, and storage and distribution. It is a broad program covering the entire U.S. fast reactor community, and interfacing closely and beneficially with the basic

nuclear data community in the U.S. elsewhere including
non-reactor nuclear applications. The effort includes
both government and privately supported participants.
The resulting data is coordinated through, and collected
in, versions of the Evaluated Nuclear Data File (ENDF/B).
The measurement program includes major programs at
national laboratories.

Some of the major goals of the latest version of
ENDF/B are given as follows:

1. To improve prediction of LMFBR power distri-
   tions, reaction rates, material worths, and
   critical masses over a wider range of fast
   spectra and to establish a way to quantitatively identify the resultant level of precision.
2. To allow calculation of reactor fission product
   worth--to permit fission product decay calculations in fast and thermal reactors to an
   identifiable precision.
3. To provide better gamma source, transport and
   heating calculations in LMFBR cores and shields.
4. To provide a consistent dosimetry file for
   radiation damage studies.
5. To update data files to reflect new and improved measurements.
6. To provide a more consistent and improved data
   file for thermal reactor application.
7. To provide error information in major evaluations of interest to the LMFBR program.

INTEGRAL EXPERIMENTS

Integral experiments are described variously but
include: "benchmarks"--designed to validate specified
data or methods; explicit mockups--to confirm design
predictions; clean assemblies--to generate basic data;
and exploratory configurations--to lead design efforts.
While the U.S. program has used all of these approaches
at times, the current program is strongly concentrated
on benchmarks and design confirmation configurations.
The current program being pursued in the integral
experiment facility, ZPPR in Idaho Falls, consists of
a program called ZPPR Assembly 3 in which control rod
interaction measurements, together with measurements of
the effect of control rods on local and global power
distributions are being performed. Moreover, reactivity
coefficients and subcritical measurements have been
carried out. In ZPPR Assembly 4, power distribution
measurements in the blankets, together with a first
attempt at breeding ratio measurement, are being carried
out.

As in the critical experiments, the shielding experiments are virtually exclusively limited to bench-mark type experiments. Some specific design features are included. The key technical areas being addressed are the shielding efficiency of the axial and radial shields, control rod streaming, and top head and support structure streaming.

CALCULATIONAL METHODS

The development of calculational methods, which permit design information to be generated, still has considerable room for research. The calculations required for the reactor design commence with the basic nuclear data and progress through the generation of energy averaged cross sections. These calculations include considerable detail as to such effects as resonance self-shielding, and spatial effects. Following the generation of these cross sections, two- and three-dimensional flux and power distributions are generated. A considerable amount of effort has been directed towards the generation of fast and accurate calculational schemes. It is not unusual for calculations employing a million space-energy points to be performed to provide essential design information.

In this area, a combination of talents is required which cover the skills of the physicist with those of a mathematician and computer programmer. We need to progress from these very complex calculational schemes to be able to perform much simpler, though equally accurate, calculational methods. A considerable contribution can be made by the physicist in the development of these empirical techniques.

COMMERCIAL REACTOR STUDIES

Many physics problems exist which must be solved in order that full exploitation of the LMFBR can be carried to the stage of development of a commercial product. In particular, the question of reactor stability, safety, control, and the refueling characteristics are of prime importance. The objective of achieving a self-sustaining fuel supply with its attendant requirement of a high breeding ratio requires considerable accuracy in the cross sections of the higher plutonium isotopes and fission products. Calculational methods to permit accurate and rapid generation of the physics characteristics are of prime importance. It is in these areas that the physicist can make his contribution.

CONCLUSION

The physicist has responded in the past to the challenge of providing information to provide a power source for today and tomorrow. It is our challenge now to respond to the "energy crisis" and to provide the next step which is development of the commercial Liquid Metal Fast Breeder Reactor. I feel that the physicist has responded admirably in reorientating himself to not only being involved in a research capacity, but also in becoming a design physicist. With this dual role, the physicist will continue to prove his value to his science.

RADIATION DAMAGE IN REACTORS*

G. H. Vineyard
Brookhaven National Laboratory, Upton, N.Y. 11934

ABSTRACT

Fuel, cladding, and structural materials, moderators, coolants, pressure vessels, and other components of nuclear reactors are bombarded by fast and thermal neutrons (with fluxes up to and above 10^{15} cm^{-2} sec^{-1}) at elevated temperatures. Fuel elements are also exposed to fission fragments; all components are exposed to high fluxes of gamma rays. As a result the physical properties of materials are modified, usually in undesirable ways. Brittleness, altered creep, and swelling which leads to complex deformations and induced stress are common problems. A broad understanding of these effects in terms of displaced atoms, point defects, dislocations, disordered regions, voids, and transmutation-induced impurities has been developed. Much empirical information has become available, and means for lessening certain effects have been found. Knowledge of the details of atomic mechanisms, particularly their quantitative aspects, in the broad range of materials of interest is still inadequate, and the body of empirical data, particularly at the higher levels of exposure, needs to be extended. Bombardment of thin specimens with charged particles from accelerators has been a useful tool for simulating high neutron exposure, but the degree to which the simulation is adequate needs further investigation.

INTRODUCTION

Among the various materials problems associated with nuclear reactors, there is one that continues to be of special interest to physicists, namely the influence on solids of nuclear radiation.✝ Beginning with the deleterious effects of fast neutrons on graphite moderator materials, which were first encountered in the early 40's, such problems have been under continuous study for about thirty years. So much has been learned in this time that all problems may appear to be solved, but this is far from true. Indeed, as will be detailed below, major developments continue to occur.[1] In this paper

---

*Supported by the Atomic Energy Commission.
✝Much of the discussion that follows applies also to materials problems of controlled thermonuclear devices, where damage to walls and structural components by energetic ions and fast neutrons is expected to present serious problems. See Refs. 14 and 15.

a short survey of effects of current interest will be given. No
attempt will be made to treat the subject exhaustively.

The principal cause of radiation effects in reactors is the
flux of neutrons throughout the interior of the reactor, and the
fission fragments which irradiate fuel and blanket materials.
Other charged particle components of the radiation exist but are
of relatively minor significance. Gamma rays, which are intense in
the core, deposit heat and, in the case of insulating materials,
may also produce a part of the radiation damage. The neutrons
originate in the fission process where they are born with a spectrum
of energies ranging from zero to somewhat above 10 MeV, with a mean
energy of about 2 MeV and most probable energy of about 1 MeV. The
actual spectrum at any point in the reactor departs from this,
generally toward lower energies, depending on the amount of moderation and absorption that has occurred. The greatest intensity of
flux, as encountered in the core of a power producing reactor, may
be somewhere between 10^{14} and 10^{16} neutrons/cm^2 sec, with a mean
energy of the order of 1 MeV. At the pressure vessel in a power
reactor the flux of fast neutrons may be in the range 10^9 to 10^{11}
neutrons/cm^{2}sec.

The interactions of neutrons with matter fall into two categories, nuclear scatterings, which produce energetic knock-on atoms
(typically with energies of tens to hundreds of keV), and nuclear
reactions which produce transmutations. The transmutation-produced
impurity atom of perhaps greatest technological significance is
helium, produced by (n,α) reactions. Numbers vary enormously with
conditions, but typically 1 to 100 atomic parts of helium per
million will be produced in stainless steel components in the core
of a reactor in a few years of operation. Reactions of neutrons
with Ni, Fe, Cr, and such impurities as nitrogen and boron all play
a role in helium production. Hydrogen is also produced in substantial quantities by (n,p) reactions, but because of its higher
diffusivity it escapes more rapidly and therefore usually is of
less significance.

The other products of neutron interaction, namely energetic
knock-on atoms, are the starting point for most of the damage to the
crystalline lattice. Such a knock-on initiates a collision cascade
in which, typically, hundreds of atoms in a region some hundreds of
Angstroms in dimensions participate. The initial product of such a
cascade is a large sprinkling of atoms that have been displaced
from their lattice sites to interstitial positions, and an equal
number of vacancies scattered over the region of the cascade. Some
of these point defects will find themselves in small clusters, such
as di-vacancies, tri-vacancies, di-interstitials, etc., which have
distinctive behavior. Immediately afterward, thermally induced
rearrangement, annihilation, and migration sets in, so that the

final state depends upon temperature, impurity content, and other variables. Fig. 1 illustrates a typical cascade, as found by Beeler[16] from a somewhat simplified computerized model.

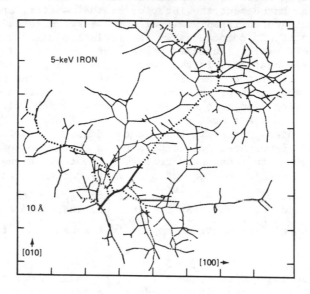

Fig. 1. Typical collision cascade in α-iron as calculated from a binary collision model by Beeler. Three-dimensional trajectories of atoms are shown projected onto a (001) plane. The short heavy line with the kink in the middle is the primary knock-on that initiated the cascade (arbitrarily assumed to start with 5 keV). Trajectories of secondary knock-ons are shown as heavy dotted lines emanating from the primary track. Thereafter alternately dashed and solid lines represent the trajectories of successively higher order knock-on atoms.

Other computerized calculations[17,18] have given realistic information on cascades in simple crystals at low or moderate energies. All of the studies together allow one to understand reasonably well the initial state of damage following a primary knock-on event in a wide range of conditions.[19,45] The interactions between defects and their thermally-activated rearrangements are less well agreed upon.

When the defects produced by the primary knock-on are numerous and close together, which is particularly true for energetic events in substances of high atomic weight, collective descriptions of the damage process, such as the so-called displacement spike models, may have limited usefulness. Such models remain highly qualitative and rather controversial.

Fission fragments, being intermediate mass ions of very high energy, produce a high density of displaced atoms along thin short paths (1 to 10 μm in length). They are responsible for intense radiation effects in fuel and blanket materials. Limitations of space prevent discussion of these rather special problems here. The interested reader is referred to the standard texts.[2-7]

MECHANICAL EFFECTS

From early days in the study of radiation effects, it has been known that heavy exposure of metals to fast neutrons produces hardening and embrittlement.[3] Typical stress-strain curves before and after irradiation are shown in Fig. 2. Irradiation raises the yield stress and ultimate stress and reduces the ductility. Generally such effects set in at lower fluences in simple annealed metals than in metals strengthened by cold working or precipitation. In ferritic steels the ductile-to-brittle transition temperature is usually raised by irradiation, and the ductility above the transition temperature is lowered. In a general way it is clear that the defects produced by irradiation will tend to increase the yield stress by locking dislocations, an effect in pure metals similar in many respects to that of alloying. The effects in the work-hardening range and beyond are more complex, and while many models have been advanced to explain observations, these

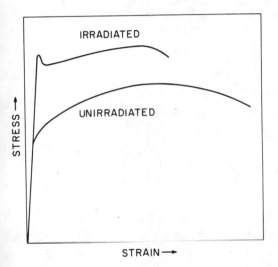

Fig. 2. Stress-strain curves typical of many structural metals before and after irradiation.

are qualitative and rather speculative. A mass of empirical data on structural alloys has been accumulated, but is not well coordinated by quantitative theoretical understanding. Generally radiation embrittlement imposes the most serious limitations on two structural components of reactors: fuel cladding, which may be caused to leak or rupture, and pressure vessels for water reactors, where the useful lifetime of the vessel may be reduced.

186

Recently a valuable development has occurred. Certain residual elements, copper and phosphorus, have been found to cause radiation brittleness in pressure vessel steels. It has been found that elimination of these elements virtually eliminates radiation embrittlement at temperature and fluences of interest in pressurized water reactors and boiling water reactors.[20,23] Fig. 3 shows brittleness before and after irradiation in three samples of steel which differ in their phosphorus and copper content.

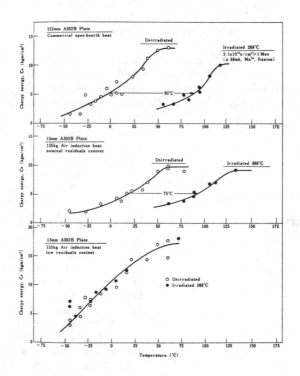

Fig. 3. Comparison of radiation embrittlement (as measured by Charpy-V notch 5.2 kgm/cm^2 transition temperature) of steels with varying amounts of residual elements (see text).

The curves show Charpy-V notch energy as a function of test temperature. The upper set of curves is for a commercial sample of A302B steel plate. The unirradiated specimen shows the typical transition from brittle behavior below about $-50°C$ to ductile behavior above about $+50°C$. After irradiation to 3.1×10^{19} fast neutrons/cm^2 at $288°C$, the curve is shifted to the right by $\sim92°C$, and the ductility at higher temperature is distinctly impaired. The middle curves show the same tests on a special heat of similar steel with a lower content of residual elements. The increase of transition temperature with irradiation is now only 75°C, and the high-temperature ductility is almost unaffected. The bottom set of curves show results on a special batch of steel with very low residuals. The irradiated and unirradiated specimens are now essentially indistinguishable. The identification of phosphorus and copper as the elements responsible has been completed in further extensive experiments with this and several other steels (summarized in ref. 23). The reasons for this effect have not been determined, although one suggestion is that copper interacts with vacancies to form defect complexes which are more numerous and more stable than in the pure steel, and that this retards the dynamic recovery process during irradiation.

Irradiation affects high temperature creep and creep rupture properties of steels, usually increasing the creep rate (although it also can decrease it) and reducing the creep rupture strength. Irradiation provides several mechanisms by which creep can be altered, but the total picture is complex. For discussions of these effects the reader is referred to Bush,[8] Reuther and Zwilsky,[23] Gilbert and Harding,[24] and King, et al.[25]

VOIDS AND SWELLING

Of the point defects introduced into metals by irradiation, some (the interstitials) are mobile at quite low temperatures. Single vacancies become mobile around 100°C to 200°C, and small clusters of vacancies have widely varying degrees of mobility.[26] While some clusters are known to be stable to higher temperatures, and voids have been seen in irradiations at lower temperature, there were grounds for believing that irradiations of structural alloys at high temperatures (e.g. 500°C) would produce little damage because of the simultaneous annealing that would occur.* Thus it was a surprise when Cawthorne and Fulton in 1966 reported pronounced swelling, with the formation of microscopic voids, in stainless steel irradiated to high fluences at elevated temperature. Since this observation, which was published[28] in 1967, an immense amount of work in many countries has been done on all aspects of the

---

*Barnes and Mazey[27] showed in the early 1960's that bombardment of copper with α-particles from a cyclotron at certain temperatures could produce voids stabilized by gaseous helium, known as bubbles. They developed a theory for nucleation, growth, coalescence, and migration of helium bubbles in copper which later proved to be of service in understanding high-temperature helium embrittlement in structural alloys and fission-product swelling in fuels.

188

phenomenon of swelling and void formation.[11, 12, 29] For the reactor designer, particularly in the case of fast reactors with their very close mechanical tolerances, this swelling phenomenon is a serious problem which requires costly modifications of design and limitations on performance. Huebotter and Bump[30] estimated that, to accommodate only 5% swelling in breeder reactors over the period 1970 to 2020 had a 1970 present worth of from $860,000,000 to $5,600,000,000 relative to the case of 15% swelling. Thus the impetus for seeking low-swelling alloys is great.

It is now known that most metals heavily irradiated at temperatures in the range approximately $0.3\ T_m$ to $0.6\ T_m$, where T_m is the absolute melting temperature, show the swelling phenomenon. Irradiations above or below this temperature range do not produce it. The swelling sets in after a threshold fluence, which depends on material and conditions, and increases thereafter, through the range 10^{20} to 10^{23} fast neutrons/cm^2. In early stages the volume is often observed to increase more rapidly than linearly with fluence above threshold, and at very high doses (experimentally achievable with ion bombardment rather than neutrons) a tendency to saturate is seen. Volume increases may ultimately reach 10 to 20%, or more. As an example, Fig. 4 shows the density change, $-\nabla\rho/\rho$, as a function of fluence in

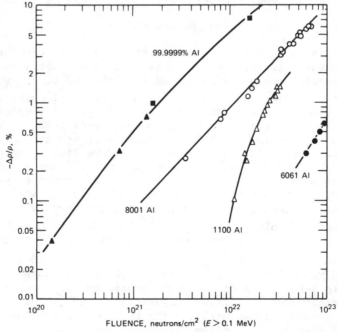

Fig. 4. Density change, $-\nabla\rho/\rho$, as a function of fluence in aluminum and several of its alloys irradiated at about 50°C (after Jostsons et al[31]).

aluminum and several of its alloys, irradiated at temperatures above 50°C.

Specimens that have swollen in this manner contain a large number of microscopic voids, clearly visible in the electron microscope. Fig. 5 is a rather typical electron micrograph of aluminum (1100 grade) irradiated to a fluence of 2.8×10^{22} neutrons/cm^2 (E > 0.1 MeV) at 50°C. The voids in this picture range from 100 Å to 1000 Å in diameter. Note that they tend to have cubic shapes and are distributed apparently at random.

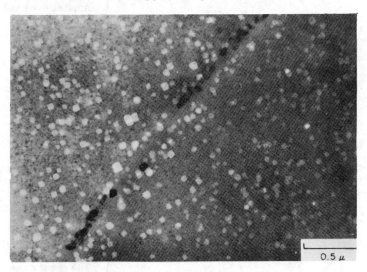

0.5 μ

Fig. 5. Electron micrograph of rather heavily irradiated aluminum showing voids within the grains and silicon precipitation at the grain boundary (after Jostsons et al[32]).

In the earliest stages of void production, the voids are very small. With further irradiation the number of voids per unit volume increases, and their mean size increases. It is typical for the voids to have geometrical shapes determined by low index (i.e. low energy) plans of the metal. At high temperatures of irradiation, for a given fluence the voids tend to be larger but less numerous. Thus, Fig. 6 shows the minimum void size observed in 304 stainless steel vs. irradiation temperature; Fig. 7 shows the void density for the same material and set of irradiation temperatures. It is observed that voids are absent or less numerous near grain boundaries, and that copious tangles of dislocations grow in at the same time as the voids.

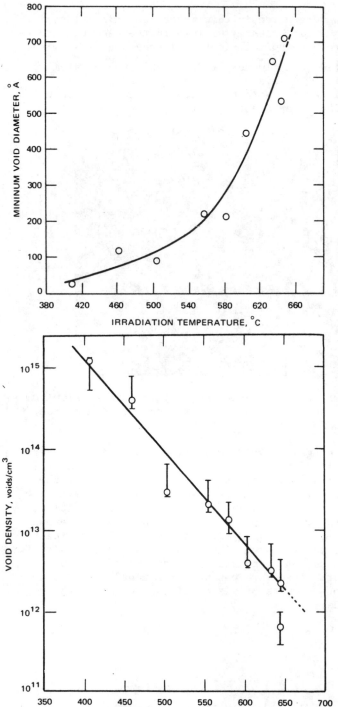

Fig. 6. Size of smallest voids observed in irradiated 304 stainless steel vs. temperature of irradiation (after Appleby et al.[42]).

Fig. 7. Density of voids in irradiated 304 stainless steel vs. temperature of irradiation (after Appleby et al.[42]).

As a result of the many investigations of void growth and
swelling in the last seven years, the qualitative nature of the
phenomenon and its scope are well known. The mechanism responsible
is widely believed to be the following: Vacancies and interstitials are produced in equal numbers in the radiation damage cascades, and at the temperatures in question migrate rapidly to
sinks. Although this process is approximately symmetrical between
vacancies and interstitials, it is not exactly so, and the
imbalance is crucial. Interstitials are attracted more strongly
to dislocations, where they are absorbed, than are vacancies (by
reason of the larger strain field around the interstitial). Moreover, even in regions far from dislocations interstitials migrate
more rapidly than vacancies.* Thus an excess of vacancies is built
up after the irradiation starts, and eventually reaches a sufficiently high supersaturation to cause nucleation and growth of
vacancy clusters, i.e., voids. The nucleation process itself
remains controversial, and indeed several different nucleation
mechanisms may operate. Homogeneous nucleation may occur, if the
supersaturation is sufficiently high; heterogeneous nucleation
probably is more common, and this may be brought about by existing
impurities or precipitates, by small heavily damaged regions
(displacement spikes, etc.) introduced by the radiation, or by
residual gases in the metal and gases, notably helium, produced
by the radiation. That displacement spikes and transmutation
gases are not essential is proved by the production of voids in
very pure metals by electron irradiation, which is incapable of
producing either spikes or transmutations. On the other hand,
injection of helium along with or prior to irradiations has been
shown to lower the threshold exposure at which voids are formed
in some metals. This shows that gases can promote void formation.
A second major issue, not yet well resolved, is the question of
why the small clusters of vacancies do not collapse into dislocation loops. The presence of enough helium (or other gas) in the

---

*If the mean residence time of a defect, before annihilation or
 removal is τ, and the rate of production of this defect, per
 unit volume, is $\dot{n}$, the mean concentration of the defect under
 steady state condition is $\dot{n}\tau$. The faster migration of interstitials, in addition to their attraction to dislocations, causes
 τ for interstitials to be less than τ for vacancies and thus
 lowers the relative concentration of the former.

nucleated cluster would prevent collapse.* In some cases there appears to be a geometric metastability of the small vacancy cluster which is sufficient. Growth of the void nucleus clearly takes place by the continued inflow of vacancies produced in the course of the irradiation. Also in the growth stage both an excess of vacancies over the equilibrium number and an excess over the number of interstitials continue to be necessary.

Several quantitative models of void production have been developed and are reasonably successful in describing the simpler observations, although a rather large number of parameters is required that are not well determined by independent means.

The observation, at first puzzling, that void swelling occurs only in a certain range of temperatures ($\sim$0.3 T_m to 0.6 T_m) can now be understood: At too low a temperature vacancies are not mobile and thus cannot aggregate. At too high a temperature the rapid motion of vacancies to existing sinks prevents the buildup of the necessary supersaturation (moreover, the equilibrium concentration, which must be exceeded, increases rapidly with temperature). The higher the flux of radiation the higher the range of temperatures in which supersaturation can be achieved. In agreement with this, experiment shows that the upper limiting temperature for swelling is flux dependent. In ion bombardments, where the damage rate is much higher than in reactors, swelling occurs in a temperature range with a substantially higher maximum.

The swelling phenomenon, since it occurs at temperatures that are common in modern power reactors, and in materials such as stainless steels which are employed for numerous structures in and near the core of the reactor, presents problems to the reactor designer.[30] In the liquid metal fast breeder reactor, fuel is contained in long cylindrical stainless steel tubes, typically 1/4 in. in diameter and about 9 ft long. These fuel "pins" are formed into bundles about 5 in. in diameter, and each bundle is contained in a larger tube of stainless steel. The bundle is inserted into the reactor and removed as a unit, and a typical large reactor will

---

*That an appreciable amount of helium is present in voids after some irradiations has been proved in ingenious experiments of Katyal, Keesom, and Cost.[33] Aluminum was irradiated with cyclotron alpha particles to produce concentrations up to 300 ppm of helium. After annealing at temperatures above 600°C, which produced voids visible in the electron microscope, the specific heat of the aluminum sample was measured as a function of temperature from 2°K upward. A pronounced peak in the specific heat centered on 4.2°K was observed. This is clearly caused by the liquid-vapor transition of the helium in the bubbles.

contain about 500 bundles. Liquid sodium, which is the heat transfer medium, flows upward through each bundle. The bundles are
supported and positioned by a pair of grids (known as grid plates)
lying in horizontal planes, one above the other. Without belabor-
ing the details, it is clear that operation of the reactor and
fuel handling both require that rather close mechanical tolerances
in this assembly be maintained. The fuel elements must retain
their integrity for the life of a core, perhaps a year, while the
grid plates and other permanent structures must retain their
integrity for the life of the reactor. Since the neutron flux is
not uniform everywhere, one side of a typical fuel assembly will be
subjected to more fluence and hence more swelling than the other.
This produces bowing of the long assembly which imposes undesirable
limits on the lifetime of the core. Unequal flux on the two sides
of each grid plate tends to cause buckling of the grid plate. For
reasons such as these it is urgent to develop alloys with minimal
swelling.

From the understanding of the mechanism sketched above, some
means of accomplishing this suggest themselves: Additional sinks
for vacancies, or increased mobility of vacancies, or reduction of
the preferential trapping of interstitials at dislocations will
reduce the excess concentration of vacancies; traps for interstitials can hold up the migrating interstitial and simultaneously
constitute a trap for vacancies; perhaps some conditions can be
found which hinder the nucleation of voids or even destabilize the
growing voids; and finally, particular alloys may be found for which
the swelling rate at desired operating conditions is lower because
of altered values of all the relevant parameters.

Adda[34] has reported that in nickel and copper the addition of
elements that reduce the stacking-fault energy minimizes swelling.
One explanation is that this reduces the strain field around dissociated dislocations, reducing their preference for interstitials.
In the case of addition of Al to Cu, one case in point, the mobility of vacancies is also increased, which would cause the maximum
swelling to occur at a lower temperature.

Certain minor changes in the composition of stainless steels
have produced substantially improved resistance to swelling.
Bramman, et al.[35] suggest that very stable, finely dispersed
coherent precipitates, such as TiC, NbC, and γ' phase, can produce
traps which promote recombination of Frenkel pairs thus reducing
the concentrations of vacancies. They also show that cold working
in such steels is beneficial, maximum swelling rates in 20% cold
worked material being some two to three times lower than those in
the same material after solution treatment. Straalsund[36] et al.
confirm that austenitic stainless steel subjected to cold work

shows negligible void formation after irradiation to 3×10^{22} neutrons/cm^2 in the temperature range $370^{\circ}C$ to $850^{\circ}C$. Bloom[37] has noted lower swelling in 10% cold-worked 304 stainless steel after similar irradiations. There is a discrepancy, however, between the microstructures reported by these various investigators, Bramman reporting that the swelling reduction from cold work is due to a reduction in size, rather than number of voids, while Straalsund and Bloom attribute the effect to a reduction by factors of 20 to 50 in the number of voids with little change in their size. The reduction in number of voids could be most readily explained on the assumption that the greater number of traps in the cold worked material reduces the concentration of vacancies attained, which inhibits the nucleation process.

Laidler[38] has investigated a niobium-stabilized austenitic stainless steel, AISI 348, in which the niobium carbide can be caused to precipitate in association with stacking faults to provide a high concentration of sinks for point defects (as well as improved tensile and creep properties at elevated temperature). It is hoped that in this material void formation, and swelling, can be strongly suppressed.

Farrell et al.[39] investigated void formation in aluminum and some aluminum alloys. They found that 6061 and 6063 aluminum alloys are much more resistant to void formation than commercial grade 99.0% purity aluminum (1100 aluminum). The former are precipitation-hardened alloys containing up to 1.2% magnesium and 0.8% silicon. They contain very finely divided coherent Mg_2Si precipitates interlaced with dislocations.

An alternative explanation for the effect of cold work is advanced by Buswell et al.[40] who argue that at high enough dislocation density, dislocations are inhibited from climbing by mutual interference which causes them to cease to perform as sinks for interstitials. They also point out that coherent precipitates will similarly inhibit dislocations from functioning as sinks for interstitials, but that incoherent precipitates provide plentiful sinks at their interfaces with the lattice. Thus coherent precipitates discourage void growth, but overaging to the stage where the precipitates become incoherent reverses the inhibition.

A curious phenomenon sometimes observed when voids are produced by radiation is the formation of a superlattice of the voids. Wiffen[41] reports such effects in molybdenum irradiated at $585^{\circ}C$

and niobium irradiated at 790°C, in both cases to a fluence of
2.5×10^{22} neutrons/cm^2 (>0.1 MeV). Two of Wiffen's electron
micrographs are shown in Fig. 8.

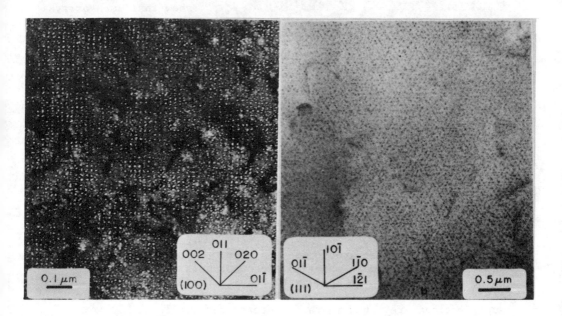

Fig. 8. Examples of superlattices of voids in (a) molybdenum
irradiated at 585°C and (b) niobium irradiated at 790° C. The
fluence for both samples was 2.5×10^{22} neutrons/cm^2, E > 0.1 MeV
(after Wiffen[41]).

Although the superlattice is not perfect, its regularity is quite
remarkable, and there is a clear alignment with the underlying
crystal lattice. Other observations of void superlattices have been
reported by Kulcinski et al.[43] and Evans et al.[44] The latter group
bombarded molybdenum foils with 2 MeV N$^+$ ions at 870°C to the rather
high dose whereby an average of 100 displacements were experienced
by each atom. Fig. 9 shows two electron micrographs at different
orientations of the resulting arrays.

196

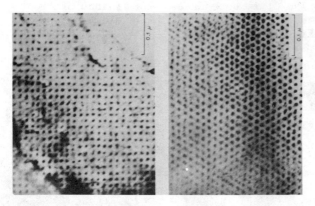

Fig. 9. Superlattice of voids in molybdenum after heavy bombardment with N^+ ions at 870°C. Left photo shows projection of void array in [010] direction, right shows projection in [111] direction. The array is b.c.c. (after Evans et al.[44]).

Considerations of elastic interactions between voids in an anisotropic medium show that such arrays minimize the interaction energy. Predictions of the superlattice constant can be made from such considerations and are in semiquantitative agreement with the observations. The process by which such a lattice is formed is not entirely clear, however, and offers an interesting field for further work that perhaps would illuminate all of the processes by which voids are formed and grow. The metals in which such superlattices have been observed are substances of high elastic anisotropy, and for the most part body-centered cubic superlattices are found in body-centered cubic alloys and face-centered cubic superlattices are found in face-centered cubic alloys.

USE OF ION BOMBARDMENT

It has already been remarked that bombardments of thin specimens with energetic ions can produce effects similar to neutron bombardment. The ion bombardment technique, in addition to its intrinsic interest, is of importance as a means of simulating heavy exposure to fast neutrons in much less time and at lower cost, with the further advantages that the irradiated specimens are not highly radioactive, as is the case after prolonged reactor irradiations, and the ambient conditions (temperature, etc.) are more easily controlled during irradiation. At least three disadvantages to this simulation should also be recognized: the spectrum of knock-on atoms is never precisely the same as with neutrons; transmutation effects are absent; and, because of the short ranges of heavy ions of appropriate energies, the specimens must be very thin, which means that disturbance from proximate surfaces must be guarded against.

As an illustration, 5 MeV nickel ions have a range of slightly less than 2 μm in metallic nickel, and 1.3 MeV protons have a range of only 10 μm in the same substance. MeV neutrons, on the other hand, penetrate a mean distance of tens of centimeters before suffering a collision. For comparing exposures it is conventional to calculate the mean number of times during a bombardment that each atom in the specimen is displaced from its lattice site to an interstitial position (from which it usually finds its way back into a lattice site by thermally activated processes before being displaced again). Kulcinsky et al.[43] have given a useful chart of this index of relative effectiveness of different types of radiation applied to nickel (Fig. 10).

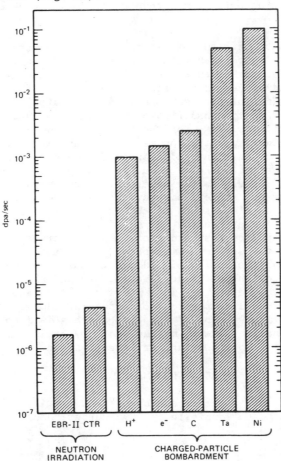

Fig. 10. Maximum damage rates available by various energetic particle bombardments of nickel, measured as displacements per atom per second (after Kulcinski et al.[43]).

The rates of production of displacements per atom per second for various kinds of radiation are shown. For the charged particle radiations the calculations are for presently attainable fluxes (in some cases limited by allowable heating of the sample). This figure shows that damage with ions can be accumulated at rates 2 to 5 orders of magnitude higher than with neutrons.

A further elaboration of the ion bombardment technique is to bombard, simultaneously or consecutively, with impurity ions and with ions of the same chemical species as the sample, thus simulating the principal effects of transmutation. This method has been employed rather frequently to introduce helium in studies of void formation.

From one point of view the ultimate in simulation would be to bombard a very thin specimen with ions of the same species provided with the same spectrum of energies as the primary knock-ons that would be produced by neutrons (allowing, of course, for the spectrum of energies of the neutrons being simulated). For elements as heavy as iron and fission neutrons, however, this presents grave difficulties, for the required ions are of such low energy that their penetration is in the sub-micron range. Thus 1 MeV neutrons on iron produce primary knock-ons in iron with a mean energy of about 30 keV. A second approach is to use protons of about the same energy as the neutrons. The energy transfer cross section for MeV protons on many nuclei is not enormously different from that of neutrons of the same energy, although it is not identical. However, unless the specimen is much thinner than the range of the proton (the order of 10 μm), degradation of energy of the proton in the specimen will occur, producing a gradient of damage and possibly retention of hydrogen. Still another approach is to bombard with ions of higher energy, use the observations to determine essential parameters in a model of the damage process, and then use the model to predict corresponding damage by neutron irradiation. Obviously none of these approaches is ideal, but the latter may eventually prove to be the most powerful. Clearly a large field remains open for further investigation.

ACKNOWLEDGMENTS

In preparing this paper I have benefitted particularly by discussions with L. C. Ianniello, T. C. Reuther, and G. W. Cunningham.

REFERENCES

1. General discussions of radiation damage will be found in
    references 2-7. Extensive treatments of materials problems
    of reactors are given in references 8-13.

2. F. Seitz and J. S. Koehler, "Displacement of Atoms During
    Irradiation," Solid State Physics (Academic Press, Inc.,
    N.Y., 1956) Vol. 2, p. 305.

3. G. J. Dienes and G. H. Vineyard, Radiation Effects in Solids
    (Interscience Pub., New York, 1957).

4. D. S. Billington and J. H. Crawford, Jr., Radiation Damage in
    Solids (Princeton University Press, Princeton, N.J., 1961).

5. J. J. Harwood, H. H. Hausner, J. G. Morse, and W. G. Rauch,
    editors, The Effects of Radiation on Materials (Rheinhold
    Pub. Corp., New York, 1958).

6. G. Leibfried, Bestrahlungseffekte in Festkörpern; eine
    Einführung in die Theorie (B. G. Teubner, Stuttgard, 1965).

7. M. W. Thompson, Defects and Radiation Damage in Metals
    (Cambridge U. Press, London, 1969).

8. S. H. Bush, Irradiation Effects in Cladding and Structural
    Materials (Rowman and Littlefield, New York, 1965).

9. "Irradiation Effects in Structural Alloys for Thermal and
    Fast Reactors," ASTM Symposium San Francisco, 1968 (Special
    Technical Publication 457, ASTM, Philadelphia, 1969).

10. Symposium on Materials Performance in Operating Nuclear
    Systems, M. S. Wechsler and W. H. Smith, editors, Ames,
    Iowa, Conference 1973 (USAEC, TN 37830).

11. "Voids Formed by Irradiation of Reactor Materials," S. F. Pugh,
    M. H. Loretto, and D. I. R. Norris, editors, Proceedings of
    the British Nuclear Engineering Society, Harwell, England,
    March, 1971).

12. "Radiation-Induced Voids in Metals," J. W. Corbett and
    L. C. Ianniello, editors, Proceedings of an International
    Conference, Albany, N.Y., June, 1971 (USAEC CONF-710601).

13. International Meeting of Fast Reactor Fuel and Fuel Elements,
    M. Dalli-Donne, K. Kummerer, and K. Schroeter, editors
    (Gesellschaft f. Kernforschung, Karlsruhe, West Germany, 1970).

14. Critical Questions in Fundamental Radiation Effects Research, L. C. Ianniello, editor (USAEC Report Wash 1240-73).

15. G. L. Kulcinski, "Materials Problems for Fusion Reactors," Materials Engineering Congress, Cleveland, 1972 (to be published).

16. J. R. Beeler, Jr., Phys. Rev. 150, 470 (1966).

17. J. B. Gibson, A. N. Goland, M. Milgram, and G. H. Vineyard, Phys. Rev. 120, 1229 (1960).

18. G. H. Vineyard, "Computer Experiments with Lattice Models," Interatomic Potentials and Simulation of Lattice Defects, P. C. Gehlen, J. R. Beeler, and R. I. Jaffe, editors (Plenum Pub. Corp., New York, 1972), pp. 3-25.

19. M. T. Robinson and I. M. Torrens, "Computer Simulation of Atomic Displacement Cascades in Solids in the Binary Collision Approximation" (to be published).

20. U. Potapous and J. R. Hawthorne, Nuclear Applications 6, 27 (1969).

21. L. E. Steele, Atomic Energy Review 3, 50 (1969).

22. J. R. Hawthorne, E. Fortner, and S. P. Grant, Welding Journal Research Supplement 35, 553 (1970).

23. T. C. Reuther and K. M. Zwilsky, "The Effect of Neutron Irradiation Toughness and Ductility of Steels," Proceedings of Kyoto International Conf., Oct. 25-26, 1971.

24. E. R. Gilbert and N. E. Harding, ref. 9, pp. 17-37.

25. R. T. King, K. Farrell, and A. E. Richt, ref. 10, pp. 133-152.

26. A. C. Damask and G. J. Dienes, Point Defects in Metals (Gordon and Breach, New York, 1963).

27. R. S. Barnes and D. J. Mazey, Proc. Roy. Soc. A275, 47 (1963).

28. C. Cawthorne and E. J. Fulton, Nature 216, 515 (1967).

29. A. L. Bement, Jr., "Void Formation in Irradiated Austenitic Stainless Steels," Advances in Nuclear Science and Technology, E. J. Henley and J. Lewins, editors (Academic Press, New York) Vol. 7, pp. 1-120.

30. P. R. Huebotter and T. R. Bump, ref. 12, pp. 84-124.

31. A. Jostsons, E. L. Long, Jr., J. E. Furguson, and K. Farrell (to be published).

32. A. Jostsons, E. L. Long, Jr., J. O. Stiegler, K. Farrell, and D. N. Braski, ref. 12, pp. 363-384.

33. O. P. Katyal, P. H. Keesom, and J. R. Cost, ref. 12, pp. 248-254.

34. Y. Adda, ref. 12, pp. 31-83.

35. J. I. Bramman, K. Q. Bagley, C. Cawthorne, and J. E. Fulton, ref. 12, pp. 125-141.

36. J. L. Straalsund, H. R. Brager, and J. J. Holmes, ref. 12, pp. 142-155.

37. E. E. Bloom, ref. 12, pp. 1-30.

38. J. J. Laidler, ref. 12, pp. 174-186.

39. K. Farrell, J. T. Houston, A. Wolfenden, R. T. King, and A. Jostsons, ref. 12, pp. 376-385.

40. J. T. Buswell, S. B. Fisher, J. E. Harbottle, D. I. R. Norris, and K. R. Williams, ref. 12, pp. 533-549.

41. F. W. Wiffen, ref. 12, pp. 386-396.

42. W. K. Appleby, D. W. Sandusky, and U. E. Wolf, ref. 12, pp. 156-173.

43. G. L. Kulcinski, J. L. Brimhall, and H. E. Kissinger, ref. 12, pp. 449-478.

44. J. H. Evans, R. Bullough, and A. M. Stoneham, ref. 12, pp. 522-532.

45. D. G. Doran, Trans. Am. Nucl. Soc. 17, 207 (1973).

HIGH-LEVEL RADIOACTIVE WASTE DISPOSAL

R. C. Liikala, R. W. McKee and W. K. Winegardner
Battelle-Northwest, Richland, Wash. 99352

ABSTRACT

The U.S. Atomic Energy Commission (AEC) is developing
additional plans and new methods for managing radioactive
wastes generated by past, present and future operations.
The objectives of these programs are to; (1) ensure the
health and safety of the public, (2) protect our environment and ecology, and (3) use methods acceptable to the
public. A brief overview is presented of the plans and
current studies for disposing of high-level radioactive
waste generated in commercial nuclear facilities. The
methodology being developed and used to assess the merits
of alternate concepts is presented. The technical areas
where the physics community might contribute to improving
this program are outlined.

INTRODUCTION

The power of the atom - a self-sustained nuclear reaction - was
demonstrated under the stands of Stagg Field in Chicago on December
2, 1942. Because the military implications of nuclear fission had
already been recognized, the United States launched a massive effort,
the Manhattan Project, to develop this energy form. This lead to
the construction of large nuclear reactors to produce nuclear weapons
materials.

The peaceful uses of splitting the atom were also recognized and
soon after World War II, the government initiated programs to realize
the benefits of producing useful energy from nuclear fission. This
effort led to the establishment of a commercial nuclear power industry in the 1960s.

Whether nuclear reactors are used to produce materials for
military purposes or to generate electrical power, a by-product of
their operation is a radioactive residue. The potential risks associated with these radioactive materials were recognized at the start.
Therefore, stringent methods have been developed to limit the release
of radioactivity and the consequences of any release so as to not
have a significant adverse effect on workers in the plant, the public, and the environment.

Because of the military priorities applied to the initial development of nuclear fission, there was little time to consider all
possible alternatives for the management of radioactive wastes. In
those early days, a prudent course of action was taken: i.e., confine the most hazardous wastes in storage tanks. The current plan
for managing radioactive wastes in AEC-owned facilities is documented
in Reference 1.

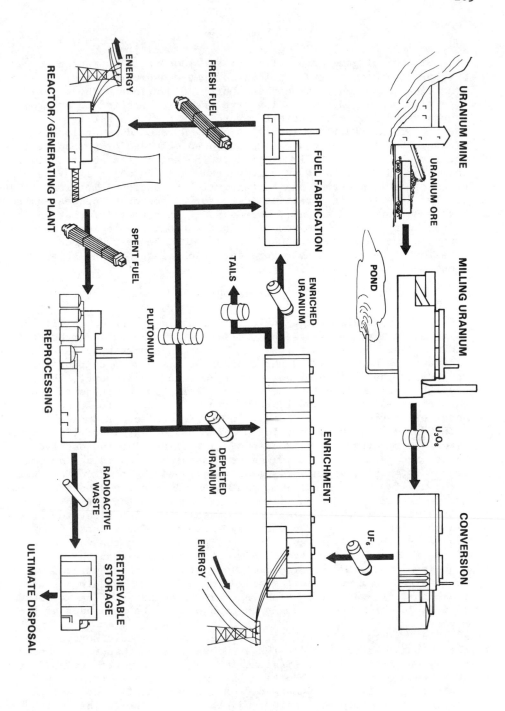

FIGURE 1. NUCLEAR FUEL LOGISTICS

204

SUMMARY

Electrical generation planners are looking to an increased use
of central nuclear power stations to meet consumer demands. Studies
are being conducted to assess the merits of various methods for
effectively managing the radioactive residues expected to be produced
by projected growth of nuclear power. The use of a Retrievable
Surface Storage Facility (RSSF) is being developed as an interim
management method while numerous advanced methods for permanent disposal are being thoroughly examined. The scientific and technical
community can contribute to solving this problem which is extremely
technically interesting and certainly in the National interest. Certainly not all of the creative ideas for disposing of nuclear waste
have been forwarded as yet. Moreover, quantitative information is
needed to provide a firm technical and economic basis for evaluating
concepts currently under study. Developing the best method for
managing high-level wastes represents a challenge to the scientific
and technical community.

BACKGROUND

The components of the nuclear fuel cycle are shown in Figure 1,
which basically represents the cycle for Light Water Power Reactors
(LWRs). The logistics of fuel in this cycle starts with the mining
of uranium, then on to the three processing steps to upgrade the
quality of the fuel (milling, conversion, and enrichment), through
the fabrication of fuel elements and utilization of these to produce
electrical energy, to the reprocessing of the spent fuel and the disposal of reprocessing residues. In the process of producing fuel
either from mining and refining uranium or in recycle of fissionable
materials recovered in reprocessing (see Figure 1), radioactive
wastes are generated. These wastes consist of such things as rags,
sweepings, boxes, piping, filters, spent resins, etc., contaminated
with small amounts of uranium, plutonium and fission products. However, the average concentration of radioactive materials in these
wastes will be low. Similar waste streams are also generated at the
reactor plants and the reprocessing facility. The volume of this
type of non-high-level waste which is generated in the various steps
of the fuel cycle is several hundred times larger than the volume of
high-level waste. Other radioactive waste streams associated with
the nuclear fuel cycle include cladding hulls, used equipment and
the high-level waste that results from reprocessing. The management
of this latter waste stream is the subject of the remainder of the
report.

In the process of producing nuclear power, radioisotopes other
than uranium are produced. These include fission products (essentially the by-product of fissions in uranium and plutonium) and other
actinide elements (by-products of neutron capture in uranium and plutonium). Spent fuel discharged from the reactor is then chemically
reprocessed to recover uranium and plutonium. During this step,

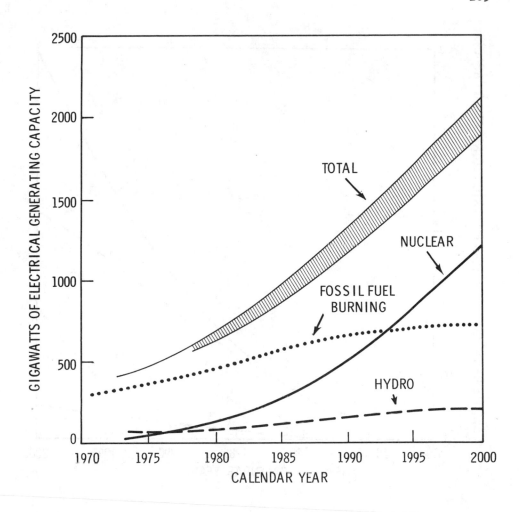

FIGURE 2. PROJECTED U.S. ELECTRICAL GENERATING CAPACITY

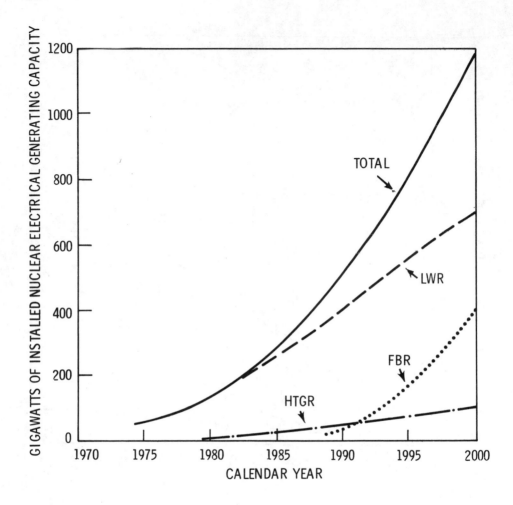

FIGURE 3. PROJECTED U.S. NUCLEAR ELECTRICAL GENERATING
CAPACITY

TABLE I   Constituents of High-Level Liquid Wastes

Constituent	Grams/metric ton (MT) from Reactor Type		
	LWR[a]	HTGR[b]	LMFBR[c]
H	400	3,800	1,300
Fe	1,100	1,500	26,200
Ni	100	400	3,300
Cr	200	300	6,700
Si	---	200	---
Li	---	150	---
B	---	1,000	---
Mo	---	50	---
Al	---	6,400	---
Cu	---	50	---
BO_3	---	---	98,000
NO_3	65,000	435,300	293,600
PO_4	900	---	---
SO_4	---	1,100	---
F	---	1,900	---
Total (Reprocessing Chemicals and Corrosion Products	68,500	452,000	379,300
U[d][e]	4,780	250	4,300
Pu[d][e]	50	1,040	470
Th[d][e]	<.01	4,200	<.01
Np[d]	480	1,440	260
Am[d]	140	30	1,250
Cm[d]	40	10	50
Other Actinides[d]	<.001	20	<.001
	5,500	7,000	6,300
Total Fission[f] Products	28,800	79,400	33,000
Total	102,800	538,400	418,600

(a) U-235 enriched PWR, using 378 liters of aqueous waste per metric ton, 33,000 MWd/MT exposure.

(b) Combined waste from separate reprocessing of "fresh" fuel and fertile particles, using 3,785 liters of aqueous waste per metric ton, 94,200 MWd/MT exposure.

(c) Mixed core and blanket, with boron as soluble poison, 10% of cladding dissolved, 1,249 liters per metric ton, 37,100 MWd/MT average equilivant exposure.

(d) At time of reprocessing.

(e) Assumes 0.5% product loss to waste.

(f) Noble gas and tritium fission products excluded.

high-level waste is formed as an acidic aqueous stream[2]. (High-level
waste is defined in Appendix F, 10 CFR 50,(Reference 2), as the
aqueous waste resulting from the operation of the first cycle solvent
extraction system, or equivalent, and the concentrated waste of subsequent extraction cycles, or equivalent, in a facility for reprocessing irradiated reactor fuels.) This high-level waste contains
most of the reactor-produced fission products and actinides except
the uranium and plutonium that are separated during reprocessing.
The high-level waste from the reprocessing of the irradiated fuels is
the most significant waste material from the standpoint of hazard and
difficulty of disposal since it contains essentially all of the
fission products and a fraction of the fuel materials. This high-
level waste generates sufficient heat to require substantial cooling.
It emits large amounts of potentially hazardous ionizing radiation
and it must be carefully isolated or contained for thousands of years
to prevent significant quantities of the more highly toxic radionuclides from entering man's environment. High-level nuclear wastes
are categorized as long-lived and short-lived. The short-lived are
defined as those with half-lives of tens of years or less, whereas
the long-lived are those with half-lives of thousands or more years.
With reprocessing, high-level waste management begins. Under the
terms of present Federal policy, liquid high-level waste from fuel
reprocessing must be converted to a stable solid material within five
years after separation in the fuel reprocessing step, and be encapsulated and shipped to a Federal repository within 10 years of its
production for long-term management by the AEC.

HIGH-LEVEL WASTE PROJECTIONS

Forecasts (3,4,5) of installed electrical generation capacity
for the United States show an increasing dependence on nuclear power.
As shown in Figure 2, installed U.S. nuclear electrical generating
capacity is projected to increase from about 15,000 megawatts in 1973
to about 1,200,000 megawatts by the year 2000. The power reactors
expected to be used include Light Water Reactors (LWRs), High-Temperature Gas Cooled Reactors (HTGRs) and Fast Breeder Reactors (FBRs)
principally the liquid metal cooled type. The nuclear projection by
reactor type is shown in Figure 3. As shown, the LWRs are the major
reactor type expected to be producing power.
The irradiated or "spent" fuel discharged from these reactors
will be chemically processed at fuel reprocessing plants to recover
the significant amount of valuable "unspent" fissile and fertile
materials for recycle. As an example, irradiated enriched uranium
Light Water Reactor (LWR) fuels typically contain about 30 percent of
the original fissile uranium and the fissile plutonium that is produced by reactor irradiation. Typical constituents of high-level
liquid wastes from the reprocessing of irradiated fuels from LWR,
LMFBR, and HTGR plants are shown in Table I. This table was developed for aqueous acidic wastes from first-cycle solvent extraction
where the addition of chemicals that could be troublesome in subsequent solidification processes is minimized. Troublesome chemicals
include the water soluble, volatile, or corrosive species or those

that result in segregation or phase separation during solidification.

Under terms of present policy[2], this high-level waste must be solidified prior to shipment to a Federal repository. The amount of solidified waste expected to be generated is shown in Figure 4. By the year 2000, about 480,000 cubic feet of solidified high-level waste will be accumulated. This amount of material will generate approximately 600 megawatts of decay heat from some 140,000 mega-curies of radioactive residue. The heat will decay to essentially insignificant levels within the first 1000 years; however, the presence of long-lived isotopes extends the safety aspects of waste management into a much longer period of time--perhaps as much as a million years.

WASTE MANAGEMENT PLANS

The near-term waste management plan that has been adopted by the AEC for high-level waste calls for AEC receiving and managing these wastes in retrievable and monitorable storage facilities until one or more of methods for ultimate disposal is selected and developed. Permanent management methods will be developed from various concepts now under study. The feasibility, technical status, safety analysis, development requirements, schedules and costs are being projected to properly assess each concept and provide data for selection of the most promising concepts.

Partitioning prior to disposal is a key element in certain of the conceptual waste management systems. By dividing the high-level waste into two fractions, one in which the major content of radio-actively toxic materials will diminish to very low levels in about a thousand years and the other, much smaller in quantity and heat generation rate, but containing long-lived materials, a substantial increase occurs in waste management options. The short-lived fractions would then decay to become radioactively non-toxic in relative-ly short times--times short enough to perhaps consider long-term storage such as in manmade structures. The long-lived fraction could be considered for treatment by other management systems. To produce a short-lived waste fraction which would decay to negligible radio-active toxicity in about 1,000 years would require removal of the actinide elements, samarium, technetium, tin, iodine, and nickel (radioactive nickel present due to dissolution of some non-core com-ponents).

Solidification of High-Level Waste

Present Federal regulations require that the liquid high-level waste from fuel reprocessing be converted to a solid material and be encapsulated prior to shipping to a Federal repository for long-term management by the AEC. The solidified high-level waste is assumed to be encased in steel canisters typically 12 inches (30 centimeters) in diameter and 10 feet (300 centimeters) long, each container holding 6.3 cubic feet of solid waste. The projected and annual accumulation

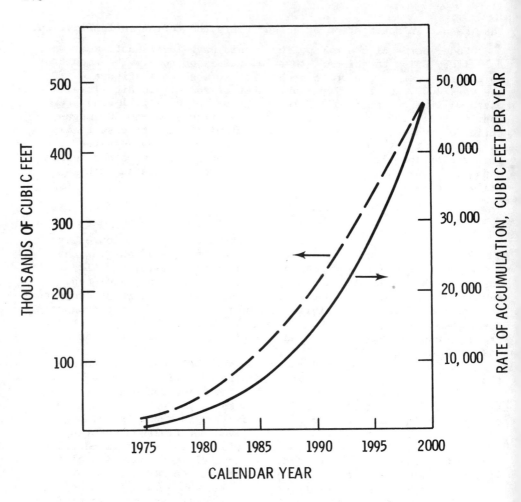

FIGURE 4. ACCUMULATED VOLUME OF HIGH-LEVEL WASTE

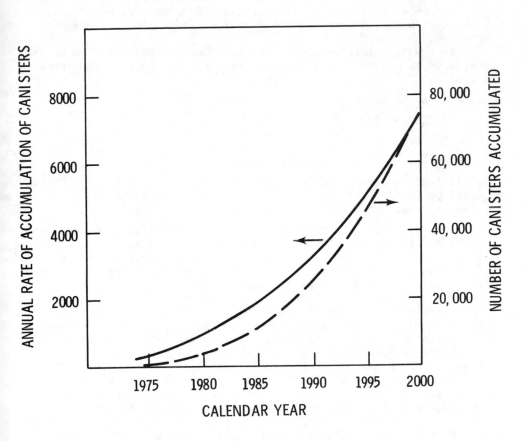

FIGURE 5. VOLUME OF SOLID HIGH-LEVEL WASTE

rates of volumes of solidified waste contained in canisters of this size are shown in Figure 5. By the year 2000, the volume of solid high-level waste in terms of the number of canisters will number about 75,000.

Four solidification processes have been developed in the United States to the point of radioactive demonstration on an engineering scale:

- Spray Solidification

- Fluidized Bed Calcination

- Pot Calcination

- Phosphate Glass Solidification

In all four processes, heat is applied to drive off volatile constituents, primarily water and nitrates, resulting in either a calcined solid or a melt that will cool to a monolithic solid.

The Waste Fixation Program (WFP)[8], (an AEC Program in progress at the Pacific Northwest Laboratory to develop and demonstrate solidification of high-level waste), has as its chief goal to provide technology for reprocessors by developing and evaluating final waste forms and developing appropriate waste solidification technology. Developed systems will be taken all the way through a radioactive demonstration phase. Solid waste forms from these demonstrations will be studied to determine the effects of time and environment. The current emphasis of the WFP is to provide early solidification technology by working with borosilicate glass or ceramic systems. As these solids have had the greatest development effort on a worldwide basis, development of acceptable systems to produce the solids should be near-term. The borosilicate solids will offer a vast improvement in waste management safety over liquids or calcined solids. In an effort parallel to the borosilicate solid development studies are aimed at determining and developing a waste form with even better confinement properties. An example of this would be a multiple-barrier material. This could involve covering small pieces of solid waste with a protective coating. The coated solids could then be dispersed in a protective matrix. Further protection could be provided by outer wrappers.

Retrievable Surface Storage Facility (RSSF)

For retrievable and monitorable storage of solidified high-level waste in a surface facility, several alternative RSSF concepts based on the enclosed basin or vault type of storage with air or water cooling of the waste are being developed by the Atlantic Richfield Hanford Company for the AEC[9,10]. For sake of providing an overview of high-level waste management, a brief summary of these concepts is given here.

The Retrievable Surface Storage Facility (RSSF) will be comprised of facilities for receiving and inspecting packaged wastes from fuel processors and facilities for safely storing these wastes. The facility will be designed to hold safely, for at least 100 years, if necessary, all of the commercial high-level waste produced in the United States through the year 2000.

Basically three RSSF concepts are under consideration by the AEC, the Water Basin Concept, the Sealed Cask Concept, and the Air-Cooled Vault Concept.

Water Basin RSSF

In the Water Basin Concept, the canisters are stored in water-filled stainless-steel-lined concrete basins. The concept consists of three major elements, namely: the waste receiving and handling facility; the storage facility, a series of water-filled concrete basins in which the wastes would be placed for cooling and long-term surveillance; and the heat rejection facilities. For the latter, a series of forced-draft cooling towers is associated with the basins, from which the waste heat would be dissipated to the atmosphere. As in all the concepts, there would also be support facilities and services. Figure 6 shows a conceptual layout of the water-cooled basin concept. Modular construction of the actual storage area would be planned so that the storage capacity would keep pace with the waste expected to be delivered to a Federal repository. Water in the basin would serve as a heat sink in case of temporary failure of mechanical cooling equipment and the water would provide both radiation shielding and a confinement barrier.

Sealed Storage Cask RSSF

In the Sealed Storage Cask Concept, the canisters would be sealed in steel casks which would be stored outdoors on concrete pads inside concrete neutron shields as shown in Figure 7 for the thick wall storage unit and in Figure 8 for the medium wall storage unit. Heat would be dissipated from the casks by natural convection air flow through the annulus between the cask and the neutron shield. The three major elements of the concept are the waste receiving and handling facility; the storage cask welding and testing facility; and the outdoor waste cask storage areas. The storage area would be surfaced with crushed rock as required to prevent wind erosion and provide access for the transport vehicle to all the storage cask locations. Area monitors and samplers would be provided to detect any radioactivity above the normal background level. A security fence would enclose the area to provide isolation and prevent inadvertent entry by personnel.

Air-Cooled Vault RSSF

In the air-cooled vault concept, the waste canisters would be sealed by welding them inside of another carbon steel container. This assembly is then placed inside concrete vaults to be cooled by natural draft convection as illustrated in Figure 9. The three major elements of the concept are the waste receiving and handling facility; the welding and testing facilities, and the canister storage cells. Expansion of facilities for handling and encapsulating (overpack)

214

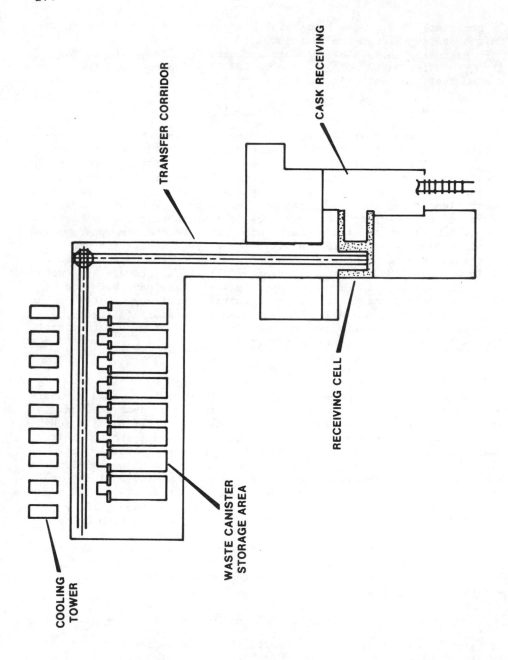

COOLING TOWER

WASTE CANISTER STORAGE AREA

TRANSFER CORRIDOR

CASK RECEIVING

RECEIVING CELL

FIGURE 6. WATER BASIN CONCEPT - FACILITY LAYOUT

FIGURE 7. STORAGE AREA - THICK WALL SEALED STORAGE CASK CONCEPT

216

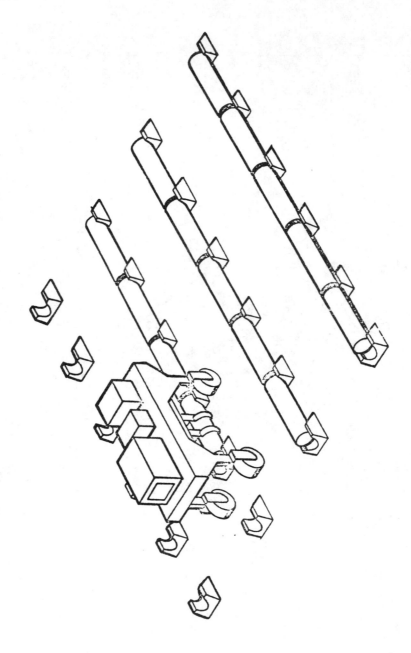

FIGURE 8. STORAGE AREA - MEDIUM WALL SEALED STORAGE CASK CONCEPT

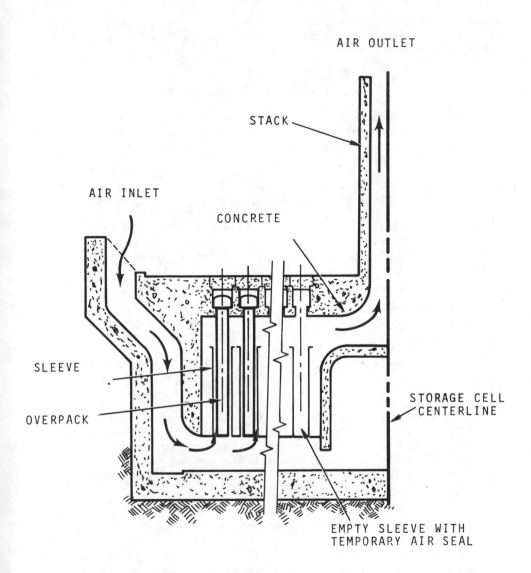

FIGURE 9. AIR-COOLED CONCEPT

would be required periodically. The receiving and welding stations could be expanded in modular components. No continuous mechanical cooling systems or emergency backup facilities would have to be provided.

Potential Future Alternatives

When the AEC made its decision to use the RSSF technique for managing the commercial high-level radioactive waste, it recognized that while this approach is simple, straightforward and safe, it does impose a long-term requirement for human surveillance and maintenance. It was noted that if management techniques can be developed which are equally safe and which would eliminate, or at least minimize this human action requirement, they should be used. For this reason, the AEC undertook a rather extensive program to identify, evaluate and possibly demonstrate feasible alternative disposal techniques for later use. Studies of disposal in bedded salt formations have been and are being conducted by Oak Ridge National Laboratory for the AEC[11,12] and studies of other alternatives for managing high-level waste are being studied by Battelle Pacific Northwest Laboratories (Battelle-Northwest) for the AEC[13]. A brief description of the bedded salt concept is given here to aid in the overview of high-level waste disposal. In the next section the advanced waste studies being conducted by Battelle-Northwest are reviewed.

As shown in Figure 10, there are fairly large deposits of rock salt in the U.S. The bedded salt concept is based upon the principle of isolating the high-level wastes in a stable underground salt formation. The handling process and emplacement operations envisioned is shown in Figure 11. Studies are underway to evaluate and demonstrate safe, competent, receiving, handling, emplacement, and retrieval operations as well as providing means to demonstrate the adequacy of analytical techniques used to predict the long-term stability of salt beds when they contain heat-producing waste.

Advanced Waste Studies Conducted by Battelle-Northwest

In addition to the RSSF and bedded salt disposal programs, the AEC has commissioned Battelle-Northwest to make an evaluation of all other potentially attractive disposal concepts. The purpose of this evaluation is to identify feasible and potentially feasible long-term waste management systems and their components, evaluate the safety of these systems, identify the research and development necessary for their establishment, and estimate the schedule and costs associated with selected systems. A synopsis of this work has been published previously[14] and details of these studies will be published in topical reports to be issued later this year[15].

Three basic types of waste management concepts are under study: (1) disposal on earth, (2) conversion by nuclear processes called transmutation, and (3) disposal in space. The earth disposal concepts involve use of geologic formations, ice sheets and the seabed. The

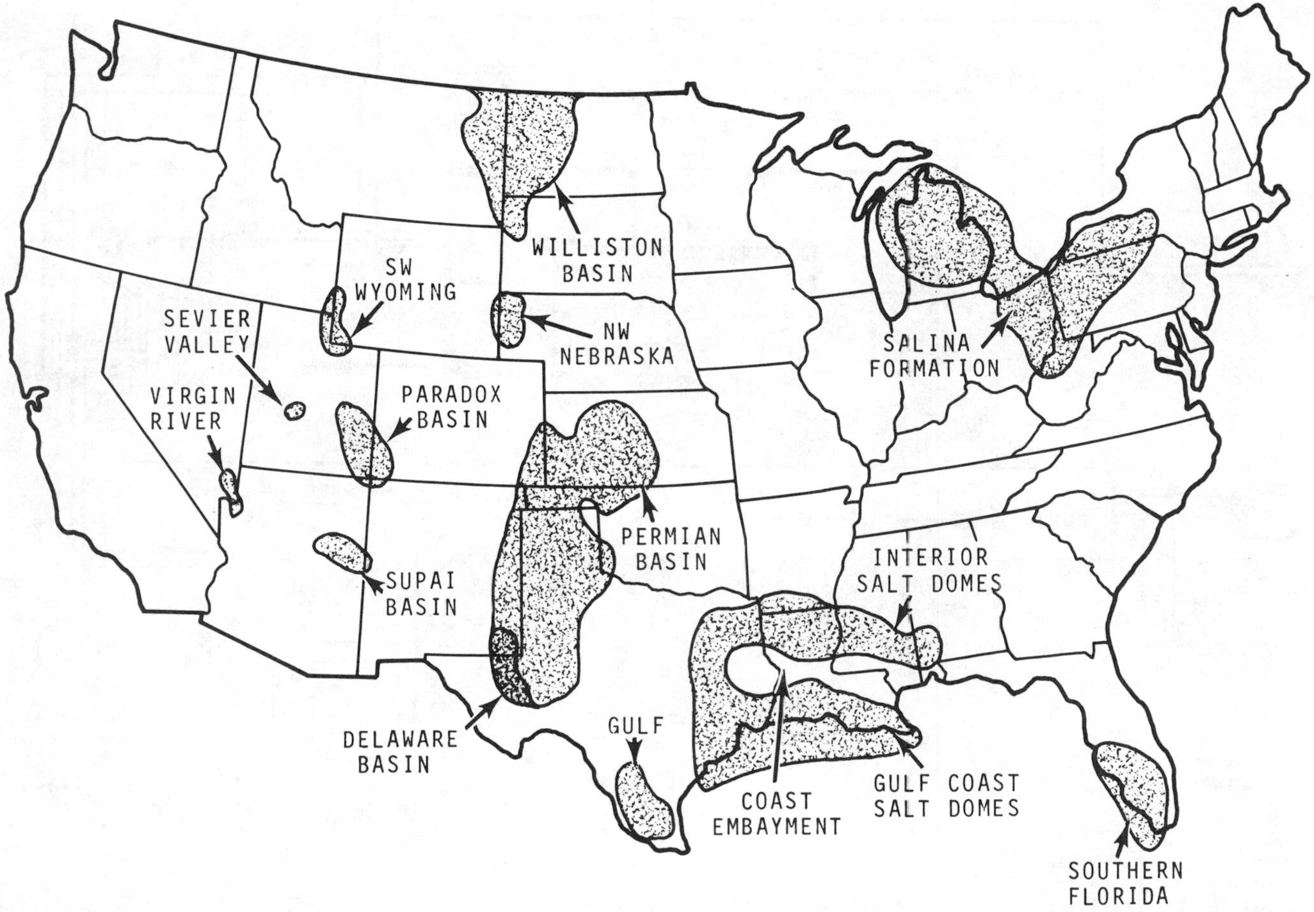
WILLISTON
BASIN
SW
WYOMING
NW
NEBRASKA
SEVIER
VALLEY
VIRGIN
RIVER
PARADOX
BASIN
SUPAI
BASIN
PERMIAN
BASIN
DELAWARE
BASIN
GULF
COAST
EMBAYMENT
GULF COAST
SALT DOMES
INTERIOR
SALT DOMES
SALINA
FORMATION
SOUTHERN
FLORIDA

FIGURE 10. ROCK SALT DEPOSITS IN THE UNITED STATES (AFTER PIERCE AND RICH, U.S.G.S. BULL 1148)

220

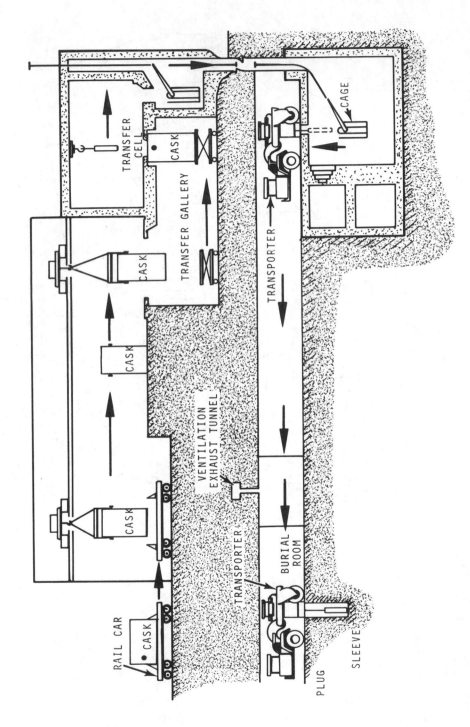

FIGURE 11. HANDLING PROCESS IN BEDDED-SALT DISPOSAL

space disposal concept involves transporting waste to various orbits
or trajectories in space. Transmutation involves elimination of some
of the more offensive waste nuclides by nuclear transition. Alternatives within these categories are listed in Table II.

Study Methodology

Briefly presented here is the overall analytical system methodology by which each disposal concept and its waste management system
elements are being studied. The factors upon which the evaluations
are being made is graphically illustrated in Figure 12.

The technical feasibility of the potential disposal concepts is
being determined in this study by answering the following questions:

1. Can the disposal concept be implemented using today's technology?
(This does not imply that additional development is not necessary to
adapt existing scientific and engineering technology to these disposal
concepts.)

2. Can the disposal concept be implemented with future technology
based upon current theory? (Is it theoretically possible?)

3. Will the disposal concept provide adequate safety for the time
period of concern? (Truly quantified answers to this point require
very extensive study, and only qualitative indications are being developed for this study using currently available data.)

4. Does the concept have a favorable energy balance? (Is the energy
consumed in the implementation of the disposal concept sufficiently
less than the electrical energy obtained from the nuclear fuel represented by the waste?)

Estimates are being made of the research and development time
and expenditures necessary for solution of the technology needs. The
date by which the concept could be in operation is also being estimated.

Capital and operating costs are being estimated, using the basic
assumption that the necessary research and development had been
successfully completed. Major legal constraints (i.e., policy con-
flicts) and expected or potential major environmental impacts are
also identified. Social psychologists are developing techniques to
measure the public's perception of the safety elements of (i.e., risk)
the various concepts.

The evaluation proceeds as follows. Since the results of these
elements are calculated in completely different units, simply adding
up the performance level by element does not necessarily lead to consistent information. Recognizing this, the technique being used is
simply one of overcoming obstacles to performance. For instance, the
technical feasibility obstacle would be of the "yes-no" type. Here
"yes" is required before analysis of R & D requirements would be
undertaken. In this way, the timing for availability of a given

TABLE II Concepts Under Study for High-Level Radioactive Waste Management

DISPOSAL

Geologic Formations

    Mined Cavity

    Nuclear Cavity

    Deep Hole

    Drilled Hole Matrix

    Manmade Structures in Geologic
      Formations

Seabed

    Stable Deep Sea Floor

    Tectonic Subsidence Areas

    Deep Trenches Other Than
      Subsidence Areas

    Rapid Sedimentation Burial

Ice Sheet

    Ice Burial - Free Flow

    Ice Burial - Anchored

    Ice Surface Facility

Extraterrestrial

    Solar Impact

    Orbiting

    Solar Escape to Deep Space

ELIMINATION

Transmutation

    Accelerator

    Fission Reactor

    Fission & Thermonuclear Explosives

    Controlled Thermonuclear Reactor
      (Fusion Reactor)

PROCESSING

Partitioning

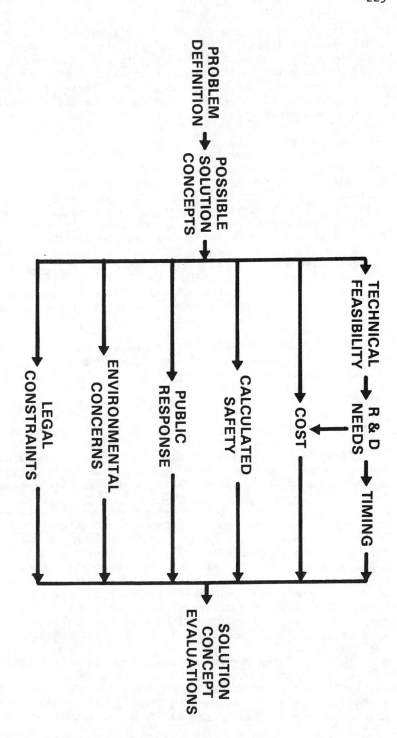

FIGURE 12. EVALUATION FACTORS

technology could be ascertained. Once the feasibility of the technology is established, a preliminary analysis of system safety is
made using the model outlined in Figure 13. Safety is, and has to be,
a major consideration in decisions on the use of any potential disposal scheme. An acceptable option must provide adequate protection
during operational phases and provide the necessary isolation during
the disposal phase. For this study "safety" is equated directly to
the potential risk to man that could result if the disposal option
was implemented.

As shown in Figure 13, the overall matrix starts with defining
the general characteristics of the disposal concept. The next step
is evaluation of the most likely sequences of failure events leading
to release of radioactive materials to man's environment and determination of the probability of these sequences taking place. The
next step follows the most likely sequences through the physical and
chemical processes required to release the waste constituents into
man's immediate environment. The characteristics of the waste must
be dealt with parametrically at the time the critical event takes
place. The generic site defines the media (granite, salt, shale,
soil, air, water, etc.) through which radionuclides must move. Finally, based on the population as indicated for the generic site and
the calculated release rate, the dose to the surrounding population
can be estimated.

The probabilistic risk to man can be determined by multiplying
the probability of the event taking place times the radiological dose
if the event happened. By comparing each of these doses with appropriate criteria, it can be determined whether or not the risk to man
exceeds acceptable criteria. If the risk level is unacceptably high,
changes could be made in the concept to improve the level of risk.
If the risk for a concept meets all criteria, the concept will be
considered to have met the safety requirements.

Analysis of system cost being made for this study considers such
costs important only if they would result in major changes in the
nuclear fuel cycle and hence alter the nuclear waste management system.

Implementation of some of the considered concepts would conflict
with present policies--such as disposal of radioactive materials in
the ocean. This policy and similar ones which are either parts of
the Federal Code of Regulations or of International agreements are
identified so that problems involved with policy changes can be
weighed against the safety and economic potentials of a particular
waste management concept. The environmental impacts, aside from the
potential release of radioactive materials, are not expected to control concept selection but will be important factors in detailed site
selection. General impacts such as land, sea or water use are listed
for evaluation.

The final area for evaluation is the potential public response
to a chosen waste management scheme. Obviously this is a nontechni-
cal subject and most difficult to evaluate. An initial study of
methodology is being designed to identify those aspects of the waste
management systems that might be deemed most important by the general

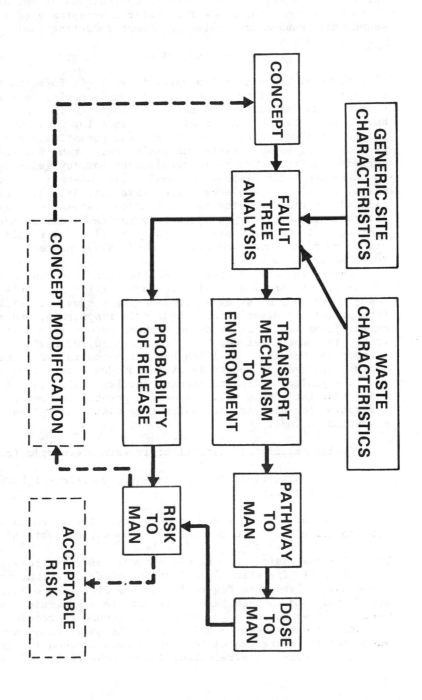

FIGURE 13. SYSTEM SAFETY EVALUATION MODEL

public. After such analysis, adequate information could be made
available on these points so that the public could better understand
and state it's opinions on overall implications of the alternative
waste management concepts. The public's acceptance of a technically
sound waste management system is a most important goal.

Geologic Disposal Concepts

Disposal of radioactive waste in geologic formations has the
potential of isolating the waste from man's environment for extended
time periods (millions of years). Geologic environments exist which
have been physically and chemically stable for millions of years, are
isolated from man's environment, and can potentially provide effective barriers between waste and man's environment for the time
periods required. The basic requirement for any geologic environment
to be suitable for disposal of radioactive waste is the capability to
safely contain the emplaced radioactive material until decay has
reduced the radioactivity to nonhazardous levels. The geologic
environment should (a) be adequately far removed from man's environment, (b) not permit waste transport readily, (c) remain relatively
stable over geologic time periods, and (d) adequately contain a highly immobile waste form.

There are several ways a geologic formation can be penetrated
and altered to provide a suitable cavity for waste emplacement purposes. The present study is considering the use of drilling, mining
(mechanical and dissolution), hydraulic fracturing, and nuclear
cavity formation. All of these methods become more difficult with
increasing depth. At depths up to about 10,000 feet, any of the
methods may be used. Drilling has the potential of going to great
depths; the present record is around 30,000 feet (about 5.6 miles).

The geologic disposal studies involve evaluation of concepts
other than the bedded salt disposal concept. The methods under consideration for disposal of radioactive waste in a given geologic
formation include:

1. Placing solidified waste directly into a geologic formation.

2. Placing solidified waste in man-made containment barriers within
a geologic formation.

3. Placing solidified waste in a geologic formation in a configuration to allow the waste to melt and form a rock-waste matrix.

Each of these basic concepts has a number of variations.

Disposal of previously solidified waste in a conventionally
mined cavity, shown in Figure 14, is one of the more basic concepts
under study and is used here to illustrate a geologic disposal alternative. It would use a building above ground to receive the waste
canisters and transfer them down into the underground area. The
waste canisters are placed in storage pods located in lined tunnels.
The storage pods are air-cooled, though other means of cooling appear

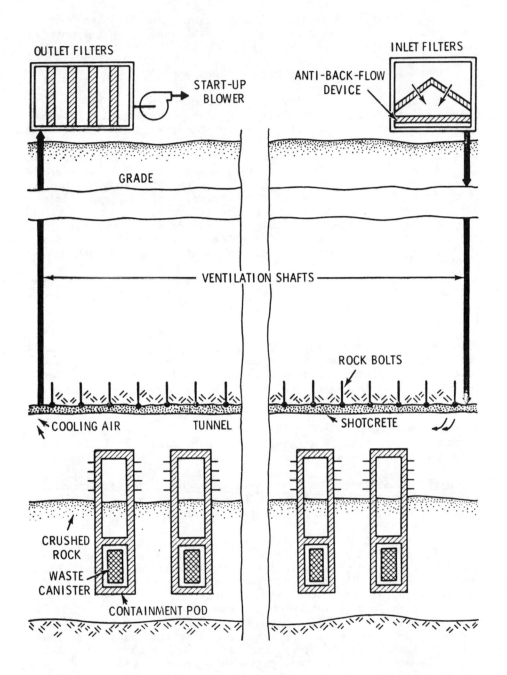

FIGURE 14. CONCEPT FOR SOLID WASTE EMPLACEMENT IN A MINED
TUNNEL WITH NATURAL CONVECTION AIR COOLING

feasible. After an appropriate time period (generally tens of years),
the cooling system can be shut down and the repository permanently
sealed.

Seabed Disposal Concepts

Disposal of high-level waste within the floors of the world's
oceans offers another possibility for permanent isolation of the
waste from man's environment. The depth to the floor or seabed would
provide isolation and safety from natural disasters such as storms,
as well as from sabotage or accidental disturbance. The large volume
of seawater could help cool the waste and effectively dilute any
material that escaped from the disposal site. The known high ion exchange capacity of the seabed sediments would aid in immobilizing
waste material if any waste escape should occur.

A number of seabed disposal concepts are being evaluated but all
are basically the same except for the site. The following geologically distinct types of sites on the seabed are being considered:

1. Stable Deep Sea Floor--areas such as deep sea basins and abyssal
plains and hills, which are considered geologically stable. The waste
would be placed in the seabed below the unconsolidated sedimentary
cover.

2. Subduction Zones--areas where, according to crustal plate tec-
tonics theory, one edge of certain crustal plates is moving under
other crustal plates and down into the earth's mantle. The waste
would be placed in these areas to be carried down, or subducted, into
the earth's mantle with the crustal plate.

3. Deep Sea Trenches other than Subduction Zones--areas where deep
trenches occur in the sea floor. The waste would be placed in the
dense seabed at the bottom of these trenches.

4. High Sedimentation Rate Areas--areas where major rivers are
building deltas into the ocean. The waste would be placed in the
seabed below the accumulating deltaic sediments.

The radioactive waste would be in a solid form and enclosed in
a durable sealed canister. To further isolate wastes, the canisters
would be placed in prepared holes in the seabed, after which the holes
would be sealed. The depth of the prepared holes would depend on the
nature of the seabed at the disposal site.

A schematic description of the seabed concept is shown in Figure
15. The previously solidified and canned bulk waste from the reprocessing plant would be transported in protective casks to special
ports of embarkation for inspection and possible short-term storage.
The waste would then be transported in protective casks by ships to
the disposal site where a number of waste canisters would be placed
in each pre-drilled hole in the basement rock from a special drilling
platform and the upper section of each hole would be filled with a
sealant.

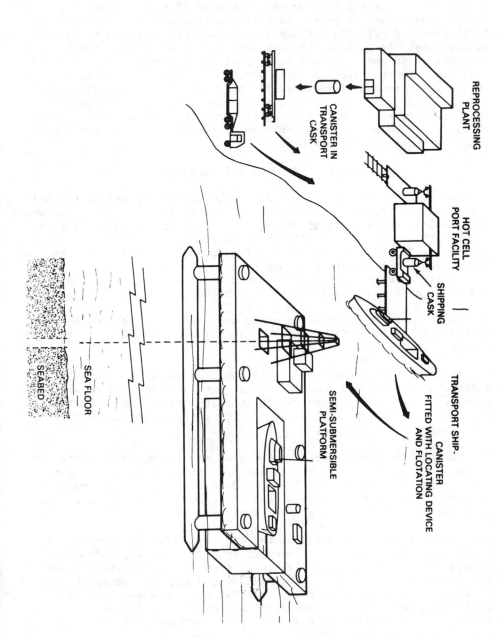

FIGURE 15. CONCEPT FOR SEABED DISPOSAL OF SOLIDIFIED WASTE

REPROCESSING PLANT

CANISTER IN TRANSPORT CASK

HOT CELL PORT FACILITY

SHIPPING CASK

TRANSPORT SHIP - CANISTER FITTED WITH LOCATING DEVICE AND FLOTATION

SEMI-SUBMERSIBLE PLATFORM

SEABED

SEA FLOOR

Implementation of the seabed disposal concepts in the stable deep sea areas and in the areas of rapid sedimentation could be attained with today's technology but a number of years would be required for development to prove the safety of the concept. Significant development of drilling and emplacement technology would be required to implement disposal in the very deep sea areas of the trenches and the subduction zones. Final sealing of the disposal holes to maintain isolation for the long time periods of concern would need to be tested (and improved if necessary) for radioactive waste disposal.

Disposal of radioactive waste in the seabed has the potential for isolating waste from man's environment for periods in the order of millions of years, depending upon confirmation of inferred knowledge by future seabed exploration.

Ice Sheet Disposal Concepts

Alternative concepts for radioactive waste disposal in the major ice sheets of the world (Greenland and Antarctica) are being evaluated. Potential advantages are great thicknesses of ice, remoteness from man's activities, and low likelihood for future development. The ice could provide effective direct cooling for the waste and, at the same time, maintain isolation from man's environment.

Three potential disposal concepts are being evaluated for the ice sheet areas such as Antarctica or Greenland.

1. Meltdown or Free Flow--the waste canister would be placed in an individual shallow drilled hole in the ice and allowed to melt down through the ice sheet to bedrock.

2. Anchored Emplacement--the waste canister would be placed in an individual shallow drilled hole in the ice but connected to surface anchors by cables or chains, which stop its descent and maintain its position (500 to 1500 feet below the ice surface) for up to about 100 years.

3. Surface Storage/Disposal--the waste canister would be placed in a temporary hot cell type of storage facility with jack-up piers on the ice sheet surface. After about 50 years, the facility would be allowed to become covered by accumulating snow and would be eventually buried in the ice sheet for final disposal.

A schematic description of the ice sheet concept is shown in Figure 16. It consists of transporting previously solidified and canned bulk waste in protective casks from the reprocessing plant to special embarkation ports. The waste would then be transported in protective casks by ships to the edge of the ice sheet when the waste canisters and casks would be off loaded to a debarkation facility near the edge of the continent. Surface vehicles would provide over-ice transport to the disposal site.

The implementation of all ice sheet disposal concepts could be done with today's technology, but a number of years would be required for development to prove out the safety of any concept. The

231

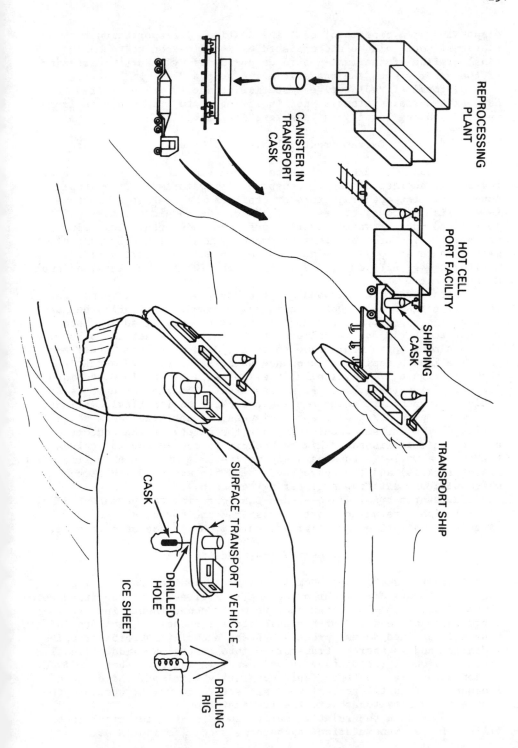

REPROCESSING PLANT

CANISTER IN TRANSPORT CASK

HOT CELL PORT FACILITY

SHIPPING CASK

TRANSPORT SHIP

SURFACE TRANSPORT VEHICLE

CASK

DRILLED HOLE

DRILLING RIG

ICE SHEET

disposal system aspects of containerization, transportation, and emplacement could all be accomplished by modification of technology. Final sealing of the waste could be performed by natural refreezing of the water around the waste in all concepts.

Disposal of radioactive waste in ice sheets is considered to have an uncertain potential for isolating waste from man's environment, depending largely on long-term ice stability.

Extraterrestrial Disposal Concepts

Disposal of radioactive waste by removing it from the earth with rockets is another disposal concept being evaluated. If a stable non-earth intercept trajectory or orbit can be guaranteed, extraterrestrial disposal offers a method for the complete removal of long-lived nuclear waste constituents from the earth. The primary unfavorable features are that the concept deals with only part of the waste, there are possible launch safety problems, retrievability and monitoring are difficult, and there is possibility for International disagreement.

Extraterrestrial disposal of the total waste constituents and of only the actinides are both considered. However, primarily because of the high space transport cost per unit of weight, space disposal of just the actinides is believed to be the most practical scheme. The remaining waste would have to be disposed of by some other means.

The launch deployment sequence using a shuttle and a tug is shown in Figure 17. Typically, the shuttle would be launched into a low circular earth orbit. From this orbit, the tugs or upper stage(s) would be launched to carry the waste package to its final destination.

The implementation of space disposal of actinide waste could be achieved with current technology, but the safety of the concept cannot yet be established. This technology is considered to include the space shuttle (with separately retrievable and reusable lift-off-assist rockets) and the space tug, which are advanced vehicles but which will use existing engineering technology.

Extraterrestrial disposal has the potential for permanent removal of radioactive waste constituents from the earth, depending largely on incentives and improved knowledge of deep space travel.

Transmutation Concepts

Another possible approach to the management of radioactive waste is the use of nuclear processes themselves to change (transmute) the hazardous long-lived radioactive waste constituents into short-lived radioactive waste or nonradioactive isotopes. Transmutation is generally defined as any process whereby a nuclide absorbs or emits radiation and is thereby transformed into another nuclide. Ideally transmutation of radioactive constituents in waste to shorter-lived or nonradioactive isotopes could completely eliminate the noxious isotopes. It is theoretically possible through use of nuclear processes themselves to achieve the transmutation.

To establish the relative merits and specific technical feasibility of the transmutation approaches, special criteria were

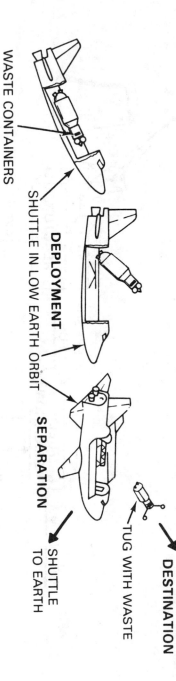

FIGURE 17. CONCEPT FOR EXTRATERRESTRIAL DISPOSAL OF SOLIDIFIED WASTE

LAUNCH

SHUTTLE, WITH TUG AND PAYLOAD

SOLID FUEL ROCKET MOTORS

EXPENDABLE EXTERNAL PROPELLENT TANK

WASTE CONTAINERS

DEPLOYMENT

SHUTTLE IN LOW EARTH ORBIT

SEPARATION

SHUTTLE TO EARTH

TUG WITH WASTE

TO FINAL SPACE DESTINATION

234

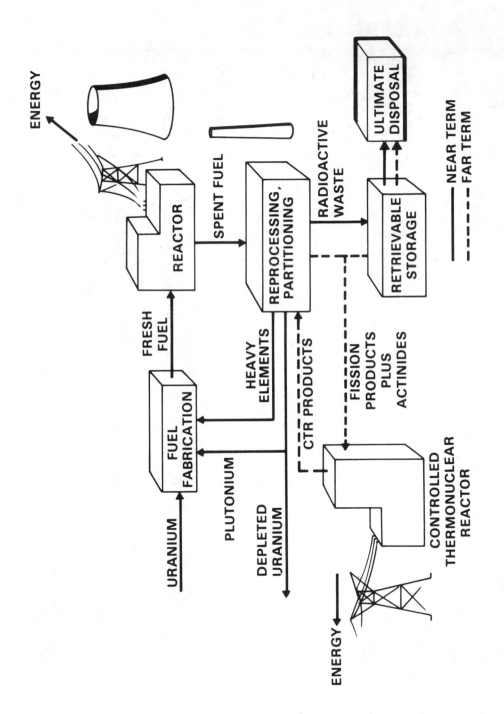

ENERGY

SPENT FUEL

REACTOR

REPROCESSING, PARTITIONING

RADIOACTIVE WASTE

RETRIEVABLE STORAGE

ULTIMATE DISPOSAL

NEAR TERM
FAR TERM

FRESH FUEL

HEAVY ELEMENTS

CTR PRODUCTS

FISSION PRODUCTS PLUS ACTINIDES

FUEL FABRICATION

PLUTONIUM

DEPLETED URANIUM

URANIUM

CONTROLLED THERMONUCLEAR REACTOR

ENERGY

TABLE III   Summary of Transmutation Device Feasibility

	Technically Feasible for Transmutation			
	Fission Products			Actinides
Device	Category 3*	Category 2*	Category 1*	Category 1
Accelerators				
• Electron Accelerator	No	No	No	No
• Proton Accelerator	No	No	No	No
• Spallation Accelerator	No	No	Possibly	No
Thermonuclear Explosives	No	No	No	No
Fission Reactors	No	No	No	Yes
Fusion Reactors	No	Possibly	Yes	Yes

*Category 1:   Storage required for >1000 years

Category 2:   Storage required for 100-1000 years

Category 3:   Storage required for <100 years

developed and applied which are unique to transmutation. These include: overall waste balance, specific transmutation rate, and total transmutation rate.

The results of the evaluation for the various transmutation alternatives are summarized in Table III. The accelerator devices failed to meet the criteria for transmutation for essentially all categories of radioactive waste. The only possible exception is the use of a spallation neutron source for transmutation of long-lived fission products. Likewise, use of neutrons from a thermonuclear explosion does not appear technically feasible. The use of fission and fusion reactors met the selection criteria for transmutation of actinides. Fusion reactors also may transmute selected fission products.

The transmutation concept of continual recycle of actinides in fission reactors appears to be particularly attractive. Calculations by Claiborne[16] at Oak Ridge National Laboratory, by Kubo[17] and Kubo and Rose[18] at the Massachusetts Institute of Technology, and at Battelle-Northwest indicate that significant reductions are possible in the cumulative toxicity index of actinides. (Toxicity index is defined as the amount of air or water required to dilute the present amount of a given isotope to levels defined in the Code of Federal Regulations, 10 CFR - Part 20.) The calculations indicate that using existing separations efficiencies with recycling of actinides in Light Water Power Reactors could achieve an order of magnitude decrease in the short-term actinide toxicity indices and about a factor of fifty decrease in the long-term toxicity index. These reduction factors may be significantly improved by achieving higher separations efficiencies, better optimization of the reactor irradiation, or by recycling in LMFBRs.

The PNL studies of the neutron-induced transmutation of actinides and fission products in the blankets of hypothetical Controlled Thermonuclear Fusion Reactors (CTRs)[19] have demonstrated that reductions of cumulative toxicity index of actinides by a factor of 10 or more below that achievable in fission reactors could be obtained in the high neutron flux levels proposed for CTRs. These studies have also shown that large reductions in the respective toxicity indices are possible for some fission product elements. For others, notably strontium and cesium, the degree of toxicity reduction is minimal, but the calculated values are uncertain by a factor of about two because of uncertainty in nuclear reaction data for these elements. All considerations of radionuclide transmutation in CTRs, of course, presuppose the successful accomplishment of controlled thermonuclear fusion.

Since it is technically feasible to transmute actinides in fission reactors and CTRs and certain fission products in CTRs, these two reactor technologies combine to form a potentially viable long-term (year 2000 or 2010), the actinides separated from the rest of the waste would continue to be recycled in fission reactors with the fission products stored in a retrievable facility. In the long-term, with the advent of CTRs, the fission products would be retrieved from storage and recycled along with the actinides in the CTR. In either strategy, some of the fission products and whatever "heel" of untransmuted waste at the end of this era must be disposed of by other means.

Summary of Advanced High-Level Waste Studies

The overall objective of this study is to prepare a comprehensive overview compendium of information pertinent to the various potential waste disposal techniques. The disposal concepts are being studied on a systematic, generic basis and are developed only to the extent necessary to perform the overall evaluations.

Initial evaluations of technical (or theoretical) feasibility for these advanced waste management concepts show that in the broad category, (i.e., geologic, seabed, ice sheet, extraterrestrial, and transmutation) all meet the criteria for judging feasibility, though certain alternatives within these categories do not (i.e., use of accelerators for transmutation).

Preliminary cost estimates have also been developed for all the principal waste management alternatives being evaluated. The results show that, although many millions of dollars may be required, the cost for even the most exotic concepts are small relative to the total cost of electric power generation. For example, the cost estimates for the disposal on earth concepts are less than 1% of the total generating costs. The cost for actinide transmutation is estimated at around 1% of generation costs, while actinide element disposal in space is less than 5% of generating costs.

Thus neither technical feasibility nor cost seems to be no-go factors in selecting a waste management system. The seabed, ice sheet, and space disposal concepts face legal constraints (i.e., International policies). The information being developed in safety, environmental concern, and public response will be important factors in determining which concepts appear most promising for further development.

CONTRIBUTIONS THE SCIENTIFIC COMMUNITY CAN MAKE

Assessment of waste disposal alternatives and selection of the most effective means of long-term management of high-level wastes requires a quantitative determination of all the evaluation factors outlined above. A quantitative assessment requires the best thinking and methodology that science and engineering can bear on this subject. To illustrate, we would like to use the safety methodology as outlined in Figure 13.

The box shown as acceptable risk precludes determining what the public's response is in relation to waste disposal options. This means that the social scientist must work in concert with the physical scientist to bring the message of the physical scientist to the public in terms of language that our society can understand and to properly evaluate the laities response. If this response is negative, then either the concept must be modified to bring it into line with the public's views or the concept must be rejected since it is an unacceptable method to the public. The social scientist is responsible for providing the tools which can be used to accurately assess the concensus view of the public and not merely the view of a vocal minority or an uninformed public.

The consequences of a release of radioactivity involves many facets which the scientific community can contribute to improving the present knowledge and understanding of. The dose to man involves the interpretation of the health effects of ionizing radiation to man. The physical chemist, the biophysicist, and the health physicist can make important contributions to the present understanding of the impact of various radiation sources on man's health. In disposal of radioactive waste, a spectrum of ionizing radiation sources exist, ranging from isotopes in the actinide element series to the species of fission products. It is important, that the mechanisms of transport, pathways to the environment and man, and the impact on man's health be thoroughly understood to assure public health and safety.

The nuclear physicist, reactor physicist, and physical chemist can contribute to improving the understanding of knowledge of the radiation characteristics of nuclear waste. The nuclear physicist is called upon to provide the best estimate of the nuclear data. He must provide the nuclear chemist the data needed to determine whether chemical and/or physical forms of high-level waste can be maintained for long periods of time. The reactor physicist, working in conjunction with the nuclear physicist, must provide the best estimates of the volumes and characteristics of wastes produced in generating electrical power with nuclear reactors.

Depending upon the concept, experts in the physical and engineering sciences must contribute to improving the knowledge of previously outlined concepts, the characteristics of the site(s) which the concept encompasses, and the known and potential modes of failure of such systems. As examples, the geophysicist must provide the most accurate information of geologic disposal sites, be they formations on land, in the seabed or in ice sheets. Moreover, the geophysicist will be required to provide accurate predictions of the behavior of these potential terrestrial disposal formations for thousands of centuries into the future. By the same token, the physicist working on extraterrestrial deployment of vehicles must provide the best estimates of the time-dependent behavior of vehicles carrying high-level waste into space.

All of science and engineering can contribute by offering ideas or methods for managing high-level waste. Central to assessment of these ideas is determining the advantages and disadvantages of proposed concepts. The fault tree (or failure mode) analysis is an important part of determining the potential advantages and disadvantages. This segment of the analysis requires the input of all kinds of experts, in science and technology, to make an overall adequate assessment.

The management of high-level radioactive waste produced by the nuclear power industry is a problem which is amenable to solution. To assure achieving the optimal solution, the science and engineering community must: (1) become aware of the problem, (2) become involved in developing and evaluating conceptual schemes for disposal, and (3) assist in communicating with the public at large.

240

REFERENCES

1. F. K. Pittman, Plan for the Management of AEC-Generated Radioactive Wastes, USAEC Report WASH-1202, January 1972.

2. Code of Federal Regulations, Appendix F to 10 CFR 50, "Policy Related to the Siting of Commercial Fuel Reprocessing Plants and Related Waste Management Facilities", U.S. Government Printing Office, Washington, D.C., p. 268, January 1, 1972.

3. Federal Power Commission, The 1970 National Power Survey, Part I, U.S. Government Printing Office, Washington, D.C., p. I-1-17, December 1971.

4. Office of Planning and Analysis, USAEC, Nuclear Power 1973-2000, Report WASH-1139 (72), Washington, D.C., December 1, 1972.

5. U.S. Energy Outlook, An Initial Appraisal 1971-1975, A Summary Report of the National Petroleum Council, December 1972.

6. J. P. Nichols et al. Projections of Fuel Reprocessing Requirements and High-Level Solidified Wastes from the U.S. Nuclear Power Industry, USAEC Report ORNL-TM-3965 Draft, to be published.

7. A. M. Platt, Editor, Quarterly Progress Report, Research and Development Activities, Waste Fixation Program, July through November 1972. USAEC Report BNWL-1699, Battelle-Northwest, Richland, WA, p. 9.

8. J. L. McElroy et al. Waste Solidification Program Summary Report, Volume 11, Evaluation of WSEP High-Level Waste Solidification Processes, USAEC Report BNWL-1667, Battelle-Northwest, Richland, WA, July 1972.

9. D. C. Woodrich, "Retrievable Surface Storage of High-Level Radioactive Wastes", Trans. Am. Nucl. Soc., 17, 326 (1973).

10. Retrievable Surface Storage Facility Alternative Concepts Engineering Studies, ARH-288, Atlantic Richfield Hanford Company and Kaiser Engineers, (December 1973).

11. Site Selection Factors for the Bedded Salt Pilot Plant, Staff of the Oak Ridge National Laboratory Salt Mine Repository Project, ORNL-TM-4219, (May 1973).

12. J. O. Blomeke, J. P. Nichols, and W. C. McClain, "Managing Radioactive Wastes", Physics Today, 26, 36 (1973)

13. K. J. Schneider and J. H. Jarrett, "Alternative Means of Ultimate High-Level Waste Management", Trans. Am. Nucl. Soc. 17, 325 (1973).

14. Overview of High-Level Radioactive Waste Management Studies,
    BNWL-1758, (August 1973)

15. A. M. Platt and K. J. Schneider, Editors, Advanced Waste Manage-
    ment Studies: High-Level Waste Disposal Alternatives, in
    preparation for the U.S. Atomic Energy Commission and to be
    published in four volumes.

16. H. C. Claiborne, Neutron-Induced Transmutation of High-Level
    Radioactive Waste, ORNL-TM-3964, Oak Ridge National Laboratory,
    December 1972.

17. A. S. Kubo, Technology Assessment of High-Level Nuclear Waste
    Management, ScD Thesis, Department of Nuclear Engineering,
    Massachusetts Institute of Technology, April 1973.

18. A. S. Kubo and D. J. Rose, "Disposal of Nuclear Wastes", Science,
    Volume 182, 21 December 1973.

19. W. C. Wolkenhauer, et al. Transmutation of High-Level Radioac-
    tive Waste with a Controlled Thermonuclear Reactor, USAEC Report
    BNWL-1772, September 1973.

REACTOR SAFETY

Herbert J. C. Kouts
U.S. Atomic Energy Commission, Washington, D.C.

It is a great pleasure to be able to discuss nuclear reactor safety before this audience. It gives me an opportunity to address the subject as it should be; that is, as an issue on which quantitative physically based questions and answers are both required and possible.

The subject is usually discussed these days in non-scientific forums. There the question is commonly put in the form, "Are nuclear power plants safe?" This question proves to be very difficult for anyone with a scientific rearing to answer in a satisfying way. We are inclined to say under these circumstances that the things we protect against are thus and such, and we take certain precautionary measures, such as the following, and all in all we feel that nuclear power plants are reasonably safe, or as safe as we think they need to be, or something like that. This produces panic. A non-scientific listener comes away with the feeling that we don't think nuclear plants are safe after all. We have just been hedging the answer.

In fact, the problem is that most people have a different view of the quantitative character of the word "safe" than a scientist does. To us, the term has an ordered and almost a continuous property, in that we would class one act or one thing as more safe than another, and we would expect that the safety of a thing could be made to vary in a continuous way as parameters that affect it are changed continuously. The layman usually thinks of safety as a two-valued function. Things are safe, or they are not. And the basis for assignment of the value is seldom a physical one, but is an emotional one.

So I want to change the subject here to two questions that make more sense. The first is, how safe are nuclear power plants now as a function of their design and operating parameters? And the second question is, how safe do we want them to be? We can shed light on these two questions when they are viewed in their scientific frames. Partial answers are coming to exist.

As a background for laying the basis on which these questions can be answered, let me take a diversion to talk about nuclear reactors as sources of electrical power, according to their use in present day technology.

First there is the physicist's view of a nuclear reactor, illustrated in my first slide (Fig. 1). Here a reactor consists

(Fig. 1)

244

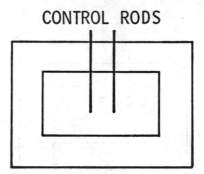

(Fig. 2)

of a central region called the core, and a surrounding region called a reflector. The core contains fissionable material such as U^{235} and other material, and it is in this region that the fissionable nuclei are induced to split on absorption of neutrons, providing the basis for a chain reaction through emitting neutrons in the act. The reflector is a region in which scattering provides a means for economizing on neutrons through returning some of the escaping ones to the core region for a further chance to induce fissions.

One further feature must be added to the physicist's version of a nuclear reactor. That is a means of balancing the neutron economy. The measure is a material that can absorb neutrons and that can be added to the core region or removed from it. This is added as control rods, shown conceptually in my next slide (Fig. 2).

That is the physicist's view. Understanding the performance of the reactor consists in being able to solve the simplified Bolzmann equation describing the time-dependent behavior of the neutron gas.

To make progress toward coping with the safety of a nuclear power plant, we must see what it looks like from an engineer's point of view. Different kinds of reactors look very different to engineers, who do not have the advantage of the simplicity of the Bolzmann equation to describe all the phenomena they are interested in. So we shall look in turn at two kinds of reactor systems used to generate almost all of the electricity produced in this country through nuclear fission.

The next slide (Fig. 3) shows a simplified engineering view of a boiling water reactor. Note that neutrons have disappeared from the field of interest. The focus is on heat. The diagram shows conceptually how water is circulated about the system, being heated to boiling in the core region, steam then being extracted to be passed through a turbine driving a generator. Removal of energy through work partially condenses the steam again, and return to the liquid state is completed through a heat exchanger called a condenser. The water is then returned to the inlet region of the reactor core for iteration of the process. The temperature of the water in a boiling water reactor is a little below 300°C, and the pressure is about 70 atmospheres.

My next slide (Fig. 4) shows a conceptual engineering view of a pressurized water reactor. The essential difference is that the water in the primary circuit is not heated past the phase transition temperature. Instead, the heat from the reactor core is transferred in a heat exchanger to water circulating in a secondary circuit. This heat exchanger is actually a steam generator. The steam produced in the secondary circuit drives a turbine, and after being condensed to water it is again returned to the steam

BOILING WATER REACTOR

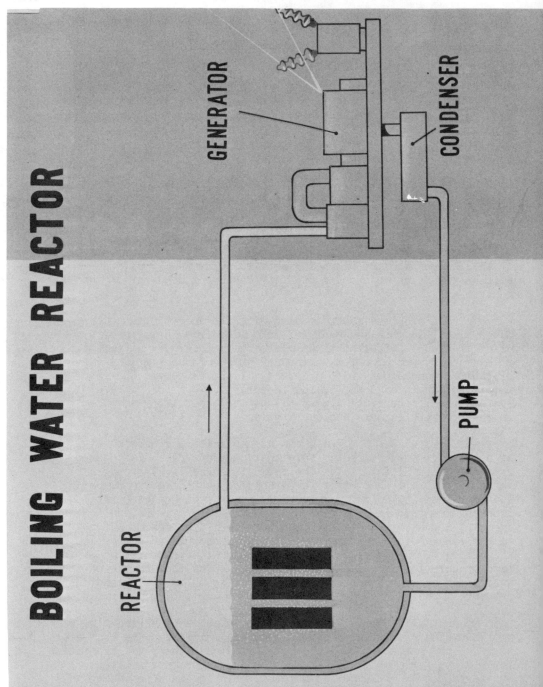

REACTOR

GENERATOR

CONDENSER

PUMP

247

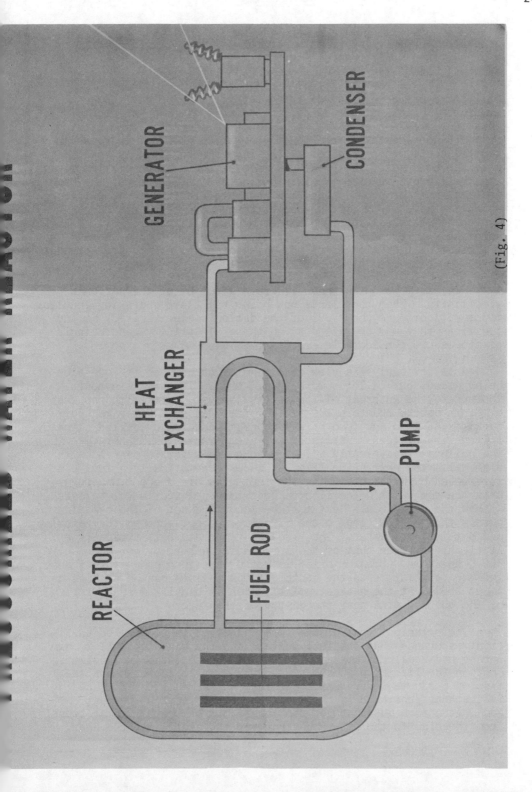

REACTOR

GENERATOR

CONDENSER

HEAT EXCHANGER

PUMP

FUEL ROD

(Fig. 4)

generator for recycle. The temperature in the primary circuit of
a pressurized water reactor is about 300°C, and the pressure is
typically about 140 atmospheres.

There has been a steady trend toward larger electrical
generating plants of all kinds, including nuclear power plants.
This increase in size makes it possible to economize on capital
costs, since increasing the power generation does not increase the
capital costs proportionately. The increase in size has been
possible because of the growth in rate of consumption of electricity
and the improved system of distributing electricity about the country.
American nuclear power plants now generate electricity at a rate
which is typically a little greater than 1,000 megawatts each.
Conventional fossil-fueled plants run even a little larger. In other
countries where the use of electricity is less, or when there is no
efficient distribution grid, the sizes of power plants are somewhat
less.

Nuclear power plants now generate about 6% of the nation's
electricity. When the plants now being built are operating, nuclear
power will produce about 20% of the nation's electricity. So any
questions that may be raised as to these electricity-generating
plants assume considerable importance.

A safety analyst's view of a reactor depends on the question
being asked, but it leans more toward the engineer's. In most
instances, the interest of the safety analyst starts with the fuel
elements of the reactors. A typical fuel pin in a water reactor
consists of a hollow tube of a zirconium-based alloy called zircaloy,
filled with pellets of slightly enriched uranium oxide ceramic.
The pellets are typically a few cm long and a little over 1 cm in
diameter. The next slide (Fig. 5) shows schematically a fuel pin
surrounded by water coolant. The fuel pins are spaced in a regular
array in the water. The fission heat is generated in the pellets
within the fuel rods. Heat must be conducted out of the ceramic,
across the physical gap to the zircaloy tubing, through the zircaloy,
and out into the water. The zircaloy temperature is typically a
little higher than that of the surrounding water, or not much above
300°C anywhere. Uranium oxide is not a very good conductor of heat,
and the temperature in the center of the pellets can be as high as
about 2000°C at the hottest point, which is usually a small region
of the hottest pins near the center of the reactor.

The principal goal of the safety analyst is to develop sufficient assurance that the fuel pellets and the cladding will remain
intact under all circumstances. This is because the fuel pellets
contain the fission products. Cladding that is intact keeps the
fission products under full control. The enemy of intact cladding
is the heat generated in the adjacent ceramic fuel. The heat
generation rate must not get too large, and the rate of heat transfer
from the clad to the water must not get too small. If the heat

249

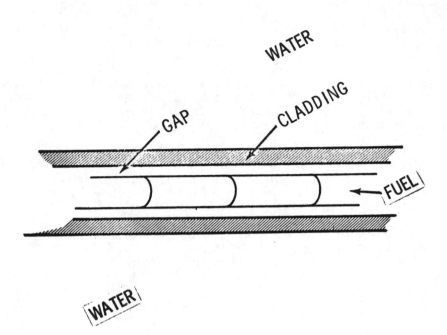

(Fig. 5)

generation rate were to become high enough, the fuel pellets might
melt or evaporate, and fission products might be liberated into the
coolant. A first protective barrier would fall. If the heat removal
rate were to diminish, the temperature of the cladding would rise.
It could exceed the melting point of zircaloy, or temperatures
could be high enough to cause extensive oxidation by the water and
resultant loss of strength of the zircaloy.

These two requirements define the two kinds of reviews that the
safety analyst must conduct. Increased heat generation would result
if the neutron balance were upset so that a positive reactivity
resulted and the power level rose. An occurrence like this if it
took place would be called a reactivity-induced accident. Reduced
cooling would be the result of a loss of pumping power or a break
in the primary system allowing the cooling water to escape. This
last occurrence would be called a loss-of-coolant accident.

The reactor designer is required to design the facility so
that conditions leading to the possibility of a reactivity-induced
accident or a loss-of-coolant accident will not develop. He is
expected to avoid these conditions through proper design and
through control of quality of fabrication. He is also expected to
incorporate into the design mechanisms such that if conditions did
develop that could lead to an accident, the conditions would be
terminated before they reached the point of true hazard. And
finally, he is expected also to include further features that would
ameliorate the effects of any release of fission products if this
did occur. This layering of protective measures is unique to the
protection afforded in connection with nuclear reactors. It is
commonly called "defense in depth."

I should now like to tell how the Atomic Energy Commission
has come to assume a regulatory role related to the safety of
nuclear power plants, and I will describe briefly the structure of
regulation.

Some important dates will be helpful as a start. (Fig. 6)
The first nuclear reactor was built under the West Stands of the
University of Chicago in 1942. The first large reactors were
built at the Hanford site near Richland, Washington in 1944.
The Manhattan Project ended and the Atomic Energy Commission came
into existence in 1947. The first commercial nuclear power plant
in the United States went into operation at Shippingport,
Pennsylvania in 1957.

In 1947, when the Atomic Energy Commission was first
established, one of its first acts was the appointment of a
committee called the Reactor Safeguards Committee. This committee
was asked to review the wartime decision to build the Hanford
reactors, to determine whether any specific measures should be
taken with respect to the agricultural operations on the Wahluke
slopes on the opposite bank of the Columbia River.

FIRST REACTOR   1942

FIRST LARGE REACTOR   1944

AEC   1947

SHIPPINGPORT   1957

(Fig. 6)

In 1950, a second committee was appointed, called the Industrial Committee on Reactor Location Problems. This group was initially asked to review plans to operate research reactors that were coming along, the Savannah River production reactors which were in the planning stages, and the land-based prototype of the first submarine reactors.

In 1952, the two committees were combined into a single Advisory Committee on Reactor Safeguards.

In 1956, the Atomic Energy Act was amended to recognize the advent of commercial generation of electrical power through fission reactors. The Advisory Committee on Reactor Safeguards was made a statutory committee, whose characteristics and duties were specifically stated by the law. Congress also wrote into the Act the establishment of a separate regulatory staff within the Atomic Energy Commission, under a Director of Regulation, who was to report directly to the five-member Commission. The Regulatory staff has an existence completely separate from the General Manager's staff of the Commission. The General Manager and the Director of Regulation are at the same level in the structure. The Commission in its direction of the Regulatory staff acts in a quasi-judicial role corresponding to that of other regulatory commissions in the Federal Government. This dual function of the Commission, to direct the very dissimilar activities of the two separate staffs under the Director of Regulation and the General Manager, is the basis of what some have termed the conflict of interest in the structure. Without making any judgment as to whether the conflict is real as well as apparent, I would like to give my own personal view that the situation has aspects that have been a real source of strength to the regulatory process. These have contributed to making the Regulatory staff of the Commission what I believe to be the most effective regulatory body in our Federal Government. We will have to be sure that the energy shortage and the proposed parting of the Atomic Energy Commission do not combine to increase the vulnerability of the regulatory body and diminish its effectiveness.

In the earlier history of nuclear reactors, the attention of safety reviews was focused predominantly on the possibility of reactivity-induced accidents. This was only natural, because memories were still heavily imprinted with the violence of nuclear weapons, and intensive programs of atmospheric tests of nuclear weapons were conducted both in the United States and the Soviet Union.

Because of the concentration on the possibility of reactivity-induced accidents, the first experimental programs on reactor safety were performed in this area. Simple research-type reactors were induced to undergo power transients through sudden reactivity addition, until an amount of reactivity large enough was added to cause the core to be destroyed. This was an extremely valuable set of experiments, showing a number of things that after the fact were almost self-evident. (Fig. 7) One conclusion was that transient behavior of the reactors is pretty well calculable over a range of reactivity additions well into values leading to heavy core damage. The significance, of course, is that safe design against reactivity-induced accidents can be based dependably on calculations, and it does not require a separate program of experiments for each new design.

A second conclusion was that certain reactivity feedback phenomena play a fundamental role in shaping the transient response to a reactivity change. Changes in atom concentrations resulting from thermal expansion and phase changes during a transient are important, tending to counteract or to augment the reactivity change, depending on the design. These effects and some others that depend on heat transfer act with some amount of time delay. Some effects which are called prompt result from direct heat deposition in the fuel. In a reactor fueled with slightly enriched uranium, such as a water reactor and a fast breeder reactor, heating the fuel leads to greater thermal excitation of the uranium oxide lattice and a doppler broadening of the capture resonances of U^{238}. This provides a negative feedback mechanism that goes a long way toward counteracting reactivity additions.

The third major conclusion of these tests was the discovery that water reactors cannot be made to explode like bombs. The obvious reason is that the time between successive generations of fissions in a water reactor is typically about 10^{-4} to 10^{-5} seconds, whereas in a bomb, the time is typically about 10^{-8} seconds. The most mechanical energy that was generated by reactivity additions made even in the most unrealistic manner never exceeded the equivalent of a very few pounds of high explosive. Any moderately well-designed reactor shield would have no difficulty containing explosive releases of these magnitudes.

The mechanisms built into reactors to protect against reactivity-induced accidents include negative reactivity feedback coefficients and safety rods activated by dependable sensors and logic circuitry. The AEC's Regulatory staff has recently established a requirement that water reactors move toward readopting an old requirement for two completely separate shutdown systems using different kinds of mechanisms. I'll come back to this.

In the early 1960's, water reactor designs began to include systems to prevent core melting that might result from fission product heating of uncooled fuel after a hypothetical break in primary system piping. At first, the Advisory Committee on Reactor Safeguards and the AEC's Regulatory staff required that such systems be provided simply as extra protection. Their function was neglected in the safety analysis. In 1966, a committee was appointed by the

1. BEHAVIOR CALCULABLE

2. FEEDBACK

3. NO EXPLOSIONS

(Fig. 7)

Director of Regulation to review the need for these so-called emergency core cooling systems, particularly for the much larger reactors for which license applications were being received. This committee was headed by William Ergen, a former member of the Advisory Committee on Reactor Safeguards from Oak Ridge. The Committee's report in 1967 said that reliable and practical methods of containing the large masses of molten fuel that would probably result from a core meltdown did not exist, and therefore it was not considered possible to assure the integrity of the containment if meltdown of large portions of the core were to occur. That is, the hot molten core might very well in time melt its way through the floor of the containment building.

The report also said that within the framework of existing types of systems, sufficient reliance can be placed on emergency core cooling following the loss of coolant, and additional steps can be taken to provide additional assurance that substantial meltdown is prevented.

This report led to AEC requirements for better emergency core cooling systems, with redundant systems provided. It also led to the start of a program to carry the analysis of loss of coolant accidents to the same stage as reactivity-induced-accidents, where confident realistic calculations of the course of events can be made. This objective will be quite difficult to achieve, because the phenomena of engineering are much more complicated than those of reactor core physics. We have confidence in our ability to calculate an upper limit to the physical consequences of a hypothetical loss-of-coolant accident, and so we can choose conditions for reactor design and operation that would ensure that no core melting at all would follow the loss of coolant. This confidence is based on experimental studies of either elementary or partial character, that cover every aspect of the sequence of events during the accident. But we do not yet have a test of the entire sequence of events, and the ability to synthesize our description so as to apply over the entire time. This test will be provided by the experimental program that will use the LOFT reactor at the National Reactor Testing Station. There we plan to conduct actual loss-of-coolant accidents, heavily instrumented to provide the data against which both realistic and conservative calculations can be checked.

Until recently, it was customary to divide conceptual accidents into two classes, called credible and incredible. The boundary between the two classes was set by a general consensus. Accidents were considered incredible if physically they were impossible or if the subjective judgment was made that the probability of occurrence was so low as to permit neglecting it. For instance, it was commonly held that a total rupture of a primary system pipe is credible, however unlikely. The rupture of a pressure vessel is so unlikely an event as to be incredible. The design of emergency core cooling systems has been based on these assumptions.

Efforts were made at times to quantify the estimates of probability that underlay the judgments as to credibility, but not very successfully. An often-quoted study in 1956 using what is now called the Delphi method concluded that the probability of a large accident might be between 10^{-5} and 10^{-7} per reactor year.

The modern approach is through reliability analysis. This procedure is being followed by a task force headed by Dr. Norman Rasmussen. The method uses event trees to estimate the probability of accidents of various kinds, estimating the overall probability through compounding probabilities of elementary events.

Let me illustrate in a simple manner how the event tree method works. My next slide (Fig. 8) shows an event tree that can be used to calculate the probability that the fire goes on when the knob controlling the supply of gas is turned on. P_1 is the probability that the gas is on. P_2 is the probability that the pilot light is on, given that the gas is on. P_3 is the probability that the burner will light when the gas is on and the pilot light is on. So the probability of failure is: (Fig. 9)

$$(1-P_1) + P_1(1-P_2) + P_1P_2(1-P_3) = 1-P_1P_2P_3$$

Event trees are being constructed for all of the conceptual accidents, and probabilities are being calculated. The initial report of the project, which will analyze the loss of coolant chain, is now being put into first draft. Some results of this and other studies of this kind have been made available. Dr. Ray in a Press Club speech recently stated that the probability of a major loss-of-coolant accident followed by failure of the ECCS to function is determined to be about 10^{-6} per reactor year, which agrees remarkably well with the old Delphi-derived value, which was derived before any ECCS system was ever invented. Dr. Ray added that the Rasmussen study is also providing calculations of consequences of hypothetical accidents, and the consequences associated with a probability of 10^{-6} per reactor year are approximately those of a large aircraft accident.

The end product of this study will finally provide us with the actuarial table we will need for a quantitative analysis of risk versus benefit, along the lines I sketched at the beginning of my talk. These will be estimates of the probability of consequences of a certain magnitude. Both probability and consequences will be functions of the design parameters and the operating parameters. This will be information that we can then use to set design and operating parameters to values that will keep the risk at whatever level the social decision makes desirable.

That final choice will not be an easy one. The criteria to be used will take some thought. Shall we require nuclear power

257

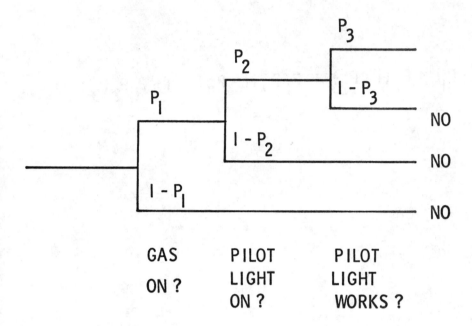

GAS	PILOT	PILOT
ON ?	LIGHT	LIGHT
	ON ?	WORKS ?

(Fig. 8)

$$(1 - P_1) + P_1(1 - P_2) + P_1 P_2 (1 - P_3) = 1 - P_1 P_2 P_3$$

(Fig. 9)

to be as safe as other methods of generating electricity? Shall we
make it some predetermined factor safer than other methods of
generating electricity? Shall we make it as safe as technologies
used for other purposes, such as transportation? Should we make it
some factor safer than these technologies? Or should we try to set
up some complicated function representing the safety of all our
technologies combined, and try to optimize through seeking a best
allocation of our resources?

I believe this final step from the scientific conclusion to
the social conclusion would be the most difficult one to make on
objective grounds.

In the meantime, a rough guideline seems appropriate, that the
probability of an accident comparable to an airliner crash should
fall in the range of 10^{-6} per reactor year over the long pull,
when as many as a thousand nuclear reactors might be operating.

Two other results of probabilistic analysis have been published
recently. The first followed a review of the reliability of control
systems for reactors and the probability of reactivity-induced
accidents. The review led the Regulatory staff to conclude that
achieving the probability of 10^{-6} per reactor year of a large
accident in the future would require two separate and non-identical
shutdown systems. I have mentioned this earlier.

The second review has considered failure probabilities for
pressure vessels for water reactors. The Advisory Committee on
Reactor Safeguards has just issued a detailed technical report
that evaluates the probability of catastrophic failure of one of
these vessels as no greater than 10^{-6} per reactor year. Other
unpublished studies conducted independently of the Committee's
are in agreement with it.

So it appears that our two objectives in conducting reliability
analyses can be achieved. First, the analysis is possible and
credible. Second, it can help us to settle on designs and on
construction and operating practices that will provide the level
of safety we are after, which is very high compared to that of
the alternatives.

LASERS FOR FUSION

Keith A. Brueckner
KMS Fusion, Inc., Ann Arbor, MI 48106

ABSTRACT

The extremely high power achievable with lasers offers the possibility of producing fusion in laser heated and inertially confined pellets of thermonuclear fuel. Special techniques of laser pulse forming properly matched to pellet size allow energy multiplication to occur with lasers which are presently technologically feasible. These techniques will be described and the current status of theory and experiments will be given. General engineering features of possible fusion reactors will be summarized.

The controlled release of fusion energy under conditions useful for power generation has been studied since the early 1950's, principally in the United States, Soviet Union, England, and West Germany. Nearly all of this effort has been directed at magnetic confinement of a very low density plasma, of deuterium and tritium under quasi-steady state conditions. This program continues very actively at present with a possible demonstration of scientific feasibility expected in the next decade.

In the early 1960's, with the development of the laser, the possibility of the use of the laser for pulsed heating of an inertially confined pellet of DT was realized. The published analyses through 1972 considered the laser to be used to heat a plasma at solid DT density (0.19 grams/cm^3) or lower, to allow the laser energy to penetrate the plasma effectively. Simple considerations show that hundreds of megajoules of laser energy delivered in a time of the order of a few nanoseconds are required under these conditions to produce fusion energy comparable to the laser energy. To reach energy multiplication sufficient for a practical application requires further orders-of-magnitude increase in laser energy, to the range of 10^{10} to 10^{11} joules of laser energy, depending on the overall efficiency of the laser system. Such lasers are not of practical interest.

Starting in the late summer of 1969 at KMS Fusion, we have studied the feasibility of using a laser to drive the hydrodynamic compression of a DT pellet to very high density, with the subsequent ignition and propagation of a burning wave in the DT giving high fuel burnup and large fusion energy release. A similar program was started in 1969 in the AEC laboratory at Livermore, either stimulated by our work or an independent effort growing out of other classified work carried out in Livermore in the 1960's.

The production of fusion energy from a pellet of thermonuclear fuel can be achieved on a level useful for power production only if the pellet is highly compressed with efficient energy transfer from the external energy source into the pellet. The simple model of a uniformly compressed DT sphere can be used to determine the fusion energy production. Figure (1) gives the ratio of fusion energy output to initial thermal energy for a uniform initial temperature of 5 keV. The energy multiplication, for an initial thermal energy of one kilojoule, is 5 at a density of 300 gm/cm^3, 16 at 600, 40 at 1000, and 80 at 2000. For high energy input or high compression, the energy multiplication levels off at about 200 corresponding to about 35% burnup of the DT. The energy multiplication can be increased if the fuel is only centrally heated to the ignition point of 5 keV, with the rest of the fuel ignited by an expanding supersonic burning front propagating outward from the fuel center. The ignition sequence is shown schematically in Figure (2). At the time of peak compression of the dense pellet core, proper timing of the shocks produces high temperature at the center of shock convergence. The shock conditions are adjusted to bring this heated region to the ignition temperature of about 5 keV over a region with dimension approximately equal to the α-particle range at the ignition temperature. The compressed DT then is heated by the energy deposition by the α-particles and the reaction rate increases very rapidly, increasing by about a factor of 100 between 5 keV and 40 keV. As the fuel heats, the α-particle range increases and the energy deposition starts to extend into the relatively cold surrounding DT, which in turn ignites. The result is a spherically expanding burning front which moves supersonically into the surrounding fuel, igniting it before appreciable hydrodynamic motion can occur.

Figure (3) shows a typical example of the propagation of a supersonic burning front. Figure (4) shows the energy multiplication with the fuel center heated to 5 keV over a few micron radius and the rest of the fuel at 500 eV. With an initial thermal energy of one kilojoule, the energy multiplication is 130 at $\rho = 600$ gm/cm^3, 400 at $\rho = 1000$ gm/cm^3, and 700 at $\rho = 2000$ gm/cm^3. The energy multiplication reaches a maximum of about 1200 for initial thermal energy of 5-10 kilojoules, independent of initial density, corresponding to about 35% fuel burnup. The effect of the centrally-initiated burning wave increases the energy multiplication by about a factor of ten over the uniformly heated case.

Calculations of the implosion of a DT sphere in spherical symmetry show that high compression can be produced by a laser pulse with proper time variation. The pressure driving the implosion is produced by the penetration of energy from the underdense laser deposition region into the dense plasma which results in ablation of the dense pellet surface. This process is shown schematically in Figure (5). The energy flow from the laser deposition region is principally by electron-thermal conduction which provides the energy flux to the conduction front at which ablation of the dense pellet is occurring. The energy flow by electron conduction also maintains nearly isothermal conditions in the expanding low density plasma

262

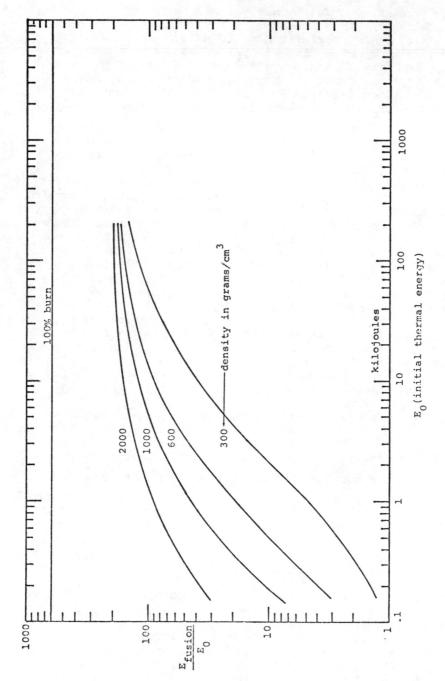

FIGURE 1 - Ratio of fusion energy to initial energy for a compressed spherical DT pellet uniformly heated to 5 keV.

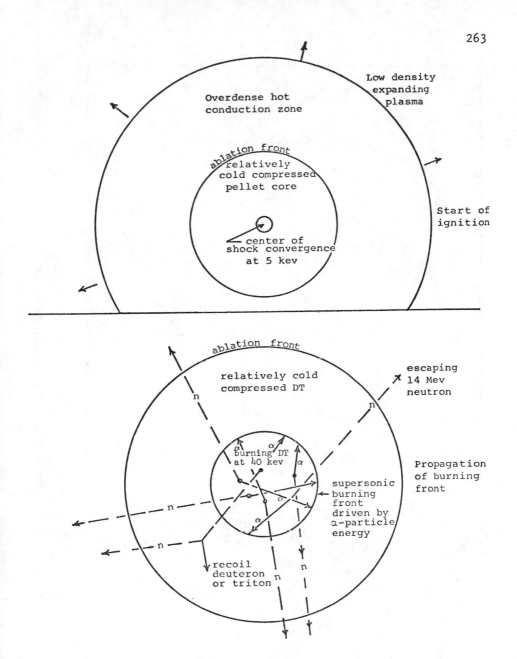

FIGURE 2 - Ignition and propagation.

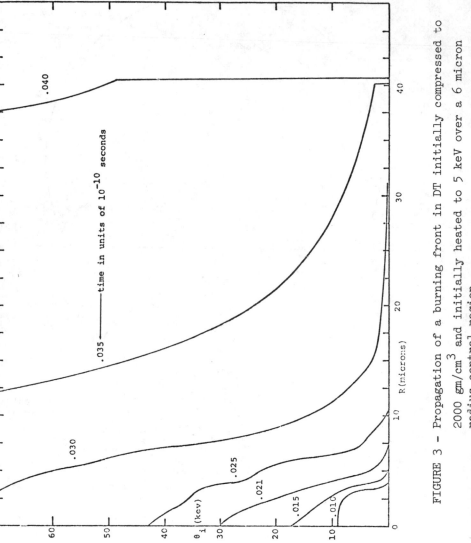

FIGURE 3 – Propagation of a burning front in DT initially compressed to 2000 gm/cm^3 and initially heated to 5 keV over a 6 micron radius central region.

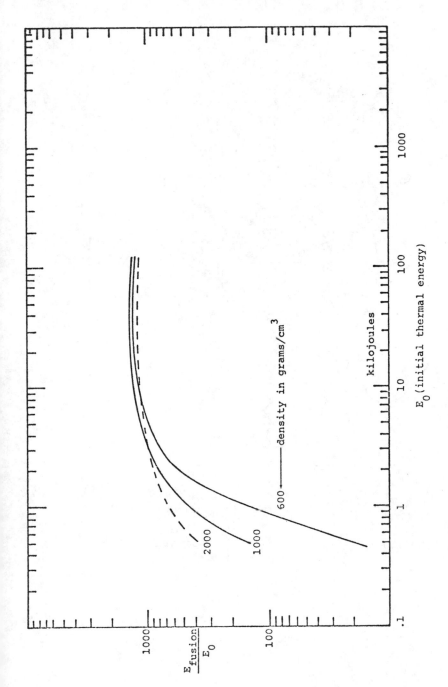

FIGURE 4 - Ratio of fusion energy to initial energy for a compressed spherical DT pellet centrally ignited by heating to 5 keV. The rest of the pellet initially was at 500 eV.

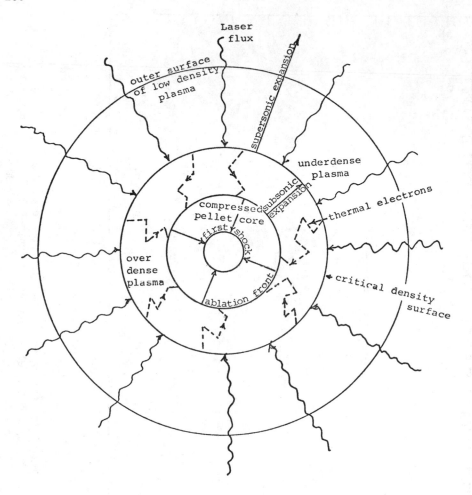

FIGURE 5 - Energy flow and hydrodynamic motion in
a laser-driven pellet.

which accelerates rapidly outward with a corresponding density drop. The motion is subsonic near the ablation front but typically becomes supersonic near the critical density surface. Inside the ablation front, the hydrodynamic flow is inward as a result of the passage of strong converging pressure waves which lead to compression of the dense pellet. During a typical implosion about 80% of the initial pellet mass is removed by ablation, 92% of the laser energy goes into the heating and expansion of the ablated pellet, 8% of the laser energy appears as energy of compression and heating of the dense pellet core, and the pellet radius is decreased by a factor of about 20 corresponding to compression by a factor of 8000 or to 1600 grams per cubic centimeter.

The efficiency of this process depends on the efficiency of energy absorption from the laser beam into the underdense plasma and on the subsequent energy partition between the compression of the dense pellet core and the energy removed in the high temperature expanding plasma produced by the ablation process. The latter partition can be estimated from a hydrodynamic model or determined by computer simulation of the energy deposition, energy transfer, and hydrodynamic processes. The result is that 6 to 10% of the absorbed energy is transferred into compression and heating of the dense pellet core.

Detailed computer simulations of the full process of laser coupling for 1.06 micron wavelength, thermal energy flow, hydrodynamics, nuclear reactions, and of the energy transport in the nuclear reaction products and radiation, give results in approximate agreement with the numbers just given. Of particular interest for the experiments now being undertaken by several groups in the USA and other countries is the prediction that the "breakeven" condition, with fusion energy equal to laser energy, can be reached with laser energy of about one kilojoule. This prediction holds, however, only if several conditions are satisfied. These are:

A. Configuration requirements
   1) laser pulse time variation properly matched to pellet configuration
   2) spherical symmetry of pellet illumination
   3) spherical symmetry of pellet configuration
B. Physics requirements
   1) stable hydrodynamic motion
   2) adequate laser-plasma coupling
   3) absence of appreciable pellet preheat.

If these requirements are not met, the breakeven energy can be markedly affected. Particularly striking is the effect of a poorly matched laser pulse. The breakeven energy for a square laser pulse and a DT sphere is several hundred megajoules. A drop in compression of a factor of ten as a result of imperfect convergence can increase the breakeven energy by a factor of ten to one hundred.

The required pellet symmetry can be provided by careful pellet fabrication and selection methods. We have studied the effect of variation in the laser and pellet parameters, using 2-dimensional computer simulation, and determined the allowable departures from complete symmetry. The illumination symmetry requirements are considerably eased by rapid energy flow by thermal conduction in the pellet surface, as shown in Figure (6). The non-uniform laser intensity gives variation in the energy deposition in the underdense plasma. The resulting lateral variation in energy flow by hot electron condition very rapidly compensates for the laser deposition non-uniformity so that the energy flow through the conduction zone to the dense pellet surface remains nearly unaltered. The rate of adjustment of the temperature by electron conduction corresponds to energy flow at much greater than sound speed so that the hydrodynamic motion is not affected during the time required for the energy flow to adjust. The presence of super-thermal electrons originating in the laser deposition region is also indicated in Figure (6). These have a large range and can penetrate the dense pellet core well in advance of the main thermal conduction front. The consequence is premature heating which increases the pressure in the dense pellet and reduces the compression which can be produced, with the consequence that pellet ignition and burning may be reduced or prevented.

The pellet implosion may be affected by hydrodynamic instabilities which perturb the spherical symmetry of the motion. The classical instabilities in an accelerated fluid are shown in Figure (7) during onset and after considerable growth. The Rayleigh-Taylor instability results from the inability of a light fluid to support a heavy fluid. Although the undisturbed arrangement is metastable, the presence of small ripples on the surface causes fluid motion which amplifies the disturbances. After growth with amplitude comparable with the wavelength of the disturbance, the motion changes to bubbles of the less dense liquid rising into the heavy liquid which falls in "spikes" between the bubbles, moving at free-fall velocity. The growth rate is most rapid for short wavelengths but is finally limited by viscosity which causes dissipation of energy comparable to the rate at which work is being done on the moving fluids by the acceleration field. The second instability, which is driven by a temperature gradient in a fluid heated from below, is the formation of Bénard cells. The heated gas forms convective columns of hot fluid. This motion is the origin of convective cloud formation in the atmosphere. The growth rate is limited by the dissipative effects of viscosity and thermal conductivity which under certain conditions can stabilize the flow. The stability of the pellet implosion has been studied analytically and by computer simulation, using a 2-dimensional code. The results show stable motion with initial disturbances not being amplified during the implosion.

The laser-plasma coupling presents difficult problems which are of a complexity very familiar for the past two decades in the controlled-fusion programs. Closely associated with the coupling problem is the effect of anomalous laser-coupling on the energy flow into the pellet. Present theories estimate that a wide range of anomalous phenomena can occur which may seriously alter the predictions of the laser-driven process. Experiments are intended to resolve these uncertainties.

The fusion yield for an optimally delivered laser pulse as a function of laser energy is given in Figure (8). The breakeven point, with fusion energy equal to laser energy, is at about one kilojoule. The only present technology which can provide this output is the neodymium glass laser at 1.06 micron wavelength.

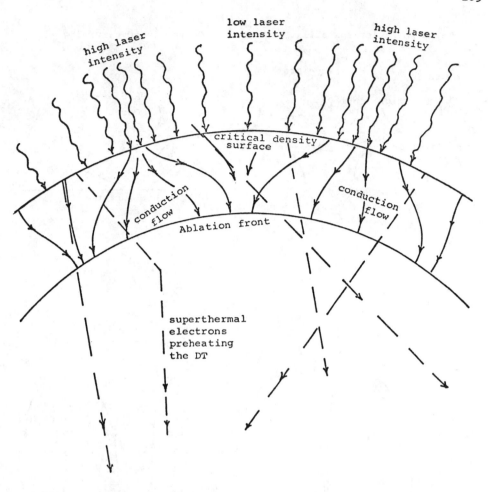

FIGURE 6 - Flow of energy on a non-uniformly
illuminated pellet.

270

Rayleigh-Taylor Instability

Convective Overturning
(Benárd Cell Formation)

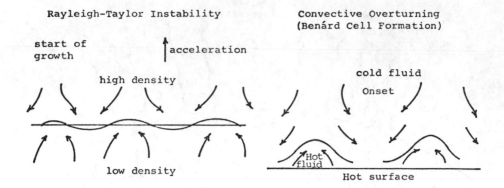

Large Amplitude Growth

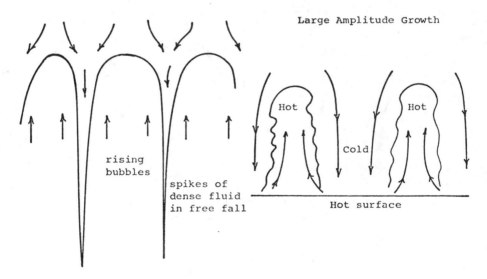

FIGURE 7 – Classical instabilities in an
accelerated fluid.

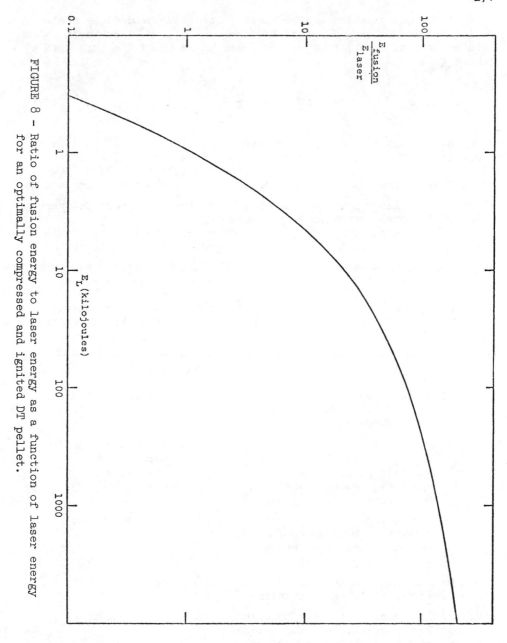

FIGURE 8 - Ratio of fusion energy to laser energy as a function of laser energy for an optimally compressed and ignited DT pellet.

The time variation of the laser pulse is determined by the pellet configuration. For a solid spherical pellet of deuterium and tritium, the time variation of the laser pulse is of the form (see Figure (9))

$$\varphi(t) \cong \varphi_0 \left(\frac{t_c}{t_c - t} \right)^2 \tag{1}$$

with t_c the time of maximum compression of the pellet. The time variation of the pulse should follow this form until $t_c - t$ is about 2% of t_c, corresponding to a variation of 2500 in the laser power. The collapse time t_c for a kilojoule pulse is about two nanoseconds, giving a peak laser power of 2.5 x 10^{13} watts. The laser delivers half of the total energy in 40 picoseconds.

For a fusion pellet in the form of a thin shell, a kilojoule pulse linearly rising in 0.8 to 1.0 nanoseconds is close to optimum, corresponding to a peak power of 2 to 2.5 x 10^{12} watts. This is about a factor of ten lower than the laser power required for a solid sphere.

The general characteristics of the laser follow from the peak power. The characteristics of neodymium-doped glass are summarized in Table I.

TABLE I - NEODYMIUM GLASS CHARACTERISTICS

I.	Safe power level Limited by surface damage or self-focusing damage	5 to 10 gigawatts/cm^2
II.	Gain (Owens-Illinois ED-2 glass) Slightly affected by beam spectral structure and pulse length	0.16 per cm per joule/cm^3 stored
III.	Energy storage Determined by pumping intensity and spontaneous emission amplification	0.3 to 0.6 joules/cm^3
IV.	Maximum glass dimension to limit parasitic amplification	50 to 60 cm
V.	Maximum glass thickness to give uniform pumping, assuming 2-sided illumination normal to surface	40 to 60 millimeters

A safe operating level for a noedymium glass laser is between 5 and 10 gigawatts (10^9 watts) per square centimeter for a beam with good spatial uniformity. We will return later to the problem of meeting

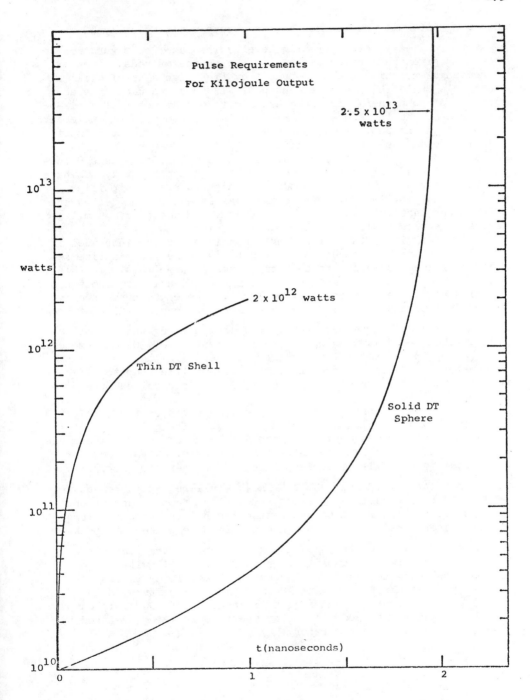

FIGURE 9 — Approximate laser pulse forms for a DT sphere and a DT shell for kilojoule fusion yield.

adequate uniformity requirements. We assume optimistically 10 gigawatts/cm^2 as a safe operating level. The laser configuration to provide a given peak power depends on the choice of rods (circular or rectangular) which are edge-pumped by flashlights or of disks at Brewster's angle which are face-pumped. If rods are used, the absorption of pump energy in the glass limits the diameter of a circular rod or the thickness of a rectangular rod to 5 to 8 centimeters, depending on the doping level of the glass. The largest rod amplifier in present use is a 80 millimeter 2% doped CD-2 rod at KMS Fusion. The large 9-beam laser at the Lebedev Institute uses 9 output rods with 45 millimeter diameter. Many lasers built by CGE have 64 millimeter diameter output rods. A rectangular rod amplifier with 40 x 240 millimeter cross-section has recently been tested at the Lebedev Institute in Moscow. We assume a diameter of 60 millimeters as a reasonable limit for rod amplifier.

The useful aperture of a single amplifier can be increased if the glass is arranged in disks at Brewster's angle, to minimize reflection losses and allow face-pumping (see Figure (10)). In this case the aperture can be increased until amplification of spontaneous emission along the disk becomes too large. Present experimentally tested designs show that an effective area in the glass disk of about 1000 cm^2 is possible corresponding to a length along the elliptical disk of about 60 cm and an open circular aperture of 30 cm diameter.

An optimized amplifier arrangement uses edge-pumped rods up to a diameter of 60 millimeters and then disk amplifiers up to a diameter of 300 millimeters. If additional energy is required, the system is paralleled.

The overall configuration of a laser depends on the gain per unit length and the final energy requirement. The gain of strongly pumped neodymium glass depends on the dimension of the glass and the details of the flash lamp geometry. A gain of 0.064 to 0.080 per centimeter in ED-2 glass, corresponding to an energy storage of 0.4 to 0.5 joules/cm^3, is reasonable. The overall gain requirement is determined by the oscillator input, typically of the order of one millijoule, and the level of saturation reached in the amplifiers. Typical laser configurations to give a kilojoule pulse with peak powers of 2×10^{12} watts and 2.5×10^{13} watts are shown in Tables II and III. The amplifiers required are well within the present state-of-the-art. The largest disk amplifier presently operating is at KMS Fusion with 100 millimeters aperture, a low signal gain of about 20, and a gain of about 8 at 1000 joule output. A disk amplifier with 300 millimeter aperture is being designed by the Lawrence Livermore Laboratory.

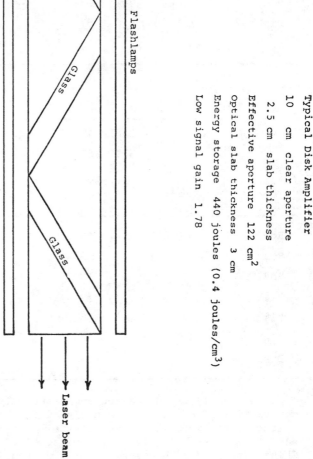

Typical Disk Amplifier

10 cm clear aperture

2.5 cm slab thickness

Effective aperture 122 cm^2

Optical slab thickness 3 cm

Energy storage 440 joules (0.4 joules/cm^3)

Low signal gain 1.78

Flashlamps

Flashlamps

Glass

Glass

Glass

Laser beam

FIGURE 10 - Typical disk amplifier.

TABLE II – LASER AMPLIFIER TO GIVE 1000 JOULES
WITH PEAK POWER OF 2×10^{12} WATTS

Amplifier Type	Energy Output (joules)	Diameter (mm)	Path Length (cm)	Low Signal Gain
Rod	0.18	4	50	24.4
Rod	3.36	16	50	24.4
Rod	16.7	33	30	6.8
Rod	76.3	64	30	6.8
Disk	324	87	30	6.8
Disk	1000	128	30	6.8

TABLE III – LASER AMPLIFIER TO GIVE 1000 JOULES
WITH PEAK POWER OF 2.5×10^{13} WATTS

Amplifier Type	Energy Output (joules)	Diameter (mm)	Path Length (cm)	Low Signal Gain
Rod	0.043	4	30	6.8
Rod	0.297	12	30	6.8
Rod	2.03	30	30	6.8
Disk	13.9	79	30	6.8
Disk	99	156	18	3.16
Disk	307	260	18	3.16
Disk(3)	3 x 333	260	18	3.16

The analysis just given indicates only approximately the characteristics of a glass laser designed for feasibility experiments. A laser must typically be substantially over-designed to provide for additional losses in mirrors, lenses, electro-optic shutters, etc. These effects are, however, readily compensated by an additional stage of amplification or increased gain per amplifier stage and hence are not the major sources of practical difficulty in achieving the desired laser performance.

As mentioned earlier, the laser pulse to be fully useful must be accurately controlled in its time variation. This cannot be achieved with electro-optic shutters of the Kerr cell or Pockel cell type. These give a pulse form which at low power is roughly Gaussian in form with a full width at half maximum of 1 to 1.5 nanoseconds. This pulse is distorted by saturation effects in the amplifiers; as a result, the amplifier output pulse has a considerably more rapid rise than the oscillator output. Other methods of pulse forming are, therefore, necessary to deliver the required pulse form at the high power and of the laser, allowing for possible distortion through the laser. A relatively simple design now operating at KMSF uses a "pulse-stacker" which starts with a pulse of about 30 picosecond duration from a mode-locked YAG oscillator, which is divided by beam splitters, selectively delayed and attenuated, and reformed to give a composite pulse of a desired form. The characteristics of the pulse-stacker are shown schematically in Figure (11). A short pulse of 20 picoseconds (one picosecond equals 10^{-12} seconds) is produced by a mode-locked oscillator. The pulse is divided into five pulses by a series of partially reflecting mirrors. Each pulse is given a predetermined time delay and attenuation and the pulses brought back to a single optical path by a second succession of partially reflecting mirrors. The "stacked" pulses consequently can be given any desired composite pulse form. The number of pulses used in practice is determined by the overall pulse length desired, by the pulse length from the oscillator, and by the acceptable intensity ripple in the stacked pulse sequence.

In operation, the pulse stacker is adjusted to compensate for the measured nonlinearity of the amplifier train, to correct for the pulse distortion which arises from saturation, possible index variation with energy depletion of the lasing medium, and relaxation effects in the lasing action.

The next requirement is the control of the laser intensity distribution through the laser amplifiers so that good spatial uniformity is maintained. This is very important since the maximum power limit is set by the onset of catastrophic self-focusing in the glass, which is initiated by the effect of intensity non-uniformity on the index of refraction of the glass. The index of refraction increases with power in the glass according to

$$n = n_0 + n_2 E^2 \qquad (2)$$

$$n_2 = 1.3 \times 10^{-13} \quad (E \text{ in ESU}) \quad . \qquad (3)$$

A simple analysis shows that a beam non-uniformity ΔP with scale b gives rises to a large field increase which in turn leads to damage, with a self-focusing length f of

$$f = 1/4 \ b \ \sqrt{n_0/n_2 E^2}$$

$$\cong 400 \ b/[\ \Delta P(\text{gigawatts/cm}^2)]^{\frac{1}{2}} \quad . \qquad (4)$$

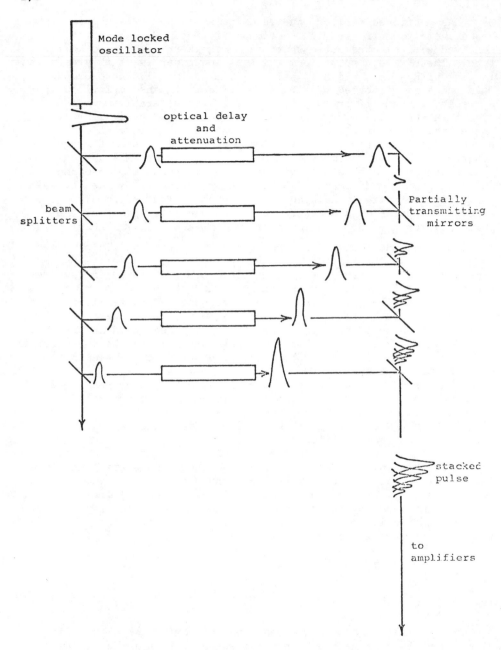

FIGURE 11 – Schematic layout of a pulse stacker.

This equation (Eq. 4) shows that self-focusing can be avoided in an amplifier with 50 cm length, with field fluctuations of 5 giga-watts/cm^2, only if the scale of the non-uniformity is greater than 2.8 millimeters. If the fluctuations are greater or the amplifier length greater, the scale of the non-uniformity must be correspondingly increased. The importance of this problem is well-known in work with high powered lasers, damage being easily produced if a laser is operated with poor control of the intensity patterns.

In practice with high quality laser glass, the intensity variations in the laser beam which initiate self-focusing are results of Fresnel rings produced by apertures in the amplifier train. Other sources of interference which can cause severe damage are improperly aligned prisms and internal reflection from rod surfaces. These problems can all be avoided if the laser is designed to have an aperture considerably larger than the useful energy producing aperture. If this is done, the laser beam is apertured only at very low intensity and the resulting Fresnel rings are very weak. This requirement is most important in the smaller laser rods which fortunately are also the lower cost elements of the amplifier train. The output disk amplifiers can be more fully filled, the diffraction patterns forming late in the system in the widely spaced disks giving relatively little self-focusing problems. The problems of Fresnel ring formation can also be alleviated by the use of "soft apertures" rather than typical apertures with sharply defined radii. We are presently testing soft apertures and designing improved apertures. Our tests show that 5 gigawatts/cm^2 is a safe operating range. We expect to reach 10 gigawatts/cm^2 with better intensity control by improved aperture design.

The next requirement is the control of the laser to deliver spatially uniform energy to the pellet. This requires good laser beam uniformity together with an optical system near the diffraction limit to focus the energy onto the target. The laser should deliver energy to the pellet with about 100 micron radius with a flux variation due to laser intensity variation of less than 20%. With an f/1 focal-ratio illuminating lens, this requires a beam divergence of less than about 10^{-4} radians. This corresponds at one micron wavelength to a diffraction limited beam at one centimeter aperture and to ten times diffraction limit at 10 centimeter aperture. These requirements are met fairly easily with uniformly pumped small diameter laser rods with a laser beam only partially filling the laser rods and hence only weakly apertured. In practice, the beam divergence can be increased by the non-uniform pumping and thermal distortion in larger diameter rods, and by edge effects from the apertures limiting the beam diameter. Closely associated with the beam divergence in non-uniformity pumped and thermally distorted rods is strong birefringence which depolarizes the laser beam. Since polarizing elements are usually present in the laser (as for example, in the disk amplifiers set at Brewster's angle), the depolarization leads to substantial energy rejection.

The problems of thermal distortion of the laser rods can be alleviated by increasing the laser cycle time, by using small diameter rods, by reducing the doping level of the rods to improve uniformity of the pumping, by filtering the flash lamp light to reject energy ineffective in pumping the rods, and by maintaining a stable firing and cooling cycle. This problem, however, remains a major source of difficulty with a glass laser system.

The laser must also be protected against damage due to amplification of laser energy reflected from the target or from other reflecting elements inadvertently placed in the laser beam. The successive laser stages must be isolated to prevent excessive amplification of spontaneously emitted light which depletes the stored energy in the glass. The target must also be protected from energy amplified through the laser before the main laser pulse, which can be the result of spontaneous emission or of oscillator energy leaking through the pulse-forming network before the main pulse. The desired isolation can be provided by both passive and active shutters. A Faraday rotator with polarizing plates before and after the rotator allows nearly loss-less forward passage of the laser beam and many tens of db reverse attenuation. Pockel and Kerr cells can be opened by short electrical pulses, allowing fairly good transmission in either direction while open. A Pockel cell can work effectively down to times of about 1.5 nanoseconds and a Kerr cell to 0.5 to 1 nanosecond. Low level isolation can also be provided by saturable dye cells which have low transmission at low power and transmit with relatively low loss at the high powers characteristic of the full laser pulse. These techniques can adequately protect the laser system.

Particular care must be taken to protect the target from very low power coming from amplified spontaneous emission ahead of the main pulse. A few millijoules of energy arriving many microseconds ahead of the main pulse can destroy the target. In addition to fast shutters open for only several nanoseconds around the main pulse, further protection can be provided with saturable dye cells. If necessary, final protection can be provided with very thin foils placed in the beam which are vaporized and made transparent by the leading edge of the main laser pulse.

With a laser designed to provide a suitable pulse, the final problem is the symmetric delivery of energy on the target. This can be done with multiple beams formed with beam splitters and mirrors combined with one or more output stages. Large diameter precision optics can also deliver the energy with adequate symmetry. Fortunately, the analysis of the physics of implosion shows that highly precise uniformity of illumination is not essential, a considerable variation being acceptable due to smoothing effects in energy flow in the pellet surface (see Figure (6)).

In summary, a kilojoule laser for fusion experiments is within the present state-of-the-art, although careful attention must be given to critical elements of the system. These are:

1) careful uniformity control to reduce self-focusing damage,
2) pulse shaping to give the required steeply rising output pulse form,

3) control of pumping levels and thermal distortion to avoid beam degradation and rod birefringence,

4) laser isolation to prevent damage from reflected energy,

5) target protection against energy arriving before the main laser pulse, and

6) symmetric target illumination.

In addition, the requirement of laser reliability should be emphasized. A kilojoule laser in a complex instrument and careful design with good engineering attention to detail is essential.

Other lasers are now being developed for experiments in the one to ten kilojoule range, for studies of laser-driven fusion feasibility. Of particular interest is the N_2-CO_2 laser being built at Los Alamos, to give a kilojoule output at 10.6 micron wavelength. The N_2-CO_2 laser is pumped by a strong electric field applied to the lasing medium in which ionization is maintained by an electron beam. This laser offers the possibility of relatively high (3 - 5%) efficiency of conversion of electrical to laser energy, compared with 0.1 to 0.3% for a glass laser. The long wavelength of this laser presents considerable theoretical problems in the laser-fusion application, which can be resolved only by experiment. The slow relaxation process of energy transfer in the lasing medium may also make pulse form control difficult.

Another interesting possibility is a lasing medium of a gaseous alklyiodide, in which flashlight pumping produces excited iodine atoms in the $5^2P_{\frac{1}{2}}$ state. This laser operates at 1.3 microns and is expected to have 0.5 to 2% efficiency, depending on the flash lamp characteristics. A kilojoule laser with a nanosecond pulse is being built at the Max Planck Institute near Munich.

We have carried out a number of experiments using the KMSF neodymium-glass laser brought to full operation during July of 1973. The laser configuration is shown in Figure (12). The laser driving the main amplifier train is a VK800 laser built by CGE, with some modifications and with Owen—Illinois ED-2 glass replacing the original French laser glass. This laser operates reliably on a six minute cycle with an energy output from the 80 millimeter output amplifier of 250 to 350 joules. A considerably higher output is possible, but has not been used because of possible glass damage from self-focusing. The output from the 80 mm amplifier is expanded to 100 mm diameter and further amplified in seven amplifier modules built by GE. Each module contains three disks of glass at Brewster's angle. The path length of the laser beam in each disk is 3 centimeters and the effective aperture is 122 cm^2. The glass stores 0.32 joules/cm^3 with 8 keV pumplamp voltage and 0.36 joules/cm^3 with 9 keV pumplamp voltage. With 200 joules input with 3 nsec pulse width (FWHM) to the GE system, the measured output from the first six modules is approximately 840 joules at 8 keV flashlamp voltage. The predicted output of seven modules at 8 keV is 990 joules and at 9 keV approximately 1200 joules. The measured gain is in good agreement with the design predictions.

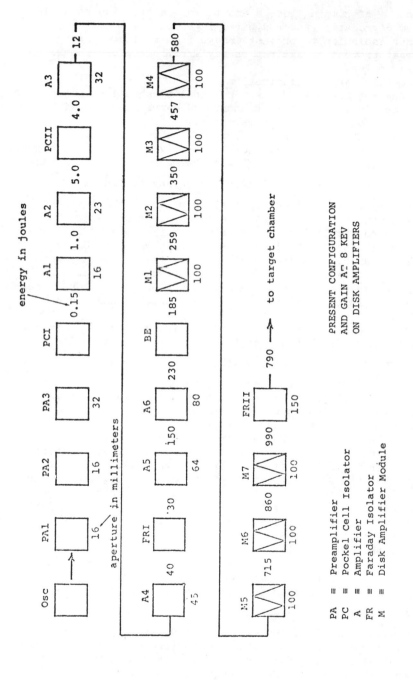

FIGURE 12 - KMSF laser arrangement.

The laser is protected against damage from reflected laser energy by Pockel cell isolators at the entrance to the 16 mm rod of the main amplifier chain, between the 23 and 32 mm rods, by a Faraday rotator between the 45 and 64 mm rods, and by a Faraday rotator at the exit of the GE amplifiers. The protection is adequate against the full output energy reentering the exit end of the rotator.

The laser pulse beam from the CGE oscillator is approximately Gaussian with a pulse width (FWHM) of 1.3 or 3 nanoseconds. The pulse is strongly distorted through the CGE and GE amplifiers due to partial saturation of the amplifiers. The pulse-stacking oscillator previously described has replaced the CGE oscillator to give a controllable pulse form.

The target area is shielded to allow breakeven experiments at the kilojoule level. The arrangement of the target area is shown in Figure (13). The chamber configuration allows the measurement of X-ray spectra by spectrometers, fast diodes, thermoluminescent dosimeters, and photographic plates. The neutron production is measured with several calibrated scintillators with large aperture and with time resolution of a few nanoseconds. In addition, the integrated neutron production is measured through silver foil activation. Provision has been made for time-of-flight measurement of the neutron energy spectrum. For measurement of the spectrum from a single neutron pulse, a DT neutron yield of about 10^8 is required.

The illumination system of the present laser configuration is shown in Figure (14). The illumination of spherical targets is uniform to 5-10%, the indicated correction plates perturbing the laser flux sufficiently to give uniform absorbed intensity on the target, after correction for the non-normal incidence. Other lens and mirror arrangements are being completed which will give further improvement in the illumination pattern.

Two-dimensional, cylindrical geometry computer simulations have been made of the response of CD_2 shells to non-uniform illumination by the KMS Fusion laser. Two configurations are shown here. The first is two-sided illumination with two f/2.6 lenses. The second is two-sided illumination with two f/1.0 lenses. The calculated energy absorption versus angle from the axis of symmetry is plotted in Figure (15) for the two cases. The calculated plasma distribution at the time of collapse of a spherical 2μm thick shell of 200μm diameter is plotted in Figures (16) and (17) for the f/2.6 and f/1.0 lenses, respectively. The grossly non-spherical convergence resulting from the f/2.6 lenses is in striking contrast to the much improved convergence resulting from using f/1.0 lenses.

The reflectivity of targets under intense laser illumination has been the subject of intensive theoretical and computational analysis and of more limited experimental study. The theory of laser deposition predicts two principal classes of instabilities resulting from coupling of ion density waves, electron plasma oscillation, and the incident and reflected laser waves. One class of instabilities resulting from the excitation of transverse ion and plasma waves with wavelength much less than the laser wavelength is expected to increase the laser energy absorption and to produce a marked increase in electron energy together with a strong departure of the energetic

284

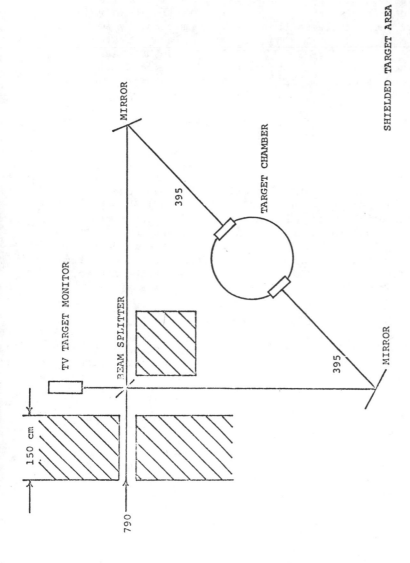

FIGURE 13 - Schematic layout of target area.

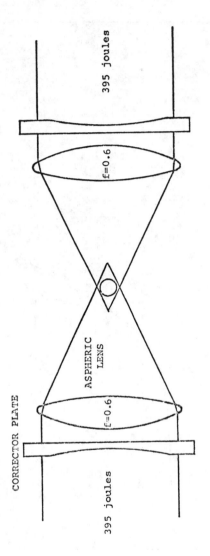

CORRECTOR PLATE

ASPHERIC LENS

f=0.6

f=0.6

395 joules

395 joules

TARGET ILLUMINATION

FIGURE 14 – Two-beam target illumination.

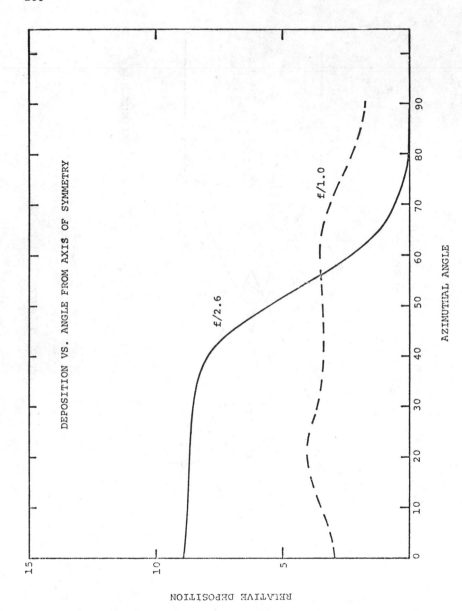

FIGURE 15 — Computed energy deposition vs. angle from axis
of symmetry.

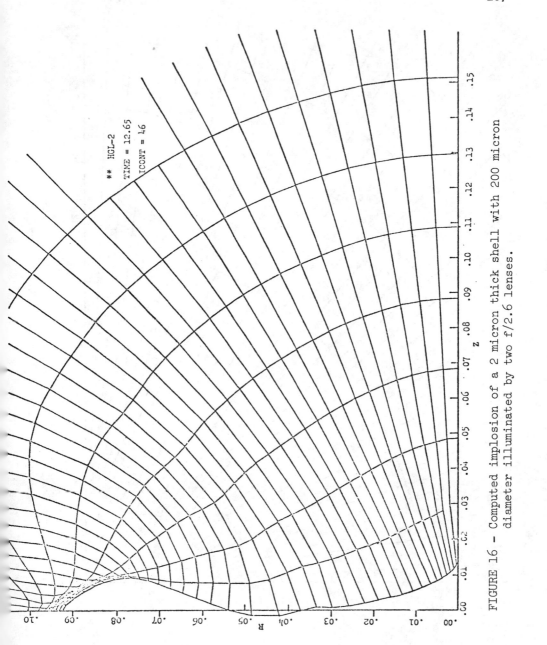

** HGL-2
TIME = 12.65
ICONT = 46

FIGURE 16 – Computed implosion of a 2 micron thick shell with 200 micron diameter illuminated by two f/2.6 lenses.

288

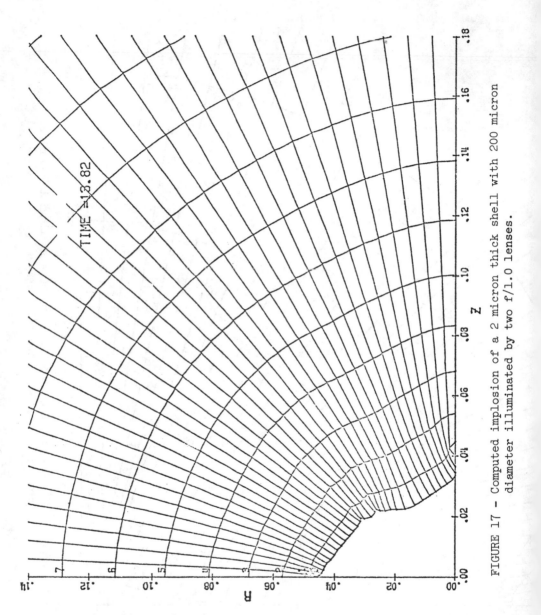

FIGURE 17 – Computed implosion of a 2 micron thick shell with 200 micron
diameter illuminated by two f/1.0 lenses.

electron distribution from Maxwellian. Another class resulting from longitudinal ion or electron density fluctuation, i.e., stimulated Brillouin or Raman scattering, is predicted to lead to a marked increase in reflectivity, possibly saturating at high laser power with reflectivity coefficient approaching unity. The threshold for the instability is expected to depend on target material and the illuminated area and pulse length, the latter since they determine the density gradients in the region of laser deposition. The thresholds are, however, expected to be reached for one-micron laser wavelength for power in the range of 10^{12} to 10^{13} watts/cm^2, for illuminated areas with characteristic dimensions of the order of 100 microns.

We have measured target reflectivity and electron temperature using an energy output up to 300 joules in a pulse with 3 nanosecond duration (full width at half maximum), delivered on plane CD_2 and CH_2 targets through a f/1.5 aspheric lens giving a measured vacuum focal spot of 60-80 microns diameter. The laser energy incident on the target chamber was measured calorimetrically; the energy on target was obtained by correction for window and focusing lens losses. The reflected energy was measured with fast photodiodes which directly compared a reference signal reflected from the incident laser pulse with the reflected energy from the target. An independent measurement was made calorimetrically of the energy from the target reflected from the 80 mm output face of the output amplifier of the laser. Relative measurements of the target reflections were also made calorimetrically at stations in the amplifier train of the laser.

The apparent electron temperature for photon energies in the 8-12 keV range was measured by a pair of fast diodes with aluminum foil attenuators. More complete measurements over a wide range of photon energies were also made on selected laser pulses using film detectors and thermoluminescent dosimeters with graduated aluminum attenuators.

The reflectivity was found to vary rapidly with target position, with the maximum reflectivity associated with the vacuum focus 50 to 75 microns above the target surface. This position also corresponded with the point of maximum hard X-ray yield and with the maximum soft X-ray flux.

The measured reflected energy was collected by the 8 cm diameter f/1.5 illuminating lens. This target reflection was also monitored at 45° and appeared to be very low, in agreement with previous measurements of the angular distribution of the reflected energy. We cannot at the present, however, exclude the possibility that some diffuse scattering occurs, increasing the true target reflectivity over our measurements. Improved reflectivity measurements will be carried out in the near future.

The variation of CH_2 and CD_2 reflectivity with target position for laser energy on target in the range of 100 to 110 joules is given in Figure (18). Figure (19) gives the measured peak reflectivity in CH_2 and CD_2 as a function of laser power on target, for a 80 micron vacuum focal diameter. The reflectivity peaks in the range of 40 to to 80 joules on target or 3 to 4 x 10^{14} watts/cm^2, and drops by a

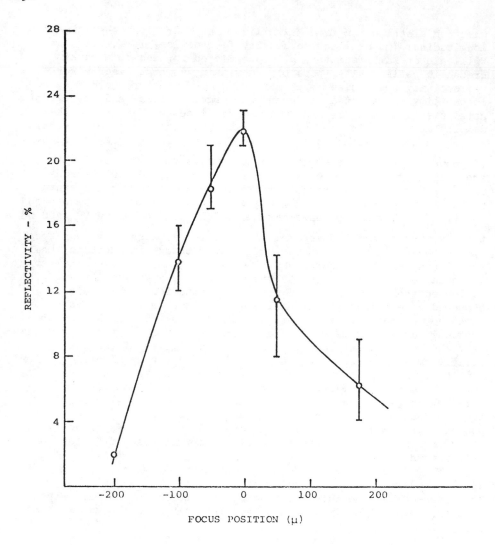

FIGURE 18 – CD_2 reflectivity vs. focus position for laser
energy of 100 to 110 joules.

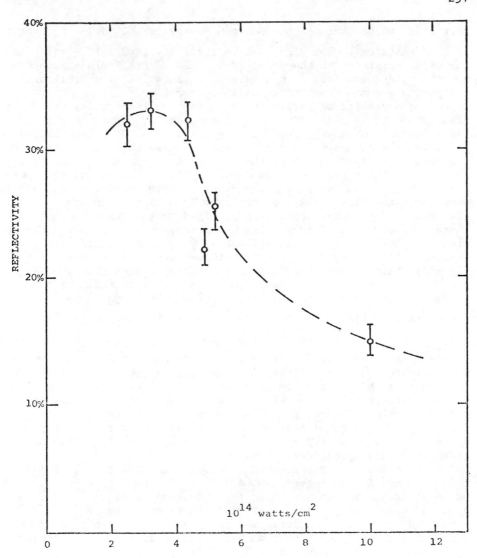

FIGURE 19 - CD_2 reflectivity vs. laser power on target.

factor of two to three at the maximum energy of 160 joules on target or 10^{15} watts/cm^2. The high peak reflectivity at 3×10^{14} watts/cm^2 may be associated with the onset of stimulated Brillouin scattering and the subsequent drop to saturation of the stimulated Brillouin scattering and the onset of anomalous absorption.

The variation of reflectivity during the laser pulse also clearly shows the onset of markedly increased absorption at high laser power. For energy less than approximately 40 joules, the reflected signal has the same time variation as the laser pulse. As the power increases, however, the reflected signal shows increased absorption, the signal becoming markedly distorted and the reflected power appearing to saturate. The details of the reflected pulse show some pulse-to-pulse variation, probably associated with small displacements of the target position relative to the laser focus. A characteristic pulse form at high power shows a strong late reflected signal following a saturated reflectivity plateau. This is an indication of the rising reflectivity with decreasing laser power for the strongly heated plasma produced by the maximum laser power.

We have also observed neutron production associated with a laser pulse directed on a solid CD_2 target. The neutron production has been observed with laser energy delivered to the CD_2 target in the range of 65 to 175 joules, corresponding to a peak power at the target in the range of 4×10^{14} to 11×10^{14} watts/cm^2. The plane target was positioned at the point of maximum reflectivity which also corresponded with the point of maximum production of hard X-rays. The neutron yield was measured with two plastic scintillators heavily shielded against X-rays, placed at 25 to 30 centimeters from the target. The time resolution of the scintillators was calibrated against a plutonium-beryllium source of known strength and against a beta source giving calibration pulses.

The characteristic signals detected consisted of an initial pulse of hard X-rays (several hundred keV) followed by a neutron pulse at 12-15 nanoseconds after the hard X-ray pulse. Delayed neutrons were also observed following the X-ray pulse by 60 to 70 nanoseconds. In several cases the delayed neutron signal was much stronger than the prompt neutron signal and occurred with a time spread of tens of nanoseconds.

The prompt neutron signals were not observed with laser energy under approximately 50 joules on target and were not always seen at the maximum energies of 175 joules on target. The strongest prompt pulses observed corresponded to several neutrons on the detector. From the detector sensitivity and geometry we infer a total neutron production of 4000 to 8000, with a weak dependence on laser energy. The time delay of 12-15 nanoseconds from the hard X-ray signal is in agreement with the transit time of the neutron from the target to the detector; we conclude that these neutrons are produced by DD reactions in the target.

The response of thick deuterated polyethylene foils to the KMS Fusion CGE laser has been computed in two-dimensional cylindrical geometry. The single fluid, two-temperature plasma model is used.

Thermal conduction and electron-ion energy exchange are included. Shocks are treated by introducing a von-Neumann-Richtmyer artificial viscosity.

The computed peak electron temperatures, θ_e, and neutron production, N, for laser energy outputs, E_L, of 100j, 200j, and 300j are presented in Table IV for various spot sizes, R_0. (The spatial distribution of the radiation is taken to be proportional to $\exp[-(r/R_0)^2]$.)

TABLE IV - CD_2 FOIL RESPONSE

Pulse Width (ns)	E_L (j)	R_0 (μm)	θ_e (keV)	N
3	100	50	1.2	$8 \cdot 10^2$
3	200	40	1.55	$6.5 \cdot 10^3$
1.2	100	30	2.0	$4.5 \cdot 10^3$
1.2	200	30	2.4	$3 \cdot 10^4$
1.2	300	30	2.56	$9 \cdot 10^4$

Although the computed peak electron temperatures range from 1.2 to 2.56 keV, the peak ion temperatures range from 0.4 to 0.7 keV. Peak ion temperatures are typically obtained in a region, between the thermal conduction front, where the plasma density is of order 0.02 gm/cm^3. The computed neutron outputs of several hundred to a few thousand neutrons for 3 nanosecond pulses containing 100 to 200 joules are in agreement with the prompt neutron measurements. The data does not indicate any need to invoke any anomalous ion heating effects. The calculated plasma distribution at peak conditions for the 200 joule, 1.2 nanosecond, 30 μm case is plotted in Figure (20).

We have also carried out a series of experiments to determine the role of a background gas on neutron production from laser heated CD_2 spheres. The targets were typically 100 microns in diameter mounted on 10 micron glass fibers. The total energy delivered on target was 300-450 joules. The gas in the chamber varied from 3-60 torr of deuterium and 30 torr of hydrogen was also used. Previous measurements on CD_2 spheres had been done with a vacuum of 10^{-5} torr. The neutrons were measured using plastic scintillators at 40 centimeters and 66 centimeters from the target. A 7" plastic scintillator was positioned at distances from 1 meter to 3.5 meters from the target. A silver activation counter at 40 centimeters was used to obtain the integrated yield. The neutron yield with no background gas present in the target chamber was 5.2×10^3 neutrons. With 10 torr of deuterium added to the chamber the yield was 1.5×10^5 neutrons.

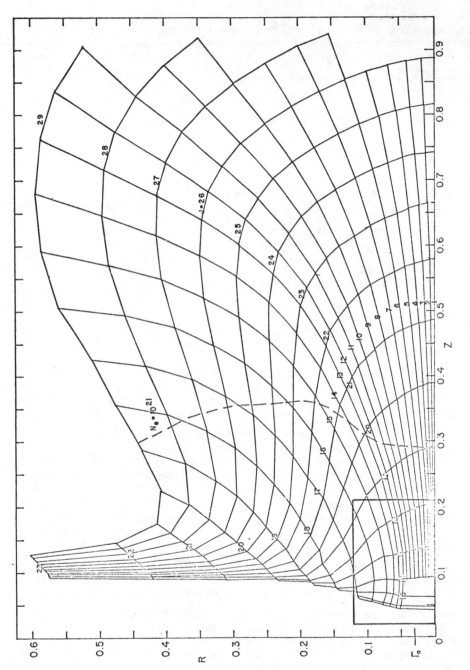

FIGURE 20 – Computed plasma motion for a plane CD_2 target irradiated by a laser beam with 30 micron radius.

The highest yield obtained using 30 torr of deuterium was 5×10^5 neutrons as observed by the three scintillators.

When 30 torr of hydrogen was used in place of the deuterium, the yield dropped to less than our detection sensitivity or less than 10^3 neutrons. An experiment using a CH_2 target in 30 torr of deuterium produced 10^5 neutrons.

There was a series of measurements in which the largest scintillator was placed at various distances from the target. The correct time of arrival of the leading edge of the neutron pulse should be 166 nanoseconds for 3.5 meters. Three of these experiments showed the leading edge of the neutron pulse arriving 11-19 nanoseconds early. The remaining 5 measurements showed that the leading edge of the neutron pulse was 5-17 nanoseconds late. The error in this measurement is less than 5 nanoseconds. The neutrons can arrive early because of the kinetic energy of the fast deuteron in the reaction. This corresponds to a deuteron energy of 20-30 keV.

With the focused surface illumination used, the spherical targets were expected to resemble foils in their response to irradiation. The yield of 5.2×10^3 neutrons in vacuum is in substantial agreement with the prediction of 10^4 neutrons from two-dimensional calculations of the yield expected from two-sided illumination of a thick CD_2 foil.

The (two-dimensional) calculations show that an approximately isothermal expansion develops, with the relatively cold ions accelerated by the electron pressure in the electrostatically neutral plasma. As a result, a small fraction of the ions have directed velocity considerably greater than thermal. The number of deuterons with energy in the 15-20 keV range is of order 10^{15}. The charge exchange cross-section for 15 keV deuteron ions traversing D_2 gas is 4×10^{-16} cm^2 per atom. This cross-section is larger than the effective cross-sections for stopping of D^+ by electron-ion collisions, and for elastic D^+ - D scattering so that a large fraction of the 15 keV deuteron ions will charge exchange and penetrate residual gas as neutrals. The D+D $\longrightarrow$ n+He^3 cross-section is 2×10^{-28} cm^2 for 15 keV deuterons. A residual gas pressure of 30 torr of D_2 corresponds to 10^{18} deuterons per cm^3. The neutron production in the region between the target and the focusing lenses, which were approximately 8 cm from the target is thus of order 10^6 neutrons. The time duration of the neutron pulse is the transit time of the deuterons to the lens, which is approximately 50 nanoseconds.

The results from the experiments with CH_2 pellets in a D_2 atmosphere indicate that a considerably smaller number of fast deuterons are produced, probably by mixing of deuterons and protons at the CH_2 - D_2 interface, leading to an acceleration similar to the CD_2 target. Fast deuterons can also be produced by recoil from coulomb collisions with the fast carbon ions. Once accelerated, the deuteron ions can in turn charge exchange and penetrate the residual gas.

In anticipation of success in the breakeven experiments, much attention has been given to the problem of producing net energy gain in a laser-driven fusion reactor. This problem is made more acute by the unavoidable inefficiency of the laser in producing high compression and heating in a pellet. To overcome this inefficiency alone, the fusion process in the pellet must multiply the energy initially supplied to the pellet in the compression process by a factor of $(0.08)^{-1} = 12.5$. Other inescapable efficiency factors, however, greatly increase the required fusion energy gain. For practical applications, the fusion energy produced in the form of kinetic energy of neutrons, plasma, and radiation must be converted into useful electrical energy. Since 80% of the fusion energy from the D-T reaction is in the kinetic energy of the neutrons, the energy can be recovered only by moderation of the neutrons and subsequent removal of the thermal energy in a thermal-electrical cycle. This process has an efficiency limited by the characteristics of materials to about 50%. The electrical energy so produced must supply the internal operating energy requirements of the fusion reactor, and provide an excess to be available for external use. The principal operating requirement is in driving the laser system which in turn provides the initiating energy for the fusion energy release.

The lasing process itself is relatively inefficient, particularly for the familiar glass or crystal lasers now widely used for commercial and scientific work. For these lasers, driven by flash lamps, the efficiency is very low, only a fraction of one percent of the electrical energy supplied to the flash lamps being released as laser energy. Such lasers would not be suitable for the driving elements in a fusion reactor. Other lasing media can be excited much more efficiently, with efficiencies above 50% being reached in some gases. There appears to be no reason in principle why a gas or chemical laser for the fusion application could not operate at reasonably high efficiency. Conservatively we assume 20% laser efficiency. The efficiency of conversion from output fusion energy to laser output energy, therefore, is 10%. Combining this with the 8% efficiency of coupling from the laser into the compressed pellet shows that the fusion process in the pellet must produce about 110 times the energy transferred from the laser into the pellet for the process to be self-sustaining, i.e., to operate a zero-power fusion reactor. For useful energy to be produced which is comparable to the energy circulated internally to drive the laser system, the energy multiplication by the fusion process must be by a factor of about 238. The details of the energy flow in the system are indicated schematically in Figure (21). The large energy multiplication required for a practical application is a very severe requirement and possible only if the energy required to initiate the fusion reaction is held as low as possible and a substantial fraction of the pellet is burned. These conditions can be met in highly compressed pellets ignited by a centrally initiated burning wave, the phenomena representing only a moderate scale-up from those occurring in the pellet in a breakeven experiment.

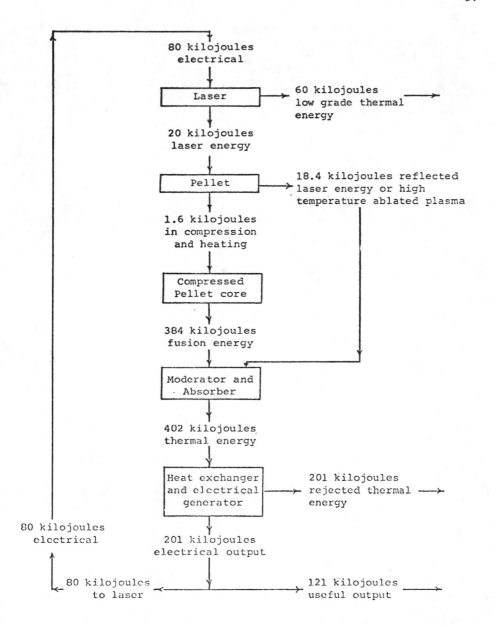

FIGURE 21 - Energy flow in a low power laser-driven fusion
reactor. The laser has been assumed to have
an efficiency of 25% in conversion of electrical
energy to laser output.

The coming experiments promise an exciting test of the ability of the scientist to analyze and predict the highly complex phenomenon of fusion burning. The experiments will bring into the laboratory a combination of densities and temperatures previously existing only in the center of stars. If successful, the experiments will be a most important first step toward the eventual production of electrical power by nuclear fusion.

ELECTRON-BEAM-INDUCED FUSION[*]

Gerold Yonas

Sandia Laboratories, Albuquerque, New Mexico 87115

ABSTRACT

The use of tightly focused relativistic electron beams has only recently been seriously considered as an alternative to lasers as a means of heating and compressing matter for achieving fusion. The potential attractiveness of this approach stems from the demonstrated high efficiency, intrinsic simplicity, and the ability to scale pulsed power and electron beam technology. In particular, the foundation has already been prepared to allow us to embark on the development of a 10^{13} W pulsed electron accelerator to deliver 10^5 J or more to a spherical target. Although such a device will in the future allow us to test more fully the feasibility of the concept, many of the fundamental questions related to the principles underlying this idea present exciting challenges to the physics community and can be tested with considerably smaller facilities. The important questions include the limitations of electron beam focusing, symmetry of irradiation of mm size pellets, and energy deposition. Vital tools in addressing these questions are (1) the existing $\sim 10^{12}$ W pulsers which are being used to determine the properties of focused beams as well as the response of materials subjected to the high energy densities which can be created, and (2) numerical methods for understanding the behavior of these beams. Results to date include the successful focusing of electron beams to provide power densities of $\sim 10^{13}$ W/cm^2, initial studies of energy absorption, and numerical studies complementing these experiments. Recent progress as well as areas for further study, particularly oriented toward the creation of reactor concepts will be delineated.

INTRODUCTION

Of the various concepts being presented at this conference as possible solutions to the world's long-range energy needs, fusion is thought by many to be the most promising, to have the greatest potential for preserving our environment, but to be the most difficult to achieve. Of the approaches to fusion which we are hearing about, most of us would agree that the one I will discuss is the most

---

[*]This work supported by the U. S. Atomic Energy Commission.

speculative. There is no question but that as one of the newest
approaches to be presented it has only begun to attract fairly widespread attention.[1-3] I will point out the known problem areas, but
since this approach is still in its infancy, it is probable that I
will have missed some difficult aspects and will have only identified
those most readily apparent. In this sense, I look forward to the
physics community for not only constructive criticisms, but also
contributions to the challenging questions raised by this approach.

Electron beam heating and compressing of DT has been proposed
as a method for achieving substantial thermonuclear yields. This
effort has been motivated by a desire to complement the laser fusion
program, to provide intense radiation sources for fusion reactor
material studies, and to permit laboratory studies of the physics of
thermonuclear ignition and burn at much higher yield levels than
would otherwise be possible. As a longer range goal, these studies
could lead to an alternate approach to a practical fusion reactor.
The reason for this optimism in pursuing such a course of study is
first, the near-term potential availability of pulse power and
electron beam technology in order to deliver beams in the 100 kJ
to 1 MJ range with efficiencies in excess of 30%, and second, the
recent theoretical and experimental breakthroughs in the physics of
beam focusing.

Although I will concentrate on the inertial confinement approach,
I should first briefly mention the possible application of pulsed
relativistic electron beams to providing the additional energy needed
to elevate the temperature of a magnetically confined plasma. If
one is to succeed with this approach then the following problems
must be solved: (1) injection of the beam into a magnetic confinement system, (2) efficient coupling of the beam to the plasma before
the beam can escape, and (3) avoiding the excitation of MHD instabilities which could destroy the confinement times which have been so
difficult to achieve.

Although this auxiliary heating approach to magnetically confined systems has been suggested fairly widely, there have been no
attempts, as yet, to heat a plasma confined within a toroidal device.
On the other hand, the recent experiments with neutral particle
sources[4] do seem to be providing promising results in this direction.
Nevertheless, since electron beam generators presently exist in
the terrawatt power range, their application should be considered
and it is clear that research into the beam plasma interaction, beam
trapping, effects of the beam on gross stability of the confined
plasma, and most important, injection, is warranted.

Although the magnetic confinement of plasmas for fusion presently
represents the most widely accepted course of study, the alternative
is to employ inertial confinement and here the application of electron
beams is in many ways very similar to the laser fusion methods already
discussed at this meeting. In a greatly over-simplified view, the
beam is to serve as the source of ignition for the DT fuel pellets,
which are injected repetively into a combustion chamber. As in the
laser approach, a lithium blanket fairly close to the reaction area
would serve as the medium for energy conversion and tritium breeding.
We will not deal with the specifics of a reactor concept, for in

fact, these specifics do not as yet exist, but will concentrate our attention primarily on energy requirements and what this implies for beam focusing, deposition, and accelerator development.

DISCUSSION

The requirements to be satisfied by relativistic electron beam technology in order to uniformly heat and ignite DT without compression are truly awesome.[5] Ignition of DT pellets with electron beams, as with lasers also, really did not get any serious attention until it became clear that initiating a fusion reaction at solid density would not be the optimum approach.

In particular, the single most important factor that changed the level of interest in inertially confined systems was the realization of the rather impressive decreases in the breakeven energy requirement that could be obtained by compressing the DT fuel many orders of magnitude beyond solid density and employing propagating burn.[6-9] In this way, the energy requirements were predicted to decrease as the square of the density increase. The problems as presented by the goal of super compression for electron beam technology are first, to produce efficient, megajoule output, 10^{13} - 10^{14} W electron accelerators and second, to find a way to couple energy efficiently into the DT fuel in such a manner as to achieve the required high degree of compression at the required ignition temperature.

In fact, a major reason for our interest in electron beams is the recent track record demonstrated by this technology which was initiated in the mid-sixties by J. C. Martin at the AWRE in Aldermaston, England.[10] In Figure 1 we see a pictorial representation of the history of the U. S. efforts to develop large pulsed electron accelerators, primarily for applications to the study of radiation effects in materials. These accelerators are now in operation on a routine basis at several laboratories in the United States and an aggressive program to develop similar capabilities is underway in the USSR.[11] Clearly, this technology has already been rather extensively developed although major improvements can still be made in energy storage (low inductance capacitor banks or inductive storage), improved properties of dielectrics in pulse forming lines, synchronized switching, and diode design. Nevertheless, the technology as depicted here has proven to be comparatively simple, inexpensive, and scalable to higher

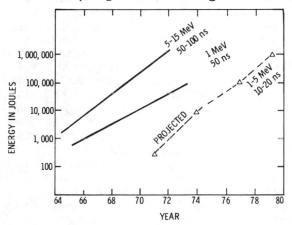

Fig. 1. History of development of relativistic electron beam generators and projected growth for shorter pulse technology.

powers. Sandia Laboratories is presently engaged in a program to apply the state-of-the-art in pulsed power to a class of accelerators which will operate in the 10-20 ns regime, and Figure 1 shows a growth projection for these accelerators as compared to previous developmental history.

The first question in any discussion of pellet fusion is to determine the specific parameters required to achieve DT burn in the super compressed fuel. One avenue to super compression which has appeared frequently in recent literature is to use a hollow high density shell to compress and heat a DT gas mixture within that shell. Based on this concept, Rudakov[12] at the Kurchatov Institute was able to predict breakeven for a constant power pulse and by considering only classical deposition in the high density shell. Other authors including Maxwell,[13] Kaliskii,[14] and Linhart,[15] have considered similar ideas. The consensus appears to be that the energy required for breakeven using electron beams lies in the 1 megajoule region.

Of even more importance to us than scientific breakeven, is the question of economic power generation. This consideration, in the case of lasers, has resulted in the requirement for a gain factor (ratio of energy released to that absorbed by the pellet) of ~ 75 in order to operate a power plant with 30% recirculating power and employing the commonly projected (although yet to be achieved) 10% efficiency of delivery of energy from the energy store to that absorbed in the pellet. This high gain then creates rather severe material problems for any reactor concept. As a result, the combustion chamber must withstand an energy release of roughly 100 million joules and this requires rather innovative approaches to design of the first wall of the chamber and the last element of the optical train delivering the power. One solution presented has been to remove the last optical element to a distance of 10 meters from the pellet[16] thus minimizing the material problem but creating a much more difficult problem for focusing and steering of the laser beams.

Similar considerations play a role in electron beam fusion. In particular, there is good reason to expect that megajoule electron beam accelerators with a 10 ns pulse duration can be developed with a more efficient conversion of stored energy to beam energy than for sub-nanosecond short wave length lasers. This optimism is warranted because of the already-mentioned fact that pulsed relativistic electron beam technology has already made significant advances in a rather short time. In addition, at least one multi-megajoule relativistic electron beam accelerator is already operating at a 50% efficiency level.[17] In principle, the 50% efficiency level is a reasonable goal thereby resulting in a required gain factor of 15.

Although there have not as yet been extensive studies of the input energy requirements to achieve such a gain factor under the constraints of electron deposition, preliminary calculations indicate that this may require a few megajoules delivered in the time period of ~ 10 ns onto a several millimeter diameter sphere.[18] This would then require a current density of several times 10^8 amps/cm^2 for a beam in the MeV range. With a projected thermonuclear energy release of tens of megajoules, this current density must be delivered to a pellet removed at some distance from the nearest materials if these materials are to survive. If the self-magnetic field of the beam can be used to constrain the pellet debris so as to not intercept with the closest material surfaces, then the most important source of damage will be the radiation flux from the pellet. If low atomic number materials are used to minimize the peak dose from the X-ray flux, and possibly by vapor resurfacing of these materials between shots, then the point of closest approach might be in the range of a fraction of a meter.

The question, then, is to find a way to transport one or more beams to the target in such a way as to deliver the required current density to the pellet in a reproducible and controlled manner. This rather formidable requirement has undoubtedly been the major drawback to electron beam fusion and has recently attracted the most attention.

There have been two basic approaches suggested to achieving this goal, either drifting the beams some distance from the accelerating diode, or to accomplish the beam convergence within the diode itself. One approach to beam drifting is to transport multiple beams at lower current densities than required and to focus the separate beams onto a target without such beams interacting.[19] Babykin and Rudakov[20] have suggested transport of an already tightly focused beam within a small plasma channel to the target. Although there has already been rather extensive research in the area of transport of high current relativistic electron beams, considerable additional effort would be needed to attempt either approach. The lack of real advances in drifting beam compression, in spite of the considerable effort already expended,[21-30] together with major recent breakthroughs in beam focusing within diodes, has led us at Sandia to concentrate our attention on the diode itself.

The first real advance in beam focusing began in 1971 when Willard Bennett and his coworkers[31] observed that a small diameter cathode when placed within a distance of less than a centimeter from an anode in a multimegavolt electron beam generator would produce a millimeter size beam. Later Condit and Pellinen[32] confirmed the earlier observation. The actual pinching mechanism was not well known at first, although extensive studies by Mesyats at the Tomsk Institute,[33,34] and later work by Parker[35] had already shown that plasma formation at electrode surfaces in vacuum diodes played an important role in impedance behavior. As a result of experimental and theoretical studies at Sandia, our understanding of this beam pinching phenomena has grown steadily during the last year. Bradley and Kuswa[36] employed streak photography and later Mix and Kelly[37] used holographic interferometry to investigate the properties of plasma generated at the electrode surfaces in a similar diode. They found that the beam pinching occurred as a result of the initial vacuum diode becoming filled with plasma at densities $> 10^{17}$ cm^{-3}. In particular they showed that the time of beam pinching corresponded to the time when a cathode plasma which had accelerated to velocities of up to 10 cm/μsec came in close contact with the more slowly moving anode plasma (Figure 2). The actual mechanism for the acceleration of the cathode plasma (which may also have been seen earlier by Mesyats' group[38]) is poorly understood but the existence of this

304

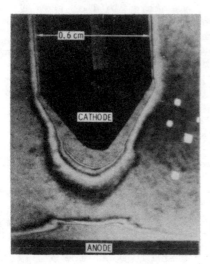

Fig. 2. Interferogram of vacuum diode showing plasma distribution after 60 ns (2 MV, 50 kA peak voltage and current).

plasma is clearly responsible for the beam pinching.

In order to model the effect of this plasma on beam dynamics, Poukey and Freeman[39] drew upon the fact that the diode impedance at the time of this tight pinching was still high ($\sim$ 10 Ω), even though the interferometry showed that the diode had become filled with plasma at densities greater than $10^{17}/cm^3$. They postulated that the plasma had become turbulent, thereby creating a uniformly resistive, charge neutralizing region between the cathode and the anode. Although no theoretical model exists to explain the plasma resistivity, they used a simplified treatment in which they assumed that the plasma provided for space charge neutralization but no current neutralization. In this way, they were able to solve numerically for the particle motion in a self-consistent manner in order to predict the existence of a tight pinch as a result of the combination of the applied electric field and the self-magnetic field of the beam (Figure 3). J. D. Lawson[40]

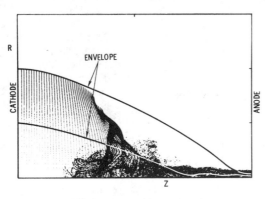

Fig. 3. Numerical simulation and envelope solution for 2.4 MeV, 54 kA beam accelerated in charge neutralizing plasma.

had earlier predicted such self-pinching of a space charge neutralized beam in a longitudinal electric field, but with far more restrictive conditions. Later Poukey and Toepfer[41] employed a model in which they ascribed an isotropic pressure tensor to the beam treated as a relativistic fluid and found good agreement with the numerical calculations. In effect, they found that the kinetic energy acquired by the electrons in acceleration across the anode-cathode gap was distributed primarily in transverse motion. As a result of this "beam heating", the magnetic field required to achieve transverse equilibrium must increase, and the beam radius therefore decreases as the beam heats up. If the beam does in fact have the behavior of a hot electron gas, then one could envision producing symmetric irradiation of a pellet by placing the target in a common anode between two cathodes (Figure 4). For the case of a target embedded in the anode

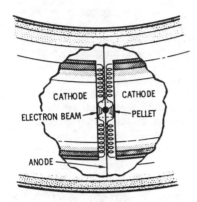

Fig. 4. Double diode configuration for symmetric irradiation of a pellet. Radial electron drift in vacuum diode is explained below.

(or to be more precise, in an electrically conducting anode plasma) one makes the problem of a fusion reactor design more difficult, but hopefully this approach will reduce the difficulty of delivering the beam in the desired manner to the target.

Let us next consider the possibility of using this double diode approach. In order to deliver a 10^{12} - 10^{14} W beam from a single generator one would choose the highest possible impedance to minimize the inductive risetime. If we consider various diode voltages of from .3 to 3 megavolts, we see that this will result in a maximum allowable diode inductance of from .01 to 100 nanohenries to achieve a pulse risetime of 5 nanoseconds assuming that the risetime limitation is the diode itself and not in the pulse forming line (Table I).

The requirement of .01 nH results from attempting to deliver 10^{14} watts with a .3 megavolt beam and will undoubtedly be prohibitively low. For such a case we would certainly have to consider multiple diodes and beam propagation. On the other hand, if we restrict ourselves to greater than .3 megavolts, and greater than 10^{13} watts, we see that the inductance must

Table I. Maximum diode inductance to achieve a 5 ns risetime

	.3 MeV	1 MeV	3 MeV
10^{12}W	1 nH	10 nH	100 nH
10^{13}W	.1 nH	1 nH	10 nH
10^{14}W	.01 nH	.1 nH	1 nH

be in the range of 0.1 to 10 nanohenries. With such extremely low inductances, we are therefore prohibited from using a single small diameter cathode within a larger diameter diode. The diameter of the diode envelope will be determined from the limitations in the power flow from the dielectric in the pulse forming line through the vacuum envelope. As a result, this provides a minimum width of the line feeding the periphery of the tube. Although there is, at this time, little data for the allowable fields in liquid dielectrics for pulses in the 10 nanosecond range, one can extrapolate existing data[42] giving at most 0.5 Ma/m of tube circumference using transformer oil and at least 3-4 times greater current per unit line width from high purity water. For accelerators delivering at least 10^{13} W, the tube radius would be roughly one meter, and as such the insulator separating the liquid dielectric from the vacuum diode would not present a substantial inductance if fields on the vacuum side of the insulator of 100-200 kV/cm could be employed.

The major difficulty of course would be the inductance of the vacuum diode itself and one way to achieve the low inductance would

be to generate one beam or multiple beams close to the insulator surface and then drift these beams to the target outside of the diode.
On the other hand, if beam transport is to be avoided, then one must
use a large diameter cathode with a small anode to cathode gap and
produce a pinched beam at the anode.

We have used the idea of the electron beam pinching within a
resistive plasma on the diode axis but have developed a hybrid concept to also achieve the requirements of low inductance and low
impedance.[39] We have proposed a large diameter cathode with plasma
injected near the diode axis to achieve the desired goals. The
concept is shown in Figure 5 where we rely on self-pinching of the
beam from the edge of the
cathode toward the axis when
the total beam current exceeds
a critical value. This critical
current level[43] corresponds to
that at which the Larmor radius
of the electrons at the edge of
the cathode is less than the
anode-cathode spacing. When
this current is reached, the
beam pinches strongly and self-
consistent numerical calculations[44] (Figure 6) show that
the electrons drift radially
after they reach a region near
the anode and focus near the
axis of the diode. An important
assumption in these calculations
is that a layer of conducting
plasma is either created by
electron bombardment of the
anode early in the pulse, or
is created beforehand by
injection of a plasma into the
anode region. In addition, by
introducing a resistive plasma

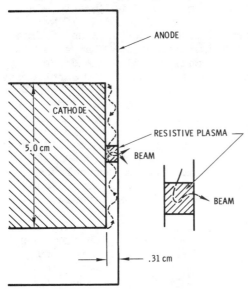

Fig. 5. Schematic representation
of beam focusing in a high current
diode using resistive plasma near
diode axis.

near the diode axis to eliminate the space-charge repulsion which
would normally restrict the size of the pinch, we have been able to
accomplish focusing to current densities in the range of 10^7 A/cm^2
with single diodes of impedances as low as 2 ohms.

The first such experiments were carried out with a 100 J beam
using a 5 centimeter diameter cathode and with an axis plasma produced by an exploding fine tungsten wire.[39] This work was then scaled
over a series of steps, first to 1000 J, and recently to the 10,000 J
level using a 1 MV accelerator (Hydra[45]). The results of these experiments indicate a current density in the range of 5-10 megamps/cm^2
with the profile shown in Figure 7.

The most promising method for reproducibly achieving the required
plasma conditions appears to be through injection of a plasma via
laser blowoff from a target inside the cathode.[46] (See Figure 8.)
Using this technique Miller and Chang[47] have recently been able to

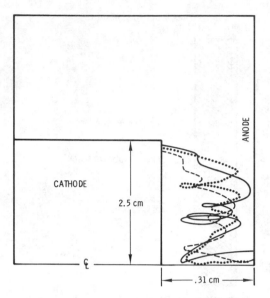

Fig. 6. Numerical simulation of electron trajectories in a 200 kV, 100 kA diode.

achieve current densities of 10^6 - 10^7 amps/cm^2 with a total delivered energy of approximately 10^4 J.

One of the greatest concerns in using this, or indeed any plasma injection technique, has been that the plasma densities required to achieve focusing of beams to current densities of $\geq 10^7$ amps/cm^2 could have resulted in a sufficiently low plasma resistivity to effectively provide a short circuit in the diode. We have found however, that plasma densities of $> 10^{15}$ cm^{-3} can be created in the diode and we can still obtain impedances of at least a few ohms. We have thus shown that the coupled phenomena of self-pinching of a high current beam in a vacuum diode, together with further compression within a plasma on axis, can lead to current densities in the 10 MA/cm^2 range. In order to study the scaling of this phenomena further to an order of magnitude higher current density, a higher power accelerator ($> 10^{12}$ W) is presently under construction and a 10^{13} W accelerator (Ripper) is being designed.

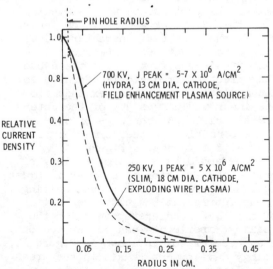

Fig. 7. Current density profiles from x-ray pinhole radiographs.

If these steps succeed, thus permitting us to produce large fusion yields, then the question still remains as to the survivability of the cathode. If a power plant must operate with an energy release per pulse of several million joules, then certainly a cathode spaced close to such an energy release would be destroyed on each shot. One must therefore find a way to separate the emitting area from the pellet itself. A possible solution to this question has arisen through our experimental and theoretical studies.

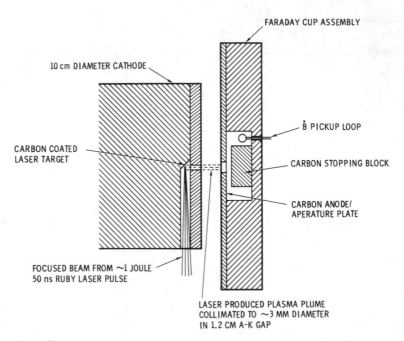

Fig. 8. Apparatus employed in ~ 700 kV, ~ 300 kA diode
to focus 10 kJ beam.

We have found that beam pinching could be accomplished even
though the cathode was rather hollow. In particular, we observed
experimentally that with a 12.7 cm outer diameter and a 7.6 cm inner
diameter cathode operating at 700 kV, that the beam characteristics
were similar to that generated with a solid cathode. Numerical
calculations showed that a virtual cathode was formed near the diode
axis by space charge giving similar electron trajectories for the solid
and hollow cases. The numerical calculations were then extended to
the case of a hollow cathode whose inner radius was 90% of the outer
radius. It was found that the radial drift from the edge of the
cathode toward the axis was still responsible for the formation of a
beam as tightly pinched as for the solid cathodes. (Figure 9)

In order to use this concept, one would have to show scaling of
the cathode outer radius to a substantial fraction of a meter in order
to survive the projected thermonuclear energy release. Such scaling
experiments are planned for the near future but the ultimate test
will come with the development of a 10^{13} W accelerator in which the
first electron beam fusion feasibility experiments can be carried out.
An early design concept for such an accelerator is shown in Figure 10.
The major requirements for technology development remain in the areas
of synchronized switching and diode design, and design concepts are
to be tested within the next year.

Thus far, we have said little of deposition, although considerations of power flow and inductance strongly urge us to employ beams
of ≥ 1 MeV kinetic energy. Because of collisions in solid targets
and the wide angular spread of highly focused beams, it is unlikely

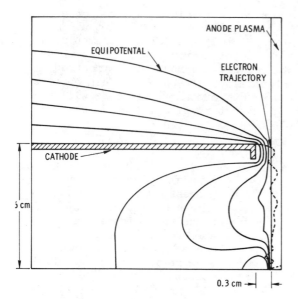

Fig. 9. Numerical simulation of
typical electron trajectory in a
700 kV, 470 kA primarily hollow diode.

that anomalous absorption
processes will be
important in the solid.
On the other hand, the
low density blowoff region
could serve as a partial
barrier to beam penetration
as a result of excitation
of two-stream or return-
current driven ion acoustic
instabilities.[48],[49]
Another effect that might
be important could be self-
magnetic stopping of the
beam in either the blowoff
or in the solid target.
This turning back of
electrons because of their
self-magnetic field will
occur only if the field
of the beam which exists
in the unneutralized
medium in the diode can
diffuse into the target
region and also if the

collisional scattering is less important than the effect of the self-
magnetic field.[50] In order for the field to penetrate to a depth of

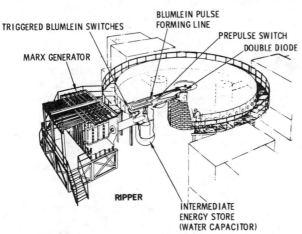

Fig. 10. Design concept for a 3 MV, 3 MA,
20 ns electron beam generator for use in
fusion feasibility studies.

one Larmor radius (the
magnetic stopping dis-
tance) the plasma
resistivity would have
to be anomalously high
as from beam-generated
plasma turbulence.

Finally, the most
significant effect will
probably be classical
scattering, and Monte
Carlo calculations have
already shown us that
rather substantial
decreases in the electron
penetration result from
having a beam of
predominantly transverse
energy. In such cases,
a large fraction of the
incident beam is

scattered from the front surface and the actual deposition profile
will, to a large extent, depend on the fate of these electrons in the
self-magnetic field of the beam and the applied longitudinal electric
field in the diode. Clearly, experiments will be needed to define the

actual deposition profile and experiments involving the response of thin films are now beginning.[51] Initial experiments have thus far consisted of comparisons of craters formed in thick aluminum plates with the predictions of a two-dimensional hydrodynamic code[52] thus allowing us to determine the total energy absorbed in the focused beam (Figure 11).[53] Although no anomalous effects have been observed, it is expected that optical measurements of target response will be required to define the deposition characteristics.

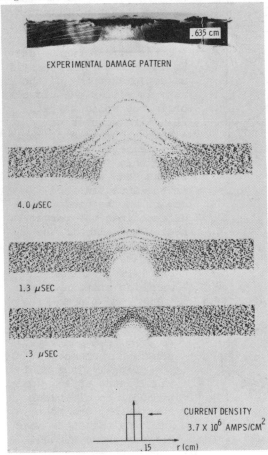

Fig. 11. Numerical simulations compared with actual damage pattern in an aluminum plate indicating 10 kJ deposited in target.

CONCLUSIONS

If the important question of scaling of the focusing process, with particular emphasis on production of low inductance diodes, is answered in a positive manner, then our degree of success in providing efficient absorption with 1 MeV beams and symmetric irradiation will determine whether or not we can achieve impressive thermonuclear yields. Further, if this concept is to proceed to a reactor prototype, then high gain pellet designs and beam focusing employing either a hollow cathode with an outer radius of a substantial fraction of a meter or by beam transport beyond the diode over a similar distance, would have to be demonstrated in order for the cathode to survive a projected yield of tens of megajoules. Further questions of course would have to deal with the matter of pellet injection and cost, efficient beam generation, high repetition rate switching and diodes, and integration of the machine concept with a lithium blanket and heat transfer system.

As one can see, the question of producing DT burn with a megajoule output does have serious problems, but solutions to these problems do present themselves. If support is provided for such an ambitious goal, then experiments in this regard could be carried out before the end of the decade. The challenge of developing a practical fusion reactor, on the other hand, presents much more severe problems, particularly in material survivability, and these problems would clearly tax the ingenuity of physicists and engineers for many years to come if the first step of scientific breakeven is demonstrated.

ACKNOWLEDGMENTS

I would like to thank my colleagues, J. Chang, M. J. Clauser, D. L. Johnson, J. R. Freeman, J. G. Kelly, G. W. Kuswa, T. H. Martin, P. A. Miller, L. P. Mix, J. W. Poukey, K. R. Prestwich, D. W. Swain, A. J. Toepfer, and M. M. Widner for their continuing creative contributions to this research and to E. H. Beckner and A. Narath for their enthusiastic support of this program.

REFERENCES

1. "Many Laboratories Try for Fusion with Electron Beams," Phys. Today, pp. 17-20, April 1973.
2. G. Yonas, J. W. Poukey, J. R. Freeman, K. R. Prestwich, A. J. Toepfer, M. J. Clauser, and E. H. Beckner, Proc. 6th European Conf. on Controlled Fusion and Plasma Phys., Moscow, Russia, July 30-August 4, 1973, pp. 483-486.
3. L. I. Rudakov and A. A. Samarsky, pp. 487-490, Ibid.
4. L. D. Stuart, et al., Bull. Am. Phys. Soc. $\underline{18}$, 1271 (1973).
5. F. Winterberg, Nucl. Fusion $\underline{12}$, 353 (1972).
6. J. Nuckolls, L. Wood, H. Thiessen, and G. Zimmerman, Nature $\underline{239}$, 139 (1972).
7. J. Nuckolls, J. Emmett, and L. Wood, "Laser-Induced Thermonuclear Fusion," Phys. Today, August 1973.
8. J. S. Clarke, H. W. Fisher, and R. J. Mason, Phys. Rev. Letters $\underline{30}$, 89 (1972).
9. K. A. Brueckner, Plasma Sci. $\underline{PS-1}$, No. 1, 13 (1973).
10. J. C. Martin, AWRE, Aldermaston, England, private communication.
11. L. I. Rudakov, et al., invited paper, 6th European Conf. on Controlled Fusion and Plasma Phys., Moscow, Russia, July 30-August 4, 1973.
12. L. I. Rudakov and A. A. Samarsky, See Ref. 3.
13. D. Maxwell, Advanced Elite Concepts, PIIR-41-67, Physics International Company, San Leandro, California (1967).
14. S. Kaliskii, Bull. de L'Academie Polanaise Des Sci. $\underline{21}$, 307 (1973).
15. J. G. Linhart, Nucl. Fusion $\underline{13}$, 321 (1973).
16. K. Boyer, Astron. and Aeron., p. 44, August 1973.
17. B. Bernstein and I. Smith, IEEE Trans. of Nucl. Sci. $\underline{3}$, 294 (1973).
18. M. J. Clauser, Sandia Laboratories, Albuquerque, New Mexico, private communication.
19. L. P. Bradley, Panel on Intense Electron Beam Application to CTR, AEC, Division of Controlled Thermonuclear Research, Washington, D.C., May 21-22, 1973.
20. M. V. Babykin, Bull. Am. Phys. Soc. $\underline{18}$, 1288 (1973).
21. S. E. Graybill and S. V. Nablo, IEEE Trans. on Nucl. Sci. $\underline{NS-14}$, 782 (1967).

22. W. T. Link, IEEE Trans. on Nucl. Sci. NS-14, 777 (1967).
23. G. Yonas and P. Spence, Proc. 10th Symp. on Electron, Ion and Laser Beam Tech., L. Martin, editor, San Francisco Press, Inc., 1970.
24. D. A. Hammer and N. Rostoker, Phys. Fluids 13, 1831 (1970).
25. M. Andrews, J. Bzura, H. Fleischmann, N. Rostoker, Phys. Fluids 13, 1322 (1970).
26. J. Benford and B. Ecker, Phys. Rev. Letters 26, 1160 (1971).
27. L. S. Levine, T. M. Vitkovitsky, D. A. Hammer, and M. C. Andrews, J. Appl. Phys. 42, 1863 (1971).
28. C. Stallings, S. Shope, and J. Guillery, Phys. Rev. Letters 28, 653 (1972).
29. P. A. Miller, J. B. Gerardo, and J. W. Poukey, J. Appl. Phys. 43, 3001 (1972).
30. C. L. Olson, Phys. Fluids 16, 529 (1973).
31. D. L. Morrow, J. D. Phillips, R. M. Stringfield, Jr., W. O. Doggett, and W. H. Bennett, Appl. Phys. Letters 19, 444 (1971).
32. W. C. Condit, Jr., and D. Pellinan, Phys. Rev. Letters 29, 263 (1972).
33. G. A. Mesyats, E. A. Littvinov, and D. I. Proshvrovsky, Proc. 3rd International Symp. on Discharges and Electrical Insulator in Vacuum, Paris, France, 1968.
34. G. A. Mesyats, Proc. 10th International Conf. on Phenomena in Ionized Gases, Oxford, England, 1971.
35. R. K. Parker and R. E. Anderson, Rec. 11th Symp. on Electron, Ion and Laser Beam Tech., Editor, F. M. Thornly, San Francisco Press, Inc., 1971.
36. L. P. Bradley and G. W. Kuswa, Phys. Rev. Letters 29, 1441 (1972).
37. L. P. Mix, J. G. Kelly, G. W. Kuswa, D. W. Swain, J. N. Olsen, J. of Vac. Sci. and Tech., November-December issue (1973).
38. F. B. Baksht, et al., Proc. V International Symp. on Discharges and Electrical Insulator in Vacuum, Poland, 1972.
39. G. Yonas, K. R. Prestwich, J. W. Poukey and J. R. Freeman, Phys. Rev. Letters 30, 164 (1973).
40. J. D. Lawson, AWRE Memo, GP/M196 (1957).
41. J. W. Poukey and A. J. Toepfer, "Theory of Super-Pinched Relativistic Electron Beams," submitted to Phys. Fluids.
42. J. C. Martin, SSWA/JCM/704/49, AWRE, Aldermaston, England (1970).
43. F. Friedlander, et al., DASA 2173, Varian Associates (1968).
44. J. W. Poukey, J. R. Freeman, and G. Yonas, "Simulation of Relativistic Electron Beam Diodes," J. of Vac. Sci. and Tech., November-December issue (1973).
45. T. H. Martin, IEEE Trans. Nucl. Sci., Volume NS-20, 289 (1973).
46. P. A. Miller, J. Chang, and G. W. Kuswa, Appl. Phys. Letters 23, 423 (1973).
47. P. A. Miller and J. Chang, Sandia Laboratories, Albuquerque, New Mexico, private communication.
48. P. A. Miller and G. W. Kuswa, Phys. Rev. Letters 30, 959 (1973).
49. R. N. Sudan, invited paper, 6th European Conf. on Plasma Phys. and Controlled Nucl. Fusion Research, Moscow, USSR, July 30-August 3, 1973.

50. G. Yonas, et al., "Applications of High Intensity Relativistic Electron Beams to Pulsed Fusion and Collective Ion Acceleration," lectures presented at International Summer School of Appl. Phys., Erice, Sicily, June 4-16, 1973.

51. M. J. Clauser, Bull. Am. Phys. Soc. <u>18</u>, 1355 (1973).

52. S. L. Thompson, <u>CSQ-2D Eulerian Code</u>, Sandia Laboratories Report, to be published.

53. M. M. Widner, J. Chang, and P. A. Miller, Sandia Laboratories, Albuquerque, New Mexico, private communication.

MAGNETIC CONFINEMENT OF THERMONUCLEAR PLASMAS

H. P. Furth
Plasma Physics Laboratory, Princeton University
Princeton, New Jersey 08540

ABSTRACT

Toroidal reactors will require a ~10-keV DT plasma
with $n\tau$ (density times confinement time) exceeding
~10^{14}cm^{-3}sec. Present-day toroidal plasmas come within
2-3 orders of magnitude in respect to $n\tau$, and within less
than an order of magnitude in respect to temperature.
Toroidal loss processes appear to be diffusive, so that $n\tau$
increases with size. In the low-pressure ("low-β") approach, typified by the tokamak, the principal physics problems are: (1) to develop scaling laws for plasma diffusion
(a problem of nonlinear microscopic instability); (2) to
minimize impurity evolution from plasma-wall interactions
(a surface physics problem); (3) to find grossly stable
configurations maximizing $n\tau$ (a MHD-stability problem).
The high-β approach, typified by the toroidal θ-pinch, is
related to item (3). Plasma pressure and density are
raised to their limits by special MHD-stabilization techniques, including feedback. Development of shock and compressional heating techniques plays an important role. An
alternative reactor approach is the open-ended magnetic
bottle, or "mirror machine." A non-Maxwellian (beam-
injected) ion population in the energy range >100 keV interacts with itself or with a plasma target. The $n\tau$-value
depends on velocity-space distribution rather than on size;
present-day $n\tau$-values need to be improved by 2-3 orders of
magnitude. The principal physics problem is anomalous
endloss — a problem of nonlinear high-frequency instability, somewhat like laser action.

I.   INTRODUCTION

The development of fusion power is an ideal occupation for
physicists. Contributions from the entire range of modern physics —
nuclear and atomic physics, the physics of fluids, electromagnetics,
optics, solid-state physics, low-temperature physics — are necessary
to achieve the objective. The social value of the objective is
clear: in the short run, fusion offers power from abundant fuel resources, at roughly conventional levels of cost and waste heat production, and at levels of radioactive waste production far below
those associated with fission power; in the long run, the evolution
of controlled fusion devices will permit a substantial reduction of
thermal pollution, as well as futher decreases in radioactive waste
production. As to the difficulty of the problem, at first it seemed

rather more than a match for the resources of Twentieth Century physics; lately, however, the impedance match has begun to look quite good. There is considerable likelihood that "break-even" conditions in deuterium-tritium can be achieved in the early 1980's.

The present talk will concentrate on the three major approaches to magnetic confinement of thermonuclear plasmas. The illustrative examples are drawn from the U.S. research program; research efforts of similar magnitude and direction are under way in the U.S.S.R., in Western Europe, and in Japan. International cooperation in control-led fusion research has been active and fruitful. It is particularly gratifying that the approach of practical success has been accompanied recently by an agreement to formalize and extend collaboration between the U.S. and U.S.S.R. controlled fusion research programs.

II. GENERAL PRINCIPLES OF MAGNETIC CONFINEMENT

The fusion reactions of principal interest for power production are listed in Table I. Their cross-sections are compared in Fig. 1 with the cross-section for ordinary Coulomb scattering. In order to achieve appreciable burn-up rates in DT mixtures, kinetic energies of $\gtrsim 10$ keV — corresponding to temperatures of $\gtrsim 10^8$ °K — are required. (At such temperatures the fuel is, of course, fully ionized: i.e., in the plasma state.) The other fusion reactions in Table I require even higher energies, and are therefore less interesting for short-range practical purposes. On the other hand, they are characterized by the release of larger fractions of the fusion energy in the form of charged particles (rather than neutrons); they will lend themselves in the future to the direct production of electric power at high efficiencies, and to the minimization of tritium inventories and neutron-activated structural materials.

TABLE I.

Fusion Reactions of Principal Interest

$$D + T \rightarrow He^4 (3.5 \text{ MeV}) + n(14.1 \text{ MeV})$$

$$D + D \rightarrow He^3 (0.82 \text{ MeV}) + n(2.45 \text{ MeV})$$

$$\rightarrow T \ (1.01 \text{ MeV}) + H(3.02 \text{ MeV})$$

$$D + He^3 \rightarrow He^4 (3.6 \text{ MeV}) + H(14.7 \text{ MeV})$$

The following discussion is specialized to the DT reaction. The achievement of a net electric power output requires a minimum product of plasma density and energy confinement time $n\tau_E$ of order 10^{14}cm^{-3} sec. The considerations involved here are discussed in greater detail by Dr. Ribe. If the density n is increased, the necessary confinement time τ_E can be reduced, but the plasma pressure then becomes larger. One must choose between very dense but freely

316

expanding plasmas, such as those discussed by Dr. Brueckner, and magnetically confined plasmas with pressures $n(T_e + T_i)$ that are limited by Maxwell's stress tensor: $\beta \equiv 8\pi n(T_e + T_i)/B^2 < 1$. Even 10^5-gauss magnetic fields can confine plasma densities of only $\sim 10^{16}\text{cm}^{-3}$ at 10 keV temperature; generally the ratio β of plasma pressure must be kept well below unity for reasons of gross plasma stability, and the maximum operating density may be reduced to $\sim 10^{14}\text{cm}^{-3}$. Thus we arrive at energy confinement times of order $10^{-2} - 1$ sec.

The necessary confinement times could be achieved in principle — and apparently also in practice — in various different types of magnetic bottle. Closed (toroidal) magnetic bottles can confine charged particles moving in all directions. The effect of a Coulomb collision is only to permit a random step by one orbit thickness (Fig. 2); hence many collisions can be tolerated during one energy confinement time, and net power can be produced at plasma temperatures as low as ~ 10 keV (cf. Fig. 1). It is also possible to use an open-ended magnetic bottle, or mirror machine, where particles moving in the direction along magnetic field lines are not confined. In that case, one large-angle Coulomb scattering is sufficient to remove a particle from confinement (Fig. 3); fortunately the Coulomb cross-section is sufficiently close to the fusion cross-sections at high energies ($\gtrsim 100$ keV) to permit production of net power. This closeness of Coulomb and fusion cross-sections has other interesting consequences, to which we shall return later.

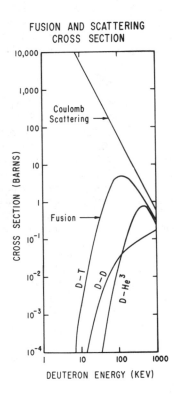

FUSION AND SCATTERING
CROSS SECTION

Figure   1.

III.   LOW-β TOROIDAL CONFINEMENT

The simplest form of closed-line magnetic field is the pure toroidal field, but this does not provide magnetohydrodynamic (MHD) equilibrium for a finite toroidal body of plasma. It is necessary to add a poloidal field component (Fig. 4), thus producing helical magnetic field lines on nested toroidal surfaces. This can be done by means of external multipole conductors, as in the stellarator, or by means of toroidal current flowing in the plasma. (A hydrogen plasma of 1 keV temperature is an electrical conductor comparable to standard copper.) The latter approach is illustrated in Fig. 4. Its special merits were first demonstrated by experiments at the Kurchatov Institute in Moscow; it goes by the name of tokamak (TOroidal KAmera MAKnetic).

CLASSICAL LOSS PROCESS OF "STRAIGHT" CLOSED SYSTEM

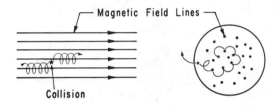

Diffusion: Time $\longrightarrow$ (Size)^2 / Diffusion Constant

Figure 2.

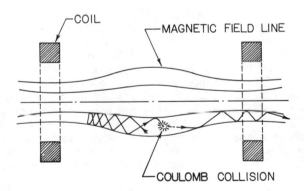

CLASSICAL LOSS PROCESS OF
OPEN SYSTEM

Figure 3.

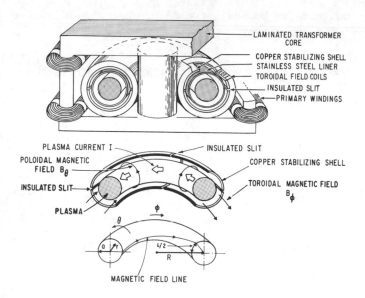

Figure 4.   Schematic of a standard tokamak.

The plasma current of the tokamak plays the incidental role of Ohmic-heating the plasma electrons to the 1-3 keV electron temperature range.   A typical time history of the plasma current and the electron temperature radial profile, as obtained on the ST tokamak (Fig. 5) at Princeton Plasma Physics Laboratory, is shown in Fig. 6. (The electron temperature is obtained by Thomson scattering a beam of ruby laser light from the plasma electrons and measuring the line broadening due to Doppler shift.)   The tokamak ions receive their energy by colliding with the electrons; in the relatively small-sized present-day tokamak experiments, the ion temperature tends to lag behind by a factor of 2-3.   In somewhat larger experiments of the near future, the electrons and ions are expected to become well-equilibrated.

Ohmic-heating alone is not expected to take the tokamak plasma temperature all the way to 10 keV, since the plasma resistivity drops as $T_e^{-3/2}$, while Bremsstrahlung increases as $T_e^{1/2}$.   Some form of auxiliary heating is required to bridge the gap from about 3 keV onward.   Heating by means of energetic neutral atom beams (Figs. 7 and 8) appears to be well suited.   The neutral beam particles pass through the walls of the magnetic bottle, are trapped in the plasma by ionization or charge-exchange, and then thermalize slowly by Coulomb collisions.   Fractional pressures of the injected ion population comparable to the levels required for reactor heating (i.e., 10-20% of the plasma thermal pressure) have already been demonstrated experimentally, with no sign of adverse side-effects.   Adiabatic compression of the plasma, following Ohmic and/or neutral-beam pre-heating (Fig. 8) has also proved successful. Several rf heating methods are in process of experimental testing and appear highly promising. The tokamak plasma heating problem is considered to be well in hand at this point.

Figure 5. The ST at Princeton.

$(n_{max} = 2 \cdot 10^{13} \ cm^{-3}$ at 40 msec$)$

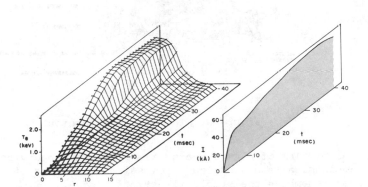

Figure 6. T_e-profiles and plasma current in the ST.

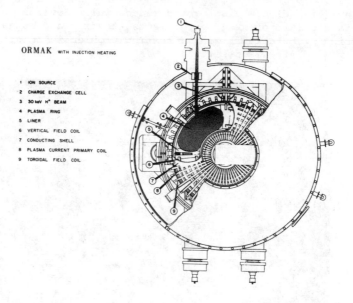

ORMAK WITH INJECTION HEATING

1 ION SOURCE
2 CHARGE EXCHANGE CELL
3 30 keV H° BEAM
4 PLASMA RING
5 LINER
6 VERTICAL FIELD COIL
7 CONDUCTING SHELL
8 PLASMA CURRENT PRIMARY COIL
9 TOROIDAL FIELD COIL

Figure 7.   The ORMAK at Oak Ridge.

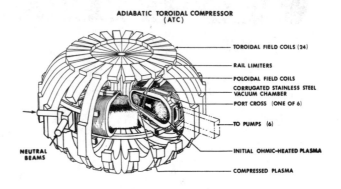

ADIABATIC TOROIDAL COMPRESSOR
(ATC)

TOROIDAL FIELD COILS (24)

RAIL LIMITERS

POLOIDAL FIELD COILS

CORRUGATED STAINLESS STEEL
VACUUM CHAMBER

PORT CROSS (ONE OF 6)

TO PUMPS (6)

NEUTRAL
BEAMS

INITIAL OHMIC-HEATED PLASMA

COMPRESSED PLASMA

Figure 8.   The ATC at Princeton.

The problem of plasma energy transport can be divided into two parts: classical and anomalous. Classical transport of particles and heat resembles the simple illustration of Fig. 2. In realistic toroidal geometry the particle orbits are more complex, and the transport coefficients are larger than in the straight cylindrical case, but the classical energy confinement for tokamaks would still be better by several orders of magnitude than actually required in a practical reactor. In present-day tokamak experiments the ions show thermal conductivities very close to classical predictions, but the electron heat conductivity and mass transport coefficients are somewhat larger than this. The anomaly is interpreted in terms of various microinstabilities; that is, collective plasma particle motions driven by temperature, density, and magnetic-field gradients. Present empirical extrapolations of the anomalous trend are favorable for tokamak reactor purposes; however, the theory of microinstabilities at very high temperatures (long Coulomb mean free paths) is not sufficiently well documented to permit precise estimates of the plasma size required to reach reactor-like $n\tau_E$-values. Next-generation tokamaks, such as the T-10 at the Kurchatov Institute and the PLT at Princeton Plasma Physics Laboratory (to become operational in 1975-76), are designed to reach the 1-MA level of tokamak current in ~3 times larger plasma size. These experiments should provide an important calibration of the energy confinement scaling laws to be used in computing tokamak reactor regimes.

A problem closely associated with plasma energy transport is the evolution of atoms from solid surfaces surrounding the plasma, with consequent plasma contamination and enhancement of radiation losses. In present experiments, radiation cooling from impurity ions is not yet dominant, but the physics of hot-plasma interaction with solid surfaces is expected to play an increasingly critical part in future tokamak research.

Tokamak confinement times increase with rising plasma current, but the maximum permissible current in a given device is limited by gross MHD instabilities. Specifically, when the helical field line of Fig. 5 "bites its own tail" on going once around the torus, a fast-growing helical perturbation (kink mode) can be anticipated; lesser disturbances occur when the line closes on going twice or three times around the torus. There are also MHD limitations on the maximum permissible β-value (cf. Section II). Present-day tokamaks, as well as future reactors, can achieve satisfactory operation within readily permitted stability limits ($\beta < 0.05$), but there would be economic advantages in increasing the MHD-stable tokamak current and β-value. The beneficial effect of noncircular minor cross-sections is being studied on the Doublet II device (Fig. 9) at General Atomic. Stronger MHD stability properties can also be obtained theoretically in the stellarator. As a low-β research device the stellarator has slipped to second place: large stellarators offer the prospect of results similar to those of tokamaks, but at greater cost and complexity of apparatus. For high-β toruses, however, a stellarator-type approach has substantial advantages, as we shall see in the next Section.

A rough idea of the remaining gap between tokamak experimental parameters and parameters for a break-even experiment is given in Fig. 10. An increase in plasma minor radius by a factor of ~10 (a factor of ~3 beyond T-10 and PLT) is expected to be accompanied by an increase in $n\tau$ of ~100 beyond the present value of $10^{12} cm^{-3} sec$. Heating power densities already demonstrated in present-day experiments would then suffice to reach the desired operating values of

322

T_e and T_i. Break-even is here defined as the generation of energy in the form of fusion-reaction products at a rate equalling the plasma heating power.

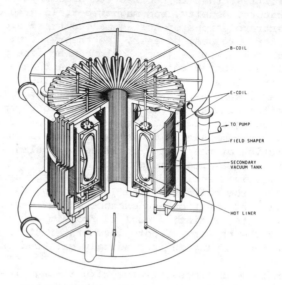

Figure 9.   The Doublet II at General Atomic.

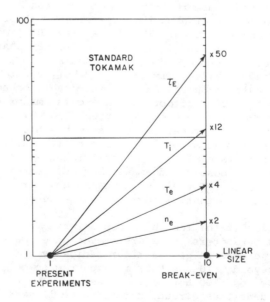

Figure 10.

IV. HIGH-β CONFINEMENT

In a long straight magnetic field configuration such as that of
Fig. 2, MHD equilibrium and stability can be attained even for the
extreme case β = 1 , where the plasma and magnetic field pressures
are equal (so that the magnetic field strength actually vanishes at
the center of the plasma). Experimental plasmas closely approxi-
mating this situation have been produced in linear θ-pinches at the
Los Alamos Scientific Laboratory.

The ideal form of θ-pinch heating is illustrated in Fig. 11.
The plasma is preheated by an intense shock wave, caused by a fast
initial rise of the axial magnetic field. The shock-wave energy is
then allowed to thermalize within the uncompressed plasma column.
In the final stage, the plasma is compressed adiabatically by a
slow-rising axial magnetic field, reaching maximum temperature and
density in a column of moderately reduced minor radius. (The pulsed
axial field shocks and compresses the plasma by inducing an azi-
muthally directed current on the plasma surface, thus exerting a
radial Lorentz force; hence the name "θ-pinch.")

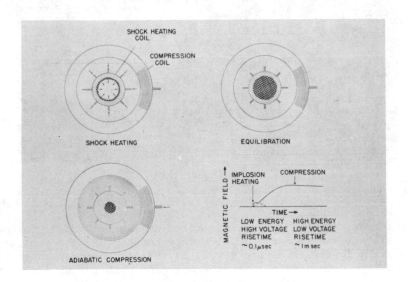

Figure 11. Ideal θ-pinch heating cycle.

A series of linear θ-pinch experiments conducted at Los Alamos
under the name Scylla have reached ion temperatures of up to 5 keV
at plasma densities of $10^{16} - 10^{17} \text{cm}^{-3}$. It is found that the transport of particles across magnetic field is very slow, and may well be
purely classical; however, the plasma confinement time has been
limited typically to ~10 μsec by the endloss of plasma moving at the
ion thermal velocity. The use of mirrors as end-stoppers has been

tested, and is found to extend the confinement time by about a factor of two. The mirror confinement effect is rather weak in this case, because of the high particle collision rates (cf. Fig. 3). Enhancements of confinement time have also been realized in proportion to increases of device length, culminating in the 5-meter linear Scyllac machine of Fig. 12. To proceed to substantially longer confinement times by further increases of length would represent a conceivable approach to the attainment of required $n\tau_E$-values; however, the realization of a closed toroidal θ-pinch is a goal of more immediate practical usefulness, and this has been the principal objective of the Scyllac effort.

Figure 12.   The 5-meter linear θ-pinch at Los Alamos.

The ideal heating cycle illustrated in Fig. 11 was approximated only qualitatively in early θ-pinch experiments. Recent experiments on the small Scylla IB device (Figs. 13 and 14), however, have verified that the objective of intensive heating with relatively little reduction of plasma radius can indeed be achieved experimentally. Thus it appears that the entire time-sequence envisaged in Fig. 11 (i.e., heating with moderate compression and subsequent long confinement time) could be carried out in an appropriately shock-heated toroidal Scyllac device.

The transition from a straight to a toroidal θ-pinch introduces problems of MHD equilibrium and stability closely related to those of the tokamak. A high-β plasma column with large ratio of major to minor radius does not lend itself to the tokamak-type of equilibrium solution. Instead, one can resort to stellarator-like, nonaxisymmetric shaping of the magnetic flux surfaces, as illustrated in Fig.

325

15.   The main idea here is to equalize the length of the inner and outer major circumferences by "scalloping" the inner one more strongly.   The basic MHD tendency for the plasma to expand outward into weaker toroidal magnetic field is then compensated.   The scalloping is achieved in the example of Fig. 15 by superposition of an "$\ell = 0$" periodic mirror and an "$\ell = 1$" sinusoidal-displacement.   The plasma remains weakly MHD-unstable against sideways excursions, so that an electronic feedback-stabilization system may be needed to maintain proper centering of the plasma column for extended times.   In plasmas that are only moderately compressed, however, a "passive" feedback, consisting of a conducting shell, may be sufficient to insure stability.   The greater technical convenience of the latter stabilization method has been a principal motivation for the development of a heating cycle that does not require large compressions in plasma column radius.

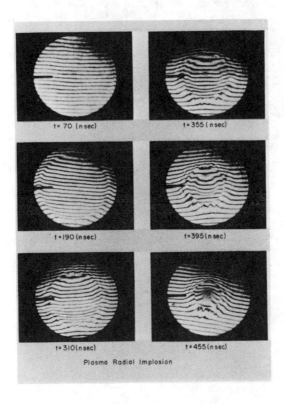

Figure 13.   Laser interferometer pattern
showing density contours in Scylla IB.

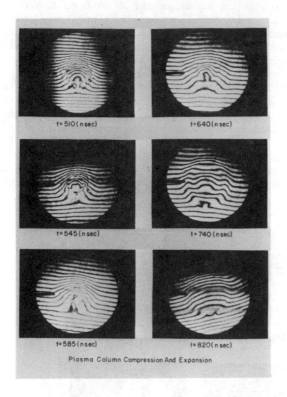

Figure 14. Continuation of Fig. 13 data.

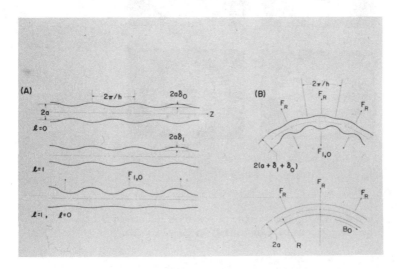

Figure 15. MHD equilibrium scheme for toroidal θ-pinch.

A Scyllac experiment studying electronic feedback stabilization on an 8-meter-long, curved θ-pinch sector of configurational design similar to that in Fig. 15 is shown in Fig. 16. The results have been encouraging for the success of the complete Scyllac torus assembly, which is to go into operation in the first part of 1974.

The remaining gap between the parameters of present-day (linear) θ-pinch experiments and the parameters of a break-even experiment is indicated in Fig. 17. A sharp improvement in confinement time should be realized simply by closing the ends of the plasma column, in the Scyllac torus geometry. Confinement times of 50-100 μsec are expected, thus raising $n\tau_E$ from the present value of $\sim 4 \cdot 10^{11} \text{cm}^{-3}$ sec to a new level of $2-4 \cdot 10^{12} \text{cm}^{-3}\text{sec}$, without increase of plasma minor radius. To attain the remaining factor of 25-50 in $n\tau_E$, a toroidal θ-pinch of roughly 5 times the linear size of Scyllac is envisaged. In such a device, present heating methods should be sufficient to raise T_i and T_e to temperatures of ~ 6 keV, where plasma heating by the alpha particles of the DT reaction would begin to take over.

Figure 16.　Scyllac 8-meter sector at Los Alamos.

V.　MIRROR CONFINEMENT

The problem of endloss can be solved in linear geometry by operating at sufficiently high temperatures (cf. Fig. 1) so that Coulomb scattering does not give rise to an excessively large plasma loss-rate (cf. Fig. 3). The natural operating regime of a mirror-type break-even experiment corresponds to ion temperatures of ~ 100 keV and densities of 10^{14}cm^{-3}. (The electrons can be considerably colder, since the reduction of Coulomb scattering in mirror experiments also permits a reduction of electron-ion thermalization rates.)

At first sight it appears that very high β-values, like those of linear θ-pinches, could not be achieved in mirror-machine geometry. The mirror confinement mechanism calls for an increase of magnetic field strength towards the open ends of the magnetic field lines; in the simple axisymmetric mirror geometry of Fig. 3, the result is a curvature of magnetic field lines that is strongly destabilizing from the MHD point of view. In a mirror device of this simple type, the plasma is expected — and found experimentally — to be unstable even at low β-values. Great progress has been made toward the feasibility of the mirror confinement approach by the application of mirror configurations like that in Fig. 18. The effectiveness of this general type of "minimum-B" or "Ioffebar" configuration was first demonstrated in 1961, by experiments at the Kurchatov Institute.

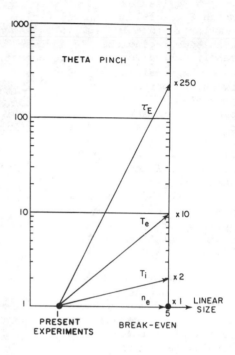

Figure 17.

Experiments at the Lawrence Livermore Laboratory, making use of the specific tetrahedral "baseball" geometry shown in Fig. 18, have now reached β-values above 0.5 in plasmas of multi-keV ion temperatures.

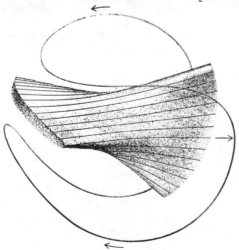

Figure 18. "Baseball"-type coil shape and magnetic field configuration.

The ideal form of hot-plasma production for mirror machines consists in building up the plasma energy distribution by direct injection of particles at the desired energy and orientation in velocity-space. Injection of neutral-atom beams (Fig. 19) is particularly convenient. Sources delivering beam currents of 5-50 amperes at energies in the range 1-30 keV, with pulse lengths of 1-100 msec, are currently in operation; further progress in source technology is being made at a rapid rate.

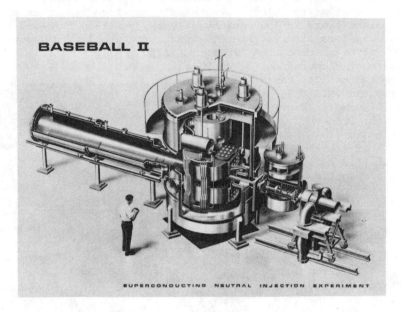

Figure 19.

A note on the difference in basic function between neutral injection into mirrors and toruses may be appropriate here. In toruses,
a small percentage of suprathermal particles is injected to heat up
the more numerous "cold" particles of the bulk plasma. In mirror
machines, the injector provides the desired plasma particles themselves, thus supplying the plasma material as well as the plasma heat
content. Mirror machines have the incidental advantage of supplying
a very pure plasma: the beam can be all hydrogen, and impurities
entering through ionization tend to scatter rapidly out the ends —
rather than concentrating in the plasma, as they tend to do in a
torus.

It is possible to build up a mirror-machine plasma starting
essentially from a vacuum, since neutral beams can be broken up and
trapped by interaction with the magnetic field itself (Lorentz
ionization). This process has been used in devices such as Baseball
II (Figs. 19 and 20), to achieve plasmas in the 1-20 keV range of ion
temperature. The build-up of density, however, is typically limited
to $\lesssim 10^{13} \text{cm}^{-3}$ by the onset of microinstabilities. These high-
frequency disturbances can arise in plasmas with non-Maxwellian
velocity distributions; they consist of various types of electromagnetic waves, amplified by the coherent relaxation of the charged-
particle distribution towards the Maxwellian — a process somewhat
resembling laser amplification. Some degree of departure from the
Maxwellian velocity distribution is inevitable in a mirror plasma,
since particles with high ratios of parallel to perpendicular
velocity (relative to the magnetic-field direction) are not mirror-
confined, and since lower-energy particles tend to Coulomb-scatter

330

most rapidly into the resultant "loss cone" in velocity-space. The problem of microinstabilities is theoretically surmountable in sufficiently dense mirror plasmas, with well-smoothed distribution functions; however, in low-density plasmas, built-up by beams that are narrowly localized in velocity-space, the problem is severe, and has not yet been surmounted experimentally.

The most encouraging experimental results have come from the 2XII device at Livermore (Fig. 21), which produces dense plasmas (10^{13}–10^{14}cm^{-3}) at ion temperatures of 1-10 keV, by adiabatic magnetic compression of colder initial plasmas injected by plasma guns. (It is in this device that β-values of >0.5 have been achieved.) The magnetic field configuration and coil structure is basically of the same type as that in Figs. 18-20. Some typical confinement results of the 2XII are shown in Fig. 22. Case A corresponds to a well-smoothed velocity-space distribution; in this case, the density falls initially at roughly the rate predicted theoretically on the basis of Coulomb scattering alone. In cases B and C, the departure from the Maxwellian is increased (this aspect can be

Figure 20.   The Baseball II device at Livermore.

controlled by changing the initial plasma production method); the result is abrupt losses of plasma, accompanied by bursts of high-frequency oscillations.

In a classical mirror plasma regime, where the endloss rate is dominated by Coulomb scattering, plasma confinement is expected to improve with rising temperature. Accordingly, the techniques of 2XII and Baseball II are currently being combined (Fig. 23) in the form of neutral-beam injection into the basic 2XII plasma. Initial results have been encouraging; a pulsed injection experiment with 12 beams of 50 A each at up to 20 keV energy will be carried out during 1974, and a roughly 5-fold increase in ion temperature is expected.

331

Figure 21.   The 2XII device at Livermore.

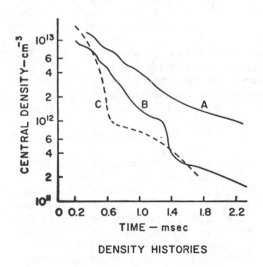

DENSITY HISTORIES

Figure 22.   Plasma confinement in the 2XII.

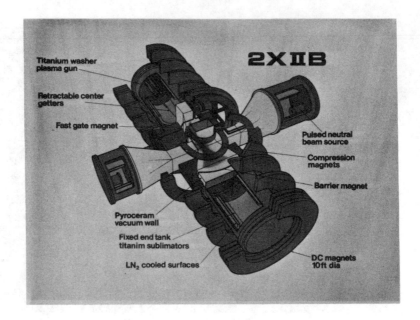

Figure 23.

The remaining gap between present-day mirror-machine plasmas and the plasma of a break-even experiment is indicated in Fig. 24. Since mirror-machine losses are related principally to diffusion in velocity-space, rather than in real space, the confinement time is relatively insensitive to plasma size: a factor of 2.5 increase in the plasma minor radius should be ample. The required temperatures can be obtained straightforwardly by choice of the injection energy of the neutral beams. The principal problem is to produce a large improvement in the plasma confinement time by appropriate control of the velocity-space distribution during plasma build-up.

VI.   TWO-COMPONENT PLASMAS

The oldest method of producing intensive fusion reactions, dating back to the early 1930's, is to strike a cold target with an energetic ion beam. This approach cannot produce a favorable overall energy balance, since the beam particles lose their energy too rapidly to the cold target electrons. If one could make a "warm" target, with ~5 keV electron temperature, however, it would be possible to reach break-even conditions.

One of the oldest ideas of controlled fusion research, dating back to the early 1950's, is to produce a warm-electron, cold-triton target plasma within a mirror machine, for interaction with a mirror-confined high-energy deuteron population. As is clear from Fig. 1, the inevitable Coulomb collisions of the energetic deuterons with the cold target tritons are not so intensive as to prevent the production of net energy from fusion. Indeed, one can calculate that substantial energy multiplication factors are easier to achieve in the two-component case than in an ordinary mirror machine. The desired electron temperature of the target plasma must be maintained by heat input from the slowing-down of the energetic-ion component; this requires an $n\tau_E$ of about $10^{13}cm^{-3}sec$ for the target plasma. (The condition on the $n\tau_E$ of the energetic component tends to be even more lenient.)

While target plasmas of sufficient $n\tau_E$ are not as yet available in linear geometry, we have noted in Section III that present-day tokamaks come close to supplying both the desired values of $n\tau_E$ and of T_e. Next-generation tokamaks such as PLT and T-10 are, in fact, hoping to exceed the minimum parameters required here. The possibility therfore arises of combining mirror-machine and torus ideas in the form of a tritium tokamak plasma with a neutral-beam-injected high-energy deuteron component, which reacts as it slows down within the "target plasma." The attainable ratios F, of fusion energy relative to injected deuteron energy, are shown in Fig. 25.

The parameter gaps that must still be bridged for attainment of break-even (F ≳ 1) in a two-component tokamak are indicated in Fig. 26. There is no requirement on the bulk plasma T_i. The

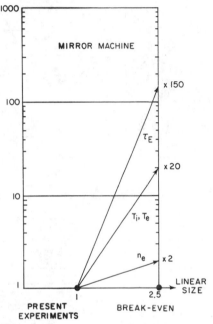

Figure 24.

most important missing item is an experimental demonstration that the energetic-ion component can be built-up to a pressure $P_{energetic}$ comparable to the plasma thermal pressure P_{plasma}, without exciting instabilities of various sorts. The outlook is somewhat favorable, both on the basis of theory and preliminary experiments, but the decisive experimental tests remain to be carried out during the next several years.

VII. PHYSICS OPPORTUNITIES

The preceding account of the state of magnetic confinement research sheds a favorable light on the question of physics opportunities — in at least two important ways. Clearly there are still many problems to be solved, so that great opportunities exist for creative

physical ideas; and clearly the problems on the way to a fusion reactor are no longer so formidable that there could be a serious
question about the prospects for success, or about the propects for
sustained growth of the international controlled-fusion research
effort.

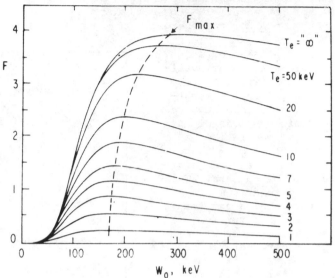

Figure 25.   Energy multiplication factor F in a
two-component tokamak with injected energy W_0.

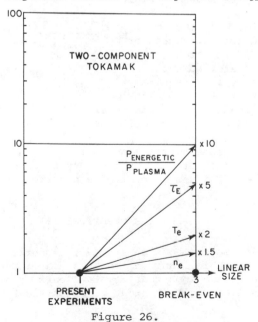

Figure 26.

In low-β toroidal research, the main plasma-physics problems have to do with microinstabilities, anomalous diffusion, and optimization of MHD-stable gross configurations. The important question of plasma-impurity control involves plasma physics, surface physics, and atomic physics. Progress in plasma diagnostics is closely paced by advances in laser and microwave technology. Computers play an important role in the on-line analysis of large data-inputs from complicated plasma-physical measurements, as well as in the development and application of plasma theory. As the experiments progress toward higher temperatures, nuclear reactions and their effects are playing an increasing role in plasma physics considerations and in the planning of experimental devices. The development of very large high-field superconducting coils will be important, not only for economical future reactors, but also for large test reactor experiments of the not-so-distant future.

The plasma physics problems peculiar to high-β research have to do mainly with shock-heating and MHD-stabilization. Special technological problems are accordingly related to the switching of very fast high-powered circuits, including feedback systems. As plasma lifetimes and sizes increase, attention turns to slower transients involving very large quantities of electrical energy, so that inductive and fast rotary energy storage systems become of key interest. In view of the high plasma densities of θ-pinches, sophisticated laser diagnostics play an even more essential role than in low-β toroidal research.

Mirror machine research concerns itself particularly with high-frequency microinstabilities due to non-Maxwellian plasmas. The special technological emphasis is on the development of intense neutral atom beams, and on the recovery of energy from particles escaping through the magnetic mirrors. The use of very large superconducting mirror coils is being pioneered already in present-day experimental devices.

ACKNOWLEDGMENT

This work was supported by U. S. Atomic Energy Commission Contract AT(11-1)-3073.

BIBLIOGRAPHY

1. World Survey of Major Facilities in Controlled Fusion Research, 1973 Edition, International Atomic Energy Agency, Vienna.
2. R. F. Post, Prospects for Fusion Power, Physics Today 26, p. 30 (April 1973).
3. H. P. Furth, Nuclear Fusion, Power Source of the Future?, 1973 Britannica Yearbook of Science and the Future, p. 110, Encyclopedia Britannica, Chicago.
4. N. A. Krall and A. W. Trivelpiece, Principles of Plasma Physics, McGraw-Hill, New York, 1973.
5. L. A. Artsimovich, Tokamak Devices, Nuclear Fusion 12, p. 215 (1972).

336

6. F. L. Ribe, Recent Developments in High-Beta Plasma Research on θ and Z Pinches, Comments on Plasma Physics and Controlled Fusion $\underline{1}$, p. 17, (January/February 1972).

PHYSICS PROBLEMS OF THERMONUCLEAR REACTORS[*]

F. L. Ribe
University of California, Los Alamos Scientific Laboratory
Los Alamos, New Mexico 87544

ABSTRACT

A problem common to all controlled fusion reactors is
that of the burning deuterium-tritium fuel under conditions
of plasma confinement which approach the ideal limit as
nearly as possible. After ignition, the balance between
alpha-particle energy deposition and plasma losses (radiation plus thermal and particle diffusion) determines the
stability or instability of the burn in toroidal systems.
Tokamak systems are described both with unstable, injection-
regulated burn cycles and stabilized steady-state burn conditions. In the theta-pinch reactor an unstable burn occurs,
somewhat regulated by high-beta plasma expansion, which is
quenched by a programmed plasma decompression. The plasma
expansion during the constant-pressure burn provides "direct"
conversion of plasma thermonuclear heat to electrical output, in addition to the electrical power derived from the
neutron energy through conventional thermal conversion equipment. The open-ended mirror reactor is characterized by a
direct conversion system for recovering end-loss plasma
energy and converting it to electrical energy for reinjection
into the plasma. This allows a favorable reactor energy balance an an amplification factor Q (= thermonuclear energy out-
put/injected plasma energy) which is compatible with classical collisional losses. For the three reactor types considered the ramifications of burn and confinement conditions for
reactor configuration, energy balance, economy, fuel handling, materials problems, and environmental-radiological
factors are considered.

I.   INTRODUCTION

The preceding paper by H. P. Furth[1] describes the basic concepts
of thermonuclear processes and magnetic confinement in fusion reactors
and gives a brief summary of the present state of research on the major
experiments. In the present paper a brief description will be given of
the properties of conceptual reactor systems based on the three major
magnetic confinement systems presently being investigated. Conceptual
fusion reactor studies have been pursued since 1965, and a summary of
the early work is to be found in the proceedings of the 1969 Culham
Conference on Nuclear Fusion Reactors.[2] The conceptual designs which
are summarized here stem from the following more recent detailed work
over the last two or three years: (a) three Tokamak reactor studies
at the Princeton Plasma Physics Laboratory (PPPL)[3], the Oak Ridge

* Work performed under the auspices of the U. S. Atomic Energy Commission.

338

National Laboratory (ORNL),[4,5] and the University of Wisconsin (UWMAK design)[6], (b) a Theta-Pinch reactor study carried out jointly by the Los Alamos Scientific Laboratory (LASL) and the Argonne National Laboratory (ANL)[7]; and (c) a Magnetic-Mirror reactor study at the Lawrence Livermore Laboratory (LLL).[8] In all of these studies plasma heating and confinement are assumed to occur according to idealized physical laws, to degrees not yet attained in the experiments, and the emphasis is on the physical problems imposed by the engineering demands of the thermal and nuclear environment and the requirements of practical power generation. The power levels and dimensions of the reactors represent compromises between economically desirable high power densities and the low power densities, and large sizes which alleviate materials problems.

II. FUSION-REACTOR CONFIGURATIONS AND POWER BALANCE

In this review we consider only the D-T fusion reaction (Table I, Reference 1) as a basis for reactor design, since its peak cross section occurs at relatively low ion temperature, allowing reactor operation at 100 to 200 × 10^6 °C (11.4 to 22.8 keV) in closed systems and 4 to 6 × 10^7 °C (45.6 to 68.4 keV) in magnetic-mirror systems. Because of this it is expected that the D-T reaction will be the basis of first-generation fusion reactors. Later power plants will probably make use of the D-D and D-He^3 reactions in which the absence of 14-MeV neutrons and their interaction with the reactor structure will be an advantage, gained at the cost of much higher plasma temperatures.

Figure 1 is a generalized cross section of the core of a magnetic

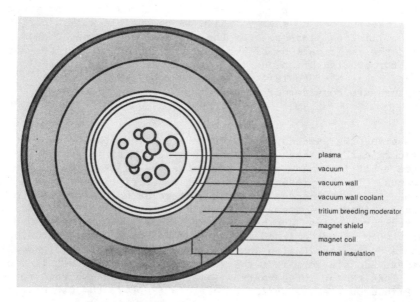

Fig. 1. Generalized cross section of the core of a nuclear fusion reactor.

confinement fusion reactor. The plasma is confined by magnetic fields (predominantly perpendicular to the plane of the figure) in excess of 6 T. Plasma ion and electron densities are in the range of 10^{20} m^{-3} for Tokamak and Mirror reactors and 10^{22} m^{-3} for the Theta-Pinch reactor. The magnetic coil producing the magnetic field is protected from the nuclear radiation of the plasma by means of a moderator and absorber of the neutrons and gamma rays, a magnet shield which absorbs neutrons and gamma rays, and thermal insulation. In the case of a deuterium-tritium plasma the moderator may breed tritium to replace that depleted from the plasma as fusion reactions take place. The plasma is surrounded by vacuum or low-density plasma, the whole being contained in a vacuum vessel with a first wall as shown. A coolant, consisting of flowing liquid metal, salt or high-pressure gas, removes heat deposited in the vacuum wall by nuclear and electromagnetic radiation from the plasma.

A quantity of basic importance to fusion physics and reactor design is the ratio β of the kinetic pressure $p = 2nkT$ of the plasma ions and electrons to the pressure of the external confining magnetic field B perpendicular to the plane of Fig. 1:

$$\beta = p/(B^2/2\mu_o) = 4\mu_o \, nkT/B^2. \tag{1}$$

In present experiments "low" beta refers to values less than 0.01 and "high" beta values lie between 0.1 and 1.0 (the maximum possible).

A quantity basic to the plasma dynamics is the reaction-rate parameter which is the product $\langle \sigma v \rangle$ of the D-T reaction cross section and relative D-T velocity, averaged over a Maxwellian velocity distribution. If we assume equal number densities n of deuterons and tritons the thermonuclear power available for each m^3 of plasma for thermal conversion to electricity or some other form of work is

$$P_T = \frac{1}{4} n^2 \langle \sigma v \rangle Q \tag{2}$$

where $Q = ME_n + E_\alpha$, and E_n and E_α are the neutron and α-particle energies of Table I, Ref. 1. The quantity M is the energy produced in the moderating blanket by each 14-MeV neutron. Substituting n from (2) we find

$$P_T = 0.78 \times 10^{23} \, \beta^2 \, B^4 \, Q \, \langle \sigma v \rangle / T^2, \qquad \text{(W/m}^3\text{)} \tag{3}$$

where Q and T are in eV. The dependence on β^2 shows the importance of the quantity beta as a measure of utilization of magnetic-field energy in fusion reactors.

The heating of the plasma is derived from the α particles which remain in the plasma, depositing a power density

$$P_\alpha = \frac{1}{4} n^2 \langle \sigma v \rangle E_\alpha = 5.1 \times 10^{-29} \, n^2 \frac{\exp(-200/T_i^{1/3})}{T_i^{2/3}}, \qquad \text{(W/m}^3\text{)} \tag{4}$$

where T_i (in eV) is the (assumed) common temperature of the deuterons and tritons, each of which is assumed to have number density $n(m^{-3})$.

The Bremsstrahlung power density for the range of n and T values appropriate to D-T reactors is given by

$$P_{Br} = 1.7 \times 10^{-38} \, n^2 \, Z^2 \, T_e^{1/2}, \qquad (W/m^3) \qquad (5)$$

where Z is the ionic charge, and n is the electron density in m^{-3}. This represents the total emission of all wavelengths of the continuum from the (optically thin) plasma (T_e in eV). In addition there is a plasma loss p_s from the synchrotron radiation emitted by the electrons in their magnetic orbits which occurs at low β, low n and high T, as well as a heat-loss power density P_Q arising from radial particle diffusion and heat conduction. This heat loss is usually characterized by an energy-loss e-folding time τ_E:

$$P_Q = 3\pi a^2 \, nT/\tau_E, \qquad (6)$$

where a is the plasma radius. Another important means of raising the energy of a plasma is adiabatic compression whereby a rate of increase $\dot{B}$ of magnetic field causes the density and temperature to rise. Assuming the plasma to be a perfect gas characterized by a specific heat ratio γ, the rate of increase of internal energy density is given by

$$P_C = 3(\gamma - 1) \, B\dot{B}/\gamma\mu_o. \qquad (7)$$

Finally, it is convenient to define a power deposition by injected beams of neutral atoms, each of energy E_o:

$$P_I = SE_o. \qquad (8)$$

In case of D-T pellet injection $E_o \approx 0$, while for energetic ions deuterium and tritium ions E_o is in the range of 500 keV.

As an example of plasma power-balance in a low-beta reactor (where β is assumed constant) consider the Tokamak case illustrated in Fig. 2. The ohmic heating rapidly declines with rising temperature and becomes ineffective in the temperature range of a few keV. The α-particle heating, on the other hand, rises rapidly in this temperature region. At the highest temperatures achieved by ohmic heating, the sum of ohmic and α-particle heating is not sufficient to overcome the losses due to ion heat conduction and Bremsstrahlung. Thus the plasma cannot get into the "ignited" regime $T \gtrsim 4$ keV where α-particle heating alone exceeds the losses and the plasma would become thermally self-sustaining. In order to achieve ignition an additional heat source of ~ 20 MW is necessary to bridge the gap between the minimum of the heat input curve and the rising curve of total heat loss. At this writing, the favored means of supplying this loss is by neutral-beam injection.

The time dependence of plasma temperature is determined by the various power input and loss quantities discussed above. Neglecting synchrotron radiation and plasma injection the rate of change of plasma internal energy is

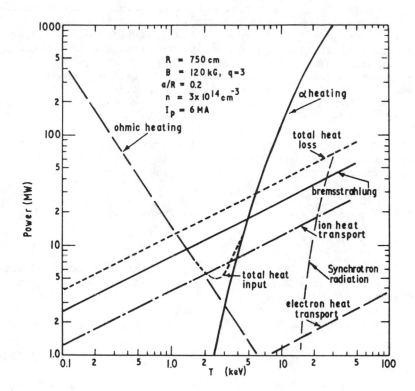

Fig. 2. Plasma power balance for a low-beta (Tokamak) reactor.[9]

$$d(3nT)dt = P_\alpha - P_{Br} - P_Q \qquad (9)$$

Neglecting particle loss, as in a low-β reactor with small fuel burn-up, we find

$$dT/dt = nf(T) - T/\tau_E, \qquad (10)$$

where $f(T) = (1/12) \langle \sigma v \rangle E_\alpha - 0.5 \times 10^{-38} T^{1/2}$. The condition for burning equilibrium $dT/dt = 0$ at same density n_o and temperature T_o is thus

$$n_o \tau_E = T_o/f(T_o). \qquad (11)$$

It is easily shown, because of the rapid variation of $\langle \sigma v \rangle$ with T, that for τ_E's which depend only weakly on T, this equilibrium is un-stable. If τ_E is adjusted for a given T_o small changes will cause an excursion of temperature upward toward the region of the $\langle \sigma v \rangle$ curve where the derivative is zero, or the temperature may decay, and the

reactor "goes out". At the unstable operating point it may be necessary to apply feedback control to the quantity τ_E or to $\langle \sigma v \rangle$.

The "$n\tau$" condition (11) is superficially similar to the famous
Lawson criterion which gives the condition that the electrical-energy
output of a given plasma volume exceed the minimum energy $3 \, nT + P_{Br} \, \tau$
necessary to produce the plasma and hold it for time τ. However, the
significances of the two conditions are greatly different. As an ex-
ample, at $T = 20$ keV $n_o \tau_E$ is about 2×10^{20} sec/m^3 and $n\tau$ (Lawson) is
about 5×10^{19} sec/m^3 for a 50%-50% D-T plasma.[10]

A quantity basic to the operation of a fusion reactor is the ratio
Q of the thermonuclear power output P_T, from the plasma core to the
thermal-conversion (turbine) equipment, to the injected power P_I necessary to sustain the plasma in a thermonuclear state:

$$Q = P_T/P_I. \tag{12}$$

The electrical power output of the thermal converter is

$$P_E = \eta_T P_T, \tag{13}$$

where η_T, related to the Carnot efficiency, is the efficiency at which
heat is converted to output electrical power P_E in the thermal converter. Its values range from 0.35 to at most about 0.6. In order to provide the injection energy at some efficiency η_I a fraction ϵ of the
plant electrical output P_E must be "recirculated". The circulating
power fraction

$$\epsilon = P_I/\eta_I P_E = 1/\eta_I \eta_T Q \tag{14}$$

is important as a cost determining factor, since it adds to the required capacity of the thermal conversion equipment. The fusion plant
output power

$$P_P = (1 - \epsilon) \, P_E = \eta_P P_T \tag{15}$$

is related to the thermal power P_T of the reactor core by the plant
efficiency

$$\eta_P = (1 - \epsilon) \, \eta_T. \tag{16}$$

III. TOKAMAK REACTORS

The essential features of a Tokamak diffuse toroidal pinch are
shown in Fig. 4 of Ref. 1 and described in the Section III of that
paper. It is useful to identify the poloidal beta β_θ, which is defined
analogously to the axial or toroidal beta of Eq. (1). Thus

$$\beta_\theta = p/(B^2/2\mu_o) \tag{17}$$

where a is the minor plasma radius. The aspect ratio $A = R/a$, where R
is the major radius. The ratio of the pitch length of the helical magnetic lines at the surface of the plasma to the major circumference
$2\pi R$ is the so-called stability margin or safety factor

$$q = aB_\phi/RB_\theta = B_\phi/AB_\theta. \tag{18}$$

The beta quantity most often referred to, which enters the thermonuclear power density (3), is the toroidal beta

$$\beta_\phi = p/(B_\phi^2/2\mu_o) = \beta_\theta/(qA)^2. \tag{19}$$

In two of the designs to be discussed divertors form an important part of the reactor core. The basic object of a divertor is to prevent particles diffusing out of the plasma from hitting the first wall and to provide a means for removing α-particle "ash" and impurities while maintaining a steady through-put of fuel from injectors. If in addition the divertor zone has low neutral-gas pressure, it is a thermally insulating layer at the edge of the hot plasma, imposing the boundary condition $dT/dr = 0$ and hence a flat temperature distribution radially across the plasma.

The magnetic divertors considered here are based on the idea of generating a null in the poloidal (r,θ) field as shown in Fig. 3. This generates a separatrix outside of which field lines are carried away from the plasma as they pass the neutral points shown in Fig. 3 as the crossing points of the zero-field lines (separatrices). Inside a separatrix the magnetic flux surfaces remain closed, and the separatrix, rather than a material wall, is the effective boundary of the plasma.

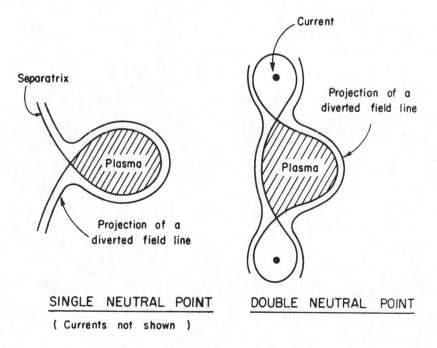

SINGLE NEUTRAL POINT   DOUBLE NEUTRAL POINT

(Currents not shown)

Fig. 3. Magnetic field configurations of poloidal field divertors.

Plasma particles diffuse across the separatrix and then follow the open field lines to particle collectors. The Tokamak design of the Princeton Group (Fig. 4) uses the single neutral-point divertor, while that of the University of Wisconsin group (Fig. 5) uses the double neutral point. The ORNL reactor (Fig. 6) has no divertor and uses a power cycle in which joule heating and neutral-beam injection produce ignition and peak power density. The plasma power density then subsides to a lower level which is sustained by the injectors. The UWMAK design (Fig. 5) operates in the unstable steady state, but in pulses whose length is determined by the decay of the poloidal field and the necessity to dispose of impurities not removed by the divertor. It uses joule heating and neutral beam injection, followed by D-T pellet injection. The PPPL design (Fig. 4) is conceived as a long-pulse, essentially steady-state system, fueled by D-T pellets.

Table I gives a comparison of the plasma and magnetic field parameters of the three conceptual Tokamak reactors. Note the large values of toroidal current (15-20 μA) and toroidal magnetic energy (100-300 GJ). In the UWMAK case the poloidal energy of 52 GJ would require a local superconducting magnetic-energy store (somewhat similar to that of a theta-pinch reactor) if the risetime of B_θ were of the order of 10 seconds or less, while for times of the order of 100 seconds the power to build up the poloidal energy [~500 MW(E)] could be taken "off the line". The

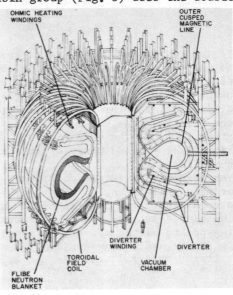

Fig. 4. PPPL Tokamak reactor.

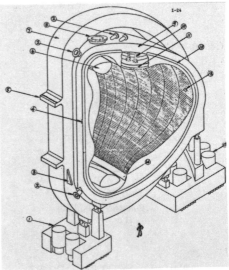

Fig. 5. UWMAK Tokamak reactor module.

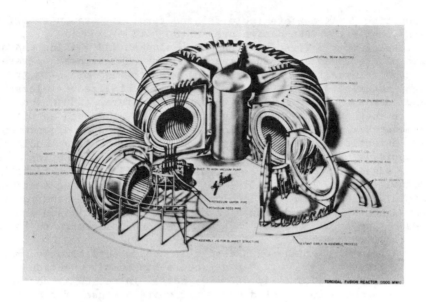

Fig. 6. General view of the ORNL conceptual Tokamak reactor.

TABLE I Plasma and magnetic field parameters of the three conceptual Tokamak reactors

	ORNL	PPPL	UWMAK
Plasma:			
Poloidal Beta β_θ (Av.)	3.0	2.3	1.08
Toroidal Beta β_ϕ	0.15	0.11	0.052
Major Radius R, m	10.5	10.5	13
Minor Radius a, m	3.3	3.3	5
Aspect Ratio A	3.2	3.2	2.6
Safety Factor q	1.4	2.1	1.75
Magnetic Fields:			
Av, Tor, B_ϕ, kG	60	60	38
Max, Tor, B_ϕ, kG	147	160	87
Pol. Field B_θ (a), kG	13	9	8.4
Plasma Current MA	21	15	21
Tor. Mag. En. GJ	120	223	290
Pol. Mag. En. GJ	~10	–	52

superconducting UWMAK magnets (poloidal and toroidal) are Cu-stabilized
NbTi (4°K), while the PPPL design uses Nb_3Sn ($\sim12^{\circ}$K).

In respect to the first-wall, neutron-blanket and fueling characteristics of the three conceptual reactors, the ORNL blanket operates
at high-temperature (1000°C), corresponding to the use of Nb-1% Zr
structural material. The PPPL and UWMAK designs utilize stainless
steel and correspondingly lower operating temperatures (680°C and
500°C). In the ORNL and UWMAK designs lithium is used both as the
neutron moderator and coolant, while the PPPL design uses a nonconduct-
ing salt, eutectic FLIBE [$(LiF)_2 BeF_2$] moderator with helium coolant,
thereby avoiding problems associated with the flow of conducting lithium across magnetic fields. The ORNL blanket utilizes segments with
longitudinal lithium flow (parallel to B_ϕ) except at turning sections
where radial currents pump the lithium electromagnetically and reverse
its flow direction. Outside the blanket regions are radiation shields
which reduce the total heat load on the superconducting magnets to very
small levels (1.5 to 15 kW).

The wall loading of the ORNL design is chosen sufficiently small
so that the first wall and blanket structure can survive neutron radiation damage for a reactor life of $\sim$ 20 years. In the UWMAK design,
first-wall changeout is expected every two years because of radiation
damage, the critical consideration being embrittlement of the stainless
steel. The tritium breeding ratios are quite adequate, resulting in
short ($\sim$100-day) doubling times which can well be lengthened by delib-
erately spoiling the breeding. The lithium inventories are of the order
of 10^6kG, and tritium inventories are of the order of 10 kG, corresponding to T_2 concentrations of $\sim$ 5 parts per million. This small
concentration is necessary to keep the radioactive tritium loss from
the plants within acceptable bounds. The tritium consumptions are
typically 0.3 kG/day.

The power-balance quantities of the three conceptual power plants
are as follows: The ORNL reactor is characterized by low thermal
(1000 MW) and electrical powers (600 MW), corresponding to small wall
loading (0.3 MW/m^2) and low plasma density (4×10^{19} m^{-3}). However
it has a large thermal conversion efficiency (56%), corresponding to
its refractory metal structure and a correspondingly high operating
temperature. The other two designs use stainless steel structure and
more conventional steam conversion plants with correspondingly low
thermal efficiencies (30 and 44%). Their thermal and electrical power
outputs are much larger (5000 and 1500-to-2300 MW, respectively). The
recirculating power of the PPPL (13%) design derives largely from the
helium circulators, while in the ORNL design the major component is
the input to the neutral-beam injectors. The UWMAK design assumes neutral-beam injection only during its 10-sec ignition phase, and it thereby makes a negligible contribution because of the 97% duty factor.
The burning cycles are quite different for the PPPL and UWMAK devices,
as opposed to the ORNL case.[*] For the assumed neoclassical diffusion

---

[*]The ORNL design used here represents a composite concept of the plasma
and burn cycle of Ref. 4 adapted by the author to the "core" and power-
conversion system of Ref. 5.

the confinement time is much too long for burning equilibrium, and the confinement time, τ_E, is "spoiled" to a value of 4 to 14 seconds. For the UWMAK and PPPL designs the long burn times τ_B (2000 to 6000 sec) are set by the L/R time of the plasma ring. Thus even though the PPPL and UWMAK reactors operate on long pulses they are essentially steady-state devices since $\tau_B \ll \tau_E$. They are sustained by injection of D-T ice pellets, coated with Argon impurity (40-μm pellets at 10^6 per sec). The burning equilibria are unstable (cf. Section II) and must be servo or feedback stabilized. The burnup fraction is 7 to 12%. In the ORNL burning cycle[5] the approach is quite different. The plasma, after ignition, is allowed to make an unstable excursion in about 20 sec to a high peak temperature (100 keV) and beta (15%), after which D-T fuel depletion (burnup fraction, 0.8) and synchrotron radiation cause a similarly sudden decrease in reaction rate, temperature (to 40 keV) and beta (5.5%). The plasma is then "driven" at an amplification factor Q between 6 and 13 by the neutral-beam injectors for about 600 sec. No confinement spoiling is used and the neutral beam injectors (30 to 60 MW at 100 to 300 A) also fuel the plasma.

In the UWMAK reactor the ohmic heating, toroidal current is excited for about 10 sec at low plasma density (3×10^{19} m^{-3}) with the plasma temperature leveling off at about 2 keV, after about 10 sec. In order to achieve ignition neutral beams of 350 to 500-keV D and T ions are injected tangentially to a major circumference. A total beam power input of 15 MW will cause ignition to occur in 11 sec. After ignition the temperature rises to its servo-stabilized equilibrium value of 11 keV ($T_e \approx T_i$), and the density is raised to its final value of 8×10^{19} m^{-3} by pellet injection. After the burn the fueling and toroidal current are decreased to cool the plasma to 500 eV, and the plasma is driven to the walls. The system is then pumped out and refueled with fresh gas. The divertors and poloidal field are then reestablished. From the end of one burn to the beginning of the next is 70 sec, giving a 97% duty factor.

IV. THE THETA-PINCH REACTOR

The basic concepts of this system are shown in Fig. 11 of Ref. 1 and described in Section IV of that paper. Unlike the other magnetic confinement systems, the theta pinch is a high-beta device ($\beta \approx 1$) in which very little penetration of the magnetic field into the plasma occurs. It is characteristic of these devices in toroidal geometry that the aspect ratio is very large, of the order of 100, roughly two orders of magnitude larger than for the Tokamaks. In the theta pinch the plasma density ($\sim 10^{22}$ m^{-3}) is also two to three orders of magnitude larger than in the Mirror and Tokamak, and confinement times are correspondingly shorter. Burning times τ_B in the theta-pinch reactor are of the order of 0.1 sec rather than ~ 1000 seconds, for the Tokamak case. This relative ordering of magnitudes follows from Eq. (3) where thermonuclear power density varies as n^2. The theta pinch is inherently a pulsed device because of its impulsive method of heating and its high instantaneous power density. For a typical cycle time $\tau_c \sim 10$ sec the duty factor $\tau_B/\tau_c \approx 10^{-3}$ results in average power

densities and wall loading which are about the same as for the other
two concepts. Total toroidal magnetic energies are of the order of
100 GJ, also comparable to those of the other concepts. However,
this energy is pulsed repetitively in and out of the compression-con-
finement coil from a superconducting magnetic energy store with rise-
times of the order of 0.03 sec. This feature and the requirement of
adapting the high voltage necessary for shock heating to the nuclear
environment of the reactor core are essential determining factors in
the design of the theta-pinch reactor.

Figure 11 of Ref. 1 shows schematically the essential elements
of a staged theta-pinch reactor. In the shock-heating stage a magnetic field B_s having a risetime of the order of one μsec and a magnitude of a few T drives the implosion of a fully ionized plasma
whose initial density is of the order of 10^{21} m^{-3}. After the ion
energy associated with the radially-directed motion of the plasma implosion has been thermalized, the plasma assumes a temperature T_E,
characteristic of collisional equilibration of the ions and electrons
(Ref. 1, Fig. 11). After a few hundred μsec the adiabatic compression
field is applied by energizing the compression coil. The plasma is
compressed to a smaller radius and its temperature is raised to a
value at or near ignition ($\sim$ 5 keV). As the D-T plasma burns for
several tens of msec, it produces 3.5-MeV α particles which partially
thermalize with the electrons and the D-T ions as the burned fraction
of plasma increases to a few percent. As a result the plasma
is further heated. Since β is approximately unity, and since B is
approximately constant, the plasma expands against the magnetic field
(not shown in Ref. 1, Fig. 11), doing work ΔW which is about 62% of
the thermonuclear energy deposited by α particles. This work produces an emf which forces magnetic energy out of the compression coil
and back into the compression magnetic-energy store. This high-β,
α-particle heating and the resulting direct conversion are important
factors in the overall reactor power balance.

A cross section of the reactor core is shown in Fig. 7. The
first wall is composed of the ends of insulator-coated blanket segments in which lithium is used both as a moderator and a coolant.
The insulator must be thin enough to allow heat extraction from the
plasma Bremsstrahlung and nuclear radiation, as well as to allow passage of heat from the plasma as it is being quenched at the end of a
burning pulse. Simultaneously the insulator must support the back-emf
electric field E_θ of the plasma implosion during the fraction of a
microsecond of shock heating. During this short time of high-voltage
stress there is no radiation present, as there is later during the
burn. Hence the insulator requirement of a resistivity of $\sim 10^8 \Omega$ - m
and a dielectric strength $E_D \sim 10^7$ V/m at a temperature in the neigh-
borhood of 1000°C need not be met in the presence of a high radiation
field. Allowing N blanket segments each coated with an insulator of
thickness X_D to support a total voltage of $2\pi b\, E_\theta$, we find $N = \pi b E_\theta /$
$X_D E_D$. Heat transfer calculations show that $X_D = 3 \times 10^{-4}$ m is adequately thin, corresponding to $N = 100$ blanket segments. With niobium metal structure as shown in Fig. 7 it is found that the blanket
provides a tritium breeding ratio of 1.11 and an energy multiplication

such that ME_n = 20.5 MeV,
while adequately protecting
the room-temperature copper
compression coil from nuclear
radiation.

The implosion-heating
coil of Fig. 7 is of an end-
fed design in which the field
B_S is admitted to the plasma
region between the blanket
segments, being excluded from
them by the skin effect during
shock-heating times and later
penetrating the segments during the time of electron-ion
equilibration.

The compression coil has
sufficient thickness (low-
current density) that joule
losses of the theta currents
which produce the compression
field B_0 are reduced to such
a level that their contribution to the plant circulating
power is acceptable. The coil
is layered in ~2-mm thicknesses to reduce to negligible
proportions the eddy-current
losses during the 0.03-sec
rise and fall of the compres-

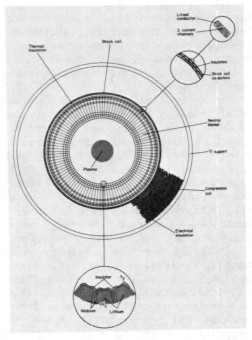

Fig. 7. Cross section of the neutron
blanket, shock-heating coil and adiabatic compression coil of the Reference
Theta-Pinch Reactor.

sion field. Similarly the segmenting of the blanket provides negligible
eddy-current losses in the lithium and graphite. An overall view of
the conceptual RTPR power plant is shown in Fig. 8. The reactor ring

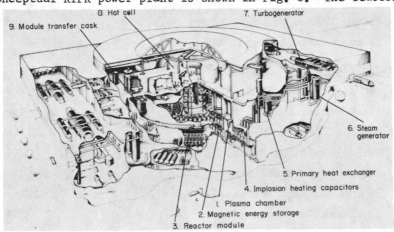

Fig. 8. Overall view of the LASL-ANL θ-pinch fusion power plant.

is below ground in an evacuated tunnel with a helium barrier to prevent escape of tritium. The transfer cask allows removal of the radioactive core modules for replacement in case of failure. During the burn cycle direct-conversion work of 9.8 MJ per meter of reactor length is produced, as compared with a thermonuclear energy $W_N =$ 93 MJ/m. The ratio of direct-conversion energy to thermally-converted electrical energy is $\Delta W/\eta_T \, W_N = 0.26$ for $\eta_T = 0.40$.

An attractive feature of a pulsed fusion reactor is the possibility of removing the alpha-particle "ash" resulting from the burnup of the deuterium-tritium fuel mixture and injecting fresh fuel between burning pulses. No divertor is required, as in the case of a steady-state, toroidal reactor. A layer of neutral gas injected between the hot central plasma and the first wall is used to cool, neutralize, and purge the partially burned D-T plasma. Preliminary calculations show that sputtering problems are alleviated because heat transfer to the wall, which would otherwise occur by energetic ions, will now occur primarily by means of low-energy neutral atoms and to a lesser extent by ultraviolet and visible radiation.

Table II provides a summary of the theta-pinch reactor plasma and magnetic-field parameters. The tritium inventory, breeding ratio and doubling time are comparable to those of the Tokamak reactors. The wall loading is about 40% higher than for the PPPL and UWMAK reactors. The toroidal magnetic energy is less for the RTPR by a factor of two to three, but, owing to its pulsed character, it should be compared to the poloidal magnetic energy of UWMAK. It is twice as large and rises

TABLE II Plasma and magnetic-field parameters for the LASL-ANL conceptual Theta-Pinch reactor.

Plasma:

Beta	0.8 - 1.0
Major Radius R, m	56
Plasma Radius a: Shock, Burn, m	0.38, 0.12
Plasma Aspect Ratio A, max	465

Magnetic Fields:

Shock, Comp. Coil Radii, m	0.91, 0.94
Shock Field, T	1.4
Compr., Burn Field, T	11.0
Helical Poloidal Field, T	0.8
Compr. Field Risetime, sec	0.031
Supercond. Mag. Energy, GJ	102
Shock-Heating Energy, GJ	0.9

much faster. The plasma burn quantities are quite different than those of the Tokamaks, with exception of the n_T and fuel-burnup parameters which are comparable.

In the power balance we have power output (2000 MWE), circulating power fraction (E = 13%), and Q value (14) comparable to those of the Tokamak systems. A unique feature of the Theta Pinch is the direct conversion power, discussed above, which offsets the compression-coil joule losses to provide the relatively low circulating power and high plant efficiency. The power level of the RTPR is inversely proportional to the cycle time τ_c since the essential plasma operation and energy balance are determined for a given pulse. The choice τ_c = 10 sec gives low wall loading and long first-wall lifetime, rather than optimum economy, which would be more favorable at the shortest cycle time allowed ($\tau_c \approx 3$ sec).

V. THE MAGNETIC-MIRROR REACTOR

The basic concepts of the Magnetic Mirror are shown in Figs. 3 and 18 of Ref. 1 and described in Section V of that report. Here, however, we point out the LLL choice of the Yin-Yang coil geometry[11], rather than the Baseball geometry, for providing the minimum-B magnetic field. The Yin-Yang coil of the LLL reactor design is shown in Fig. 11.

The toroidal reactors described in the previous sections allow the possibility, under ideal confinement conditions, of plasma ignition, i.e., the plasma can be self-sustaining, requiring little or no injected energy. Under these conditions the amplification factor Q Eq. (12) of the plasma itself becomes very large. The magnetic-mirror, because it is an open-ended device with an intrinsic loss of plasma, allows no such operation. Under ideal collisional circumstances the theoretical value of Q is only slightly greater than unity (about 1.2 for the mirror ratios considered here). The magnetic-mirror plasma is therefore a driven power amplifier whose power output is a factor Q times its injected power. In order to achieve economical net output with such low values of Q a magnetic-mirror reactor must make use of the plasma energy which escapes from its mirror in order to power the injectors. The means by which this is accomplished at high efficiency is called direct conversion. This leads to a large recirculating power fraction, of order unity. However, the power circulation loop does not pass through the thermal-conversion equipment and can be shown to be economical if its efficiency is sufficiently high.

The method by which the injection power is supplied directly from the energy of plasma ions which escape out the mirrors is illustrated in Fig. 9. The fraction (1-h) P_N of thermonuclear power which occurs as α particles remains in the plasma and, along with $\eta_I P_I$, is converted to useful electrical output through a direct converter of efficiency η_D. The thermal converter provides useful output and that recirculating power necessary to sustain itself. According to Fig. 9 net power output occurs when

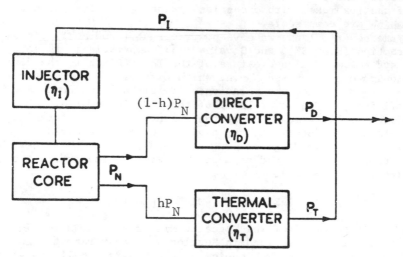

Fig. 9. Power flow diagram showing the main features of a Magnetic-Mirror power plant with direct conversion.

$$\eta_T h P_N + \eta_D[(1-h)\ P_N + \eta_I P_I] - P_I > 0. \tag{20}$$

Using $Q = P_N/\eta_I P_I$, this corresponds to the condition

$$Q > \frac{1 - \eta_D \eta_I}{\eta_I[h\eta_T + (1-h)\eta_D]} . \tag{21}$$

R. F. Post and his colleagues have shown that values of η_D as large as 0.8 can be attained. For $\eta_T = 0.8$, inequality (21) shows that reactor breakeven can occur for Q values as low as 0.8, thus allowing the system to accommodate the attainable value of about 1.2.

The method by which end-loss plasma energy from a magnetic mirror is converted to useful electrical power is illustrated in Fig. 10. This shows a vertical section of one mirror of a minimum-B mirror system like that of Fig. 11 and a typical escaping ion orbit. First the escaping plasma is expanded in the horizontal fan-shaped magnetic field which extends 76 m from the mirror. In this process the plasma density is reduced, and the ion motion is converted, by means of an inverse of the mirroring process, into motion parallel to the field lines. After expansion, the plasma density is sufficiently low ($\sim 10^{13}$ ions/m^3) that charge separation of ions and electrons can occur. The electrons (whose energy is negligible) are diverted away vertically in the separator along the magnetic lines, but the ions escape across the lines and continue horizontally to a collector. Here, depending on their energy, the ions are decelerated in a periodic set of charge-collecting electrodes which collect them as they are brought to rest by the retarding potentials. There results a distribution of high voltages on the collector electrodes which store the energy of the slowed-down

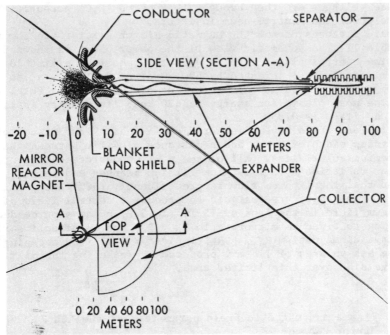

Fig. 10. Illustrating the main components of an apparatus for converter end-loss ion energy from a mirror reactor into direct-current
electrical output.

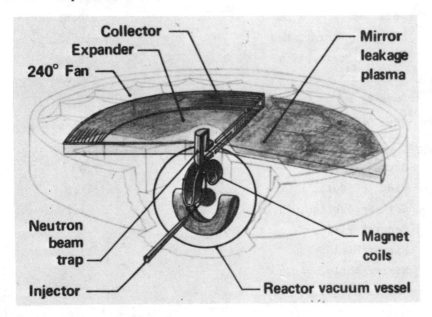

Fig. 11. Illustrating the main components of the LLL Magnetic-Mirror
Reactor with direct conversion.

ions. These voltages are then brought to a common D.C. potential V_D and used to power the neutral-beam injectors.

The main features of the LLL Magnetic-Mirror Reactor are summarized in Table III. A schematic view of the power plant is shown in Fig. 11. The central plasma core is a nearly spherical ellipsoid of about 3.3-m mean radius, injected by 490 MW of neutral beam at an average ion energy of 550 keV and a current of 890 A over a total area of 5 m^2. The mean plasma ion energy is 620 keV; its density is 1.2 X 10^{16} m^{-3}; and its β is 0.85.

The Yin-Yang coils have a mean radius in their "fan" planes of 10 m, a winding width of 6.6 m and a distance of 4.2 m between the parallel semicircular sides. All of the coil conductor is protected from neutron radiation by a 1-m-thick lithium blanket and shield, arranged so that the lithium flow is predominantly (90%) along magnetic lines. The coils are designed to produce a central field of 5 T and a maximum field in the fans of 15 T, (16.5 T on the superconductor) corresponding to a vacuum mirror ratio R_V = 3.0 and an effective ratio due to plasma diamagnetism $R_{eff} = R_V/(1-β)^{1/2}$ = 7.7. The escaping plasma beam has an area of 1.5 m^2, produced by weakening the mirror field on one side over this limited area.

TABLE III Plasma and magnetic field parameters of the LLL 200 MWe DT Mirror Reactor

Plasma:

Shape	Ellipsoidal, Min. B, Vol.
Beta	0.85
Mean Ion Energy, MeV	0.62
Z Axis Intercept, m	3.5
Volume m^3	130
Density m^{-3}	1.2 x 10^{20}
Beam Area at Mirror, m^2	1.5

Magnet Coil:

Radius, m	10
Width of Coil, m	6.6
Coil Separation, 2h, m	4.2
Central Field, T	5
Mirror Field, T	15
Mirror Ratio, Vac.	3
Mirror Ratio, β = 0.85	7.7

The plasma beam escaping from the vertical mirror of Fig. 11 into the expander is first deflected by 90 degrees along the mirror magnetic lines which are bent by means of "steering" coils. The neutrons emerging from the mirror are buried in a neutron trap. From the mirror to the end of the expander the magnetic field decreases from 15 T to 0.15 T, decreasing the perpendicular ion energy by a factor of 100.

The expander has a horizontal radial extent of 76 m, accepting a beam height of 0.87 m over a total horizontal angle of 240°. Past the expander the electrons are deflected away by separator coils, and the ion energy is converted over the 22-m radial length of the collector. The expander and collector are enclosed in a containment vessel of three radial sections separated by 1.7 m, composed of dished hexagonal modules, each with a horizontal dimension of about 32 m. Support columns intercept about 3% of the plasma beam. In addition to this flat containment vessel there is a spherical vacuum vessel, mostly below ground level, in order to limit tritium leakage to the atmosphere.

The thermonuclear power carried by neutrons is 470 MW, and the escaping power of charged particles is 610 MW. Emerging from the direct converter (η_D = 0.70) are 430 MW and from the blanket and thermal conductor (η_T = 0.45) are 360 MW. Accounting for 560 MW of injection energy and 20 MW to auxiliaries, the net electrical output is 170 MW, giving an overall efficiency of 27%. The system Q is chosen to be 1.2.

REFERENCES

1. H. P. Furth, Magnetic Confinement of Thermonuclear Plasmas, Bull. Am. Phys. Soc., Ser. II, 19, No. 1, 85 (1974). Preceding paper of these proceedings.

2. Proceedings, British Nuclear Energy Society Nuclear Fusion Reactor Conference, September, 1969. (UKAEA, Culham Laboratory, Abingdon, Berks., 1970).

3. R. G. Mills, et al, in Proceedings of the Texas Symposium on Controlled Thermonuclear Experiments and Engineering Aspects of Fusion Reactors, November 20-22, 1972, Paper II/B.6. Also private communication (1973).

4. A. P. Fraas, Oak Ridge National Laboratory Report, ORNL-TM-3096 (May 1973).

5. J. F. Etzweiler, J. F. Clarke, and R. H. Fowler, Oak Ridge National Laboratory Report ORNL-TM-4083 (June, 1973).

6. B. Badger, et al., Wisconsin Tokamak Reactor Design, University of Wisconsin Nuclear Engineering Department Report UWFDM-68 (November, 1973) Vol. 1.

7. R. A. Krakowski, et al., Joint Los Alamos Scientific Laboratory and Argonne National Laboratory Report, LA-5336/ANL-8019 (December 1973).

8. R. W. Werner, et al., in Proceedings of the Texas Symposium on Controlled Thermonuclear Experiments and Engineering Aspects of

356

Fusion Reactors, November 20-22, 1972. Paper 1, Session IV. Also private communication (1973).

9. D. R. Sweetman, Nuclear Fusion 13, 157 (1973).
10. R. G. Mills, Proceedings, British Nuclear Energy Society Nuclear Fusion Reactor Conference, September 1969. (UKAEA, Culham Laboratory, Abingdon, Berkshire, 1970). p. 322.
11. R. W. Moir and R. F. Post, Nuclear Fusion 9, 253 (1969).

MHD POWER GENERATION

A. Kantrowitz and R.J. Rosa
Avco Everett Research Laboratory, Inc.
Everett, MA 02149

ABSTRACT

The MHD Generator is now widely recognized as one of the more promising new methods for large-scale electric power generation. Its primary function, in the terminology of thermodynamics, is that of an "expansion engine" like the reciprocating piston engine and the turbine; and although premature, it is tempting to suppose that it represents the next and perhaps the ultimate step in the refinement of such engines.

The potential advantages of the MHD generator over the turbine are to a degree analogous to the advantages of the turbine over the piston engine. As a result of their conceptual simplicity, turbines outclass piston engines in power handling capacity and reliability. Likewise, the MHD generator, which represents a still further increase in simplicity, also represents a still further increase in intrinsic power handling capacity and potential reliability.

But beyond this, the MHD generator also represents a large step forward in temperature handling capability and thus efficiency. With respect to temperature the turbine is actually a step backwards, whereas the MHD generator, in principle, can handle even the very high temperature that may be produced someday in the plasma of a fusion reactor.

While the bulk of the present MHD development effort is directed toward fossil-fired plants, the advance of reactor technology toward higher temperature could eventually make it possible for nuclear plants to take advantage of MHD also.

INTRODUCTION

The Magnetohydrodynamic (MHD) Generator is now widely recognized as one of the more promising new methods for large-scale electric power generation. It represents an important application of plasma physics and in the long run could have a major impact upon technology. To understand why this is so it may be helpful to review what the role of the MHD generator is.

The term "generator," although correctly implying that this is a device for producing electric power, is misleading in the sense that it is not truly descriptive of its primary function. This primary function, in the terminology of thermodynamics, is that of an "expansion engine" like the reciprocating piston engine and the turbine. That is to say, it is a device for converting heat (or more specifically the enthalpy of a gas) into a more useful form of energy. Since expansion engines are the prime movers of our modern technological society, the refinement of such engines is potentially of far-reaching importance.

Although premature, it is tempting to suppose that the MHD generator represents the next important step in the refinement of such engines. One might view the development of expansion engines as having proceeded from the piston engine with its multiplicity of valves, cams, and sliding seals, through the turbine, which is a windmill inside of a pipe, to the MHD generator in which the windmill too is removed, the pipe alone remaining--with the aerofoils of the windmill replaced by lines of magnetic force.

The potential advantages of the MHD generator over the turbine are to a degree analogous to the advantages of the turbine over the piston engine. As a result of their conceptual simplicity, turbines out-class piston engines in power handling capacity and reliability. Likewise, the MHD generator, which represents a still further increase in simplicity, also represents a still further increase in intrinsic power handling capacity and potential reliability.

But beyond this, the MHD generator also represents a large step forward in temperature handling capability and thus efficiency. With respect to temperature the turbine is actually a step backwards, whereas the MHD generator is a large step forward and, in principle, can handle even the very high temperature that may be produced someday in the plasma of a fusion reactor.

While the bulk of the present MHD development effort is directed toward fossil-fired plants, the advance of reactor technology toward higher temperature could eventually make it possible for nuclear plants to take advantage of MHD also.

THE MHD ENERGY CONVERSION PROCESS

The magnetohydrodynamic (MHD)generator transforms the internal energy of a gas into electric power in much the same way that a turbogenerator does, and the basic physical phenomena that are employed are the same in both cases. Figure 1 illustrates this point and also illustrates the essential difference. In a turbogenerator, the energy of a gas is converted into the motion of a solid conductor by means of turbine blades and a connecting mechanical linkage. In the MHD generator the gas itself is a conductor and by expanding through a nozzle moves itself. In either case, the motion of a conductor through a magnetic field gives rise to an electromotive force and a flow of current in accordance with Faraday's law of induction. In a conventional generator, the current is carried to the external load circuit through "brushes." In the MHD generator, the same function is performed by "electrodes."

In order to conduct electricity, a gas must of course be partially ionized. So far, only thermal ionization has proven

TURBO GENERATOR   MHD GENERATOR

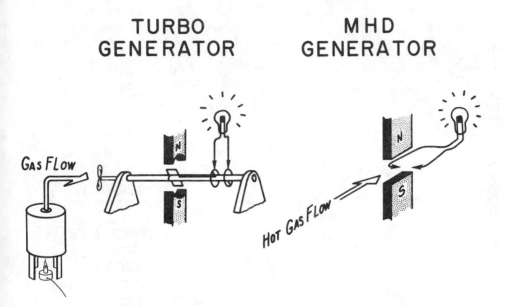

GAS FLOW

HOT GAS FLOW

Fig. 1. Comparison of Basic Principles.

practical for this purpose because instabilities associated with other
methods of ionization tend to produce a plasma that is not sufficiently uniform for use in an MHD generator. Most common gases, such
as air, CO, CO_2, or the noble gases, do not ionize appreciably until temperatures in excess of 4000°K are reached. However, if a
common alkali metal-bearing compound such as potassium carbon-
ate is added to a gas in small amounts (on the order of one part per
hundred or less), sufficient thermal ionization can be obtained at
temperatures of 2000° to 2500°K. This process, called seeding,
thus results in temperatures low enough to be withstood by some
solid materials and to be produced in furnaces. Figure 2 shows a
curve of conductivity vs. temperature, typical of combustion product gases.

At this point it may be helpful to summarize the major differences between the plasma in a fusion reactor and the plasma in
an MHD generator. The latter will consist primarily of neutral
particles, only about one in 10^4 being ionized. The energy distributions of all particles will be essentially maxwellian and--under most
circumstances--all will have the same mean energy which will
correspond to only a few thousand degrees Kelvin. The gas pressure
will range from one-half atmosphere to several atmospheres, hence
even though magnetic fields of up to 8 tesla are contemplated, the
ratio of electron cyclotron frequency to collision frequency ($\omega\tau$)
will not exceed four or five. The magnetic Reynolds number
($\mu_0 \sigma u L$) will be less than unity--generally much less. That is to
say, the magnetic field distribution will not be appreciably altered

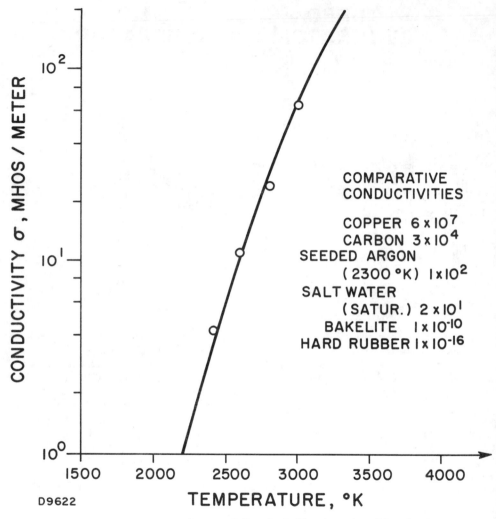

Fig. 2. Conductivity of Seeded Combustion Products.

by the motion of the plasma.

It will be recognized that these conditions are very much less complex than those in a fusion plasma. Indeed the MHD generator problem is essentially solved by combining the gas-dynamics of wind tunnels with the equations of current flow pertaining to ordinary solid or liquid conductors--with one notable exception, namely that the gas exhibits a much stronger Hall effect than all but a very few solids, i.e., the conductivity is a tensor. This fact does significantly complicate the design of MHD generators, and it does lead to some potential instabilities. These are mild, however, compared with those with which the fusion researcher wrestles. This rather drab phenomenological picture doubtless accounts for

the relatively slight interest that fusion researchers have traditionally exhibited toward MHD. It is worth remarking, however, that
this has not been the case in the U.S.S.R. where fusion researchers
have made some notable contributions to MHD generator theory.

An MHD generator in its most typical form (Figure 3)

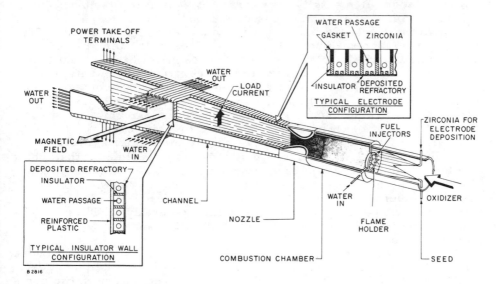

Fig. 3. MHD Generator Construction - Faraday Configuration.

consists of a duct down which the gas flows owing to an applied
pressure gradient, an arrangement of coils which produce a magnetic field across the duct, and electrodes on either side which
carry off the current. Because of the absence of hot, highly stressed moving parts, or of any solid parts at all which are not readily
accessible for external cooling, and because there are no close
tolerances to be maintained, the device can handle gas conditions
which would quickly destroy a conventional turbine. Experimental
generators have been built which have withstood gas temperatures
of 3000°K and the erosive and corrosive effects of ash for many
hours.

CONSTRUCTION

It was concluded at an early stage in MHD development at
Avco that it would not be practical to employ thick refractory duct
liners operating at the gas temperature, especially if ash-bearing
fuels such as residual oil or coal were to be used. This meant that
the walls had to be cooled. However, the properties of high thermal
conductivity and low electrical conductivity are not found in any
single material. Therefore, the composite channel structure illustrated in Figure 3 was developed. In this drawing the top and bottom walls are electrodes, while the side walls have a voltage

gradient along them, and therefore must be insulators. In both cases, the walls are made up of water-cooled metal elements with a thin layer of refractory insulation in between. Such a wall looks to the gas like an electrical insulator so long as the voltage between adjacent elements is less than that required to support an arc (about 50 volts under typical gas conditions).

On the electrode walls an additional feature is present. This is a groove filled with refractory in the face of the metal element. The depth and width of the groove are chosen so that the heat transfer from the gas raises the surface temperature of the refractory to approximately $2000^{\circ}K$. At this temperature, it becomes a good electron emitter and electrical conductor. The refractory most often used for this purpose has been stabilized zirconia. However, there is evidence that many other refractory mixtures, including coal slag, would serve the purpose nearly or equally as well.

The wall structure described above is capable of withstanding the high-temperature combustion gases with ease. That is to say, the handling of high temperature per se is not a problem. The major problem has been rather a gradual degradation of the resistivity of the inter-element ceramic (generally MgO) due to penetration by seed compounds (KOH, K_2SO_4, and K_2CO_3) dissolved in water.

There appear to be several solutions to this problem. These are illustrated in Figure 4. One is to use insulators of pure, dense MgO. A second is to add a small amount of refractory in the combustion chamber along with the fuel. This refractory coats the entire wall and, because its surface temperature is above the dew point of the seed, prevents the seed from condensing and penetrating the insulators. In the case of a direct coal-fired generator the ash contained in the fuel may serve for this purpose.

A third solution, favored in the U.S.S.R., is to allow a coating of seed (presumably K_2CO_3) to form while maintaining a wall temperature cold enough so this layer is solid but hot enough to prevent the layer from absorbing water vapor from the gas. The validity of this solution is open to question if direct coal firing is contemplated. However, it seems adequate if the fuel is gas or light oil.

MACHINE EFFICIENCY

An MHD generator uses a thermodynamic cycle similar to that used by a turbine, i.e., a Rankine cycle if the working fluid is condensed at some point, or a Brayton cycle if it remains a gas. The efficiency of an MHD generator as a cycle component is defined as the ratio of the energy actually extracted to the energy that would be extracted by a perfectly reversible or isentropic expansion between the same two pressures. This is the same definition as that used for a turbine. The most important dissipative processes are joule dissipation and wall friction, but heat loss to the walls and excitation power consumed by the magnet (if it is not superconducting) can also be important.

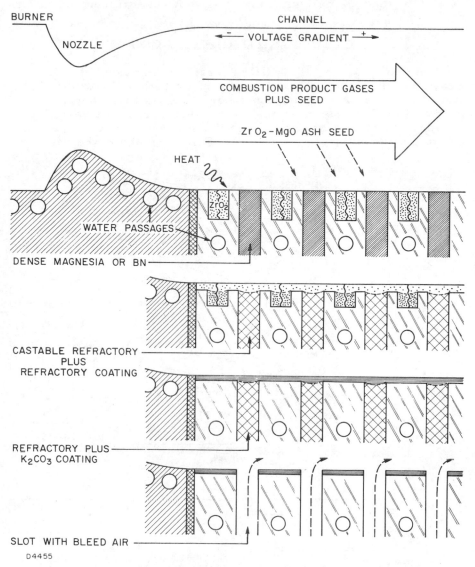

BURNER

NOZZLE

CHANNEL

← − VOLTAGE GRADIENT + →

COMBUSTION PRODUCT GASES
PLUS SEED

ZrO_2-MgO ASH SEED

HEAT

ZrO_2

WATER PASSAGES

DENSE MAGNESIA OR BN

CASTABLE REFRACTORY
PLUS
REFRACTORY COATING

REFRACTORY PLUS
K_2CO_3 COATING

SLOT WITH BLEED AIR

D4455

Fig. 4. Methods of Wall Insulation.

All losses tend to become smaller compared with the total output as the device becomes larger, or as the power output per unit volume goes up. The latter requires either increased magnetic field strength and/or increased conductivity resulting from increased flame temperatures. Present practical limits are represented, on the one hand, by a fossil fuel burned with pure oxygen and, on the other hand, a fossil fuel burned with air preheated to 2000°F. In the first case, power density as high as $1000 \ MW/m^3$ and an efficient generator as small as approximately 1 MW is possible. In the second case, power density is on the order of $25 \ MW/m^3$ and a smallest efficient size of several hundred MW results.

CYCLES AND CYCLE EFFICIENCY

As noted above, the MHD generator may be used in either a Brayton or Rankine cycle depending upon whether the working substance is a gas or a vapor. In either case, its thermodynamic role is that of an "expansion engine" like a turbine. For a variety of reasons, primarily greater suitability for high temperature, the Brayton or "gas" cycle is favored. Figure 5a shows the basic

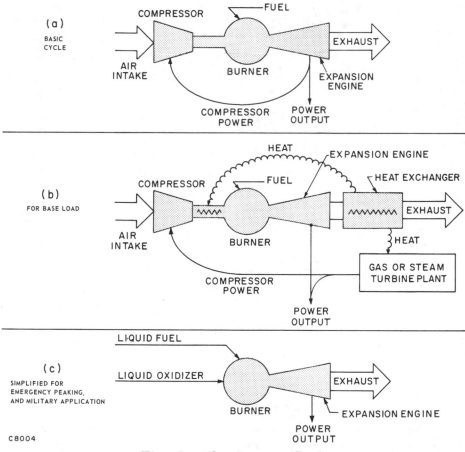

Fig. 5. The Brayton Cycle.

Brayton cycle. Figure 5b shows the cycle with embellishments added for maximizing its efficiency. This is the cycle proposed for base-load use. Finally, Figure 5c shows a simplified cycle proposed for emergency, peaking, and military applications.

Thermodynamically, it is only a partial cycle since it depends upon an outside source for liquefied (or compressed) oxidizer. It
could be called a "rocket cycle." (As might be imagined, there is
also a spectrum of cycles lying between the extremes represented
by b and c and exhibiting a corresponding spectrum of characteristics. However, these need not be considered here.)
      Detailed cycle efficiency calculations performed according to
the standards of and with the aid of utility power engineers are
summarized in Figure 6 and compared with calculations performed

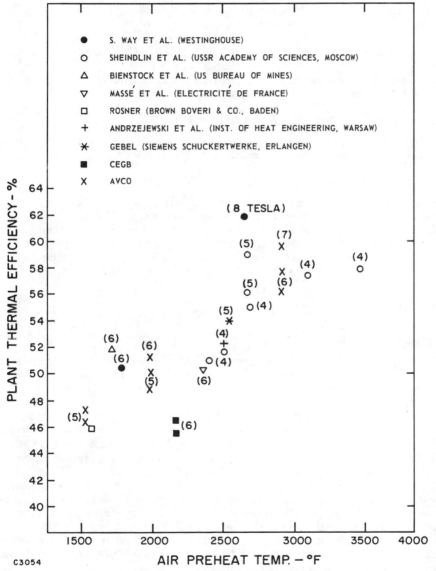

C3054

Fig. 6. Baseload MHD Plant Efficiency.

by other U. S. and foreign groups. Efficiency is here plotted against air preheat temperature since it is one of the most important practical parameters controlling top flame temperature and cycle efficiency. A considerable amount of experimental testing, both here and in Russia, indicates that 3000°F preheat is practical in gas or oil fired plants. Much less testing with fuel having a high ash content, i.e., coal, has been done, but the indications are that at least 2000°F is feasible. Schemes for raising this figure have been advanced, but none has yet been tested. In any case, it is now widely accepted that first-generation MHD plants ought to reach 50 percent efficiency and with further development attain an efficiency of 60 percent.

EMERGENCY AND PEAKING APPLICATIONS

An MHD plant using the simplified cycle shown in Figure 5c has a number of striking features. These are:

1. A virtually unlimited single unit rating.

2. Extremely rapid start-up (one to five seconds), and even more rapid response to load variation (milliseconds).

3. Low capital cost.

These have led to its consideration for utility applications where the objectives are to replace spinning reserve and to handle the daily and seasonal peaks of the power demand curve. A significant fraction of installed utility capacity must be assigned to meet such needs. Since peaking calls for a relatively simple plant that is used intermittently, it is attractive as an early application for MHD.

POLLUTION

Thermal

High efficiency means low thermal pollution. The amount of heat rejected by a power plant per kilowatt of useful output is proportional to $(1-\eta)/\eta$ where η is the plant efficiency. This relation is plotted in Figure 7.

Another important consideration is the type of cycle used. The closed Rankine cycle used in conventional steam plants must reject most of the heat through a heat exchanger, and its performance in terms of both cost and efficiency is very dependent upon having the lowest possible heat rejection temperature. Thus, such a plant is always sited on a large body of water, or else must employ large cooling towers to transfer the waste heat to the atmosphere. The open Brayton cycle, on the other hand, intrinsically rejects all heat to the atmosphere. Therefore, an MHD Brayton cycle, if bottomed by a gas turbine Brayton cycle as shown in Figure 5b, eliminates thermal water pollution entirely, and without the expense of cooling towers.

COMPARISON OF HEAT REJECTION FROM THERMAL POWER PLANTS

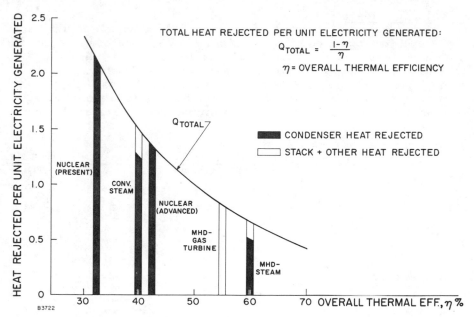

Fig. 7. Heat Rejection vs. Cycle Efficiency.

Particulate Emission

A large fraction of the particulate emission from conventional plants is due to unburned fuel. This will be greatly reduced in an MHD plant due to higher combustion chamber temperature and pressure. Furthermore, as an integral part of the process, particulate matter in the effluent gas will be removed by -- for example -- electrostatic precipitation. This is done because of the economic necessity for recovery of an alkali seed impurity added to enhance the electrical conductivity of the combustion gases. Highly efficient removal of particulate matter (in excess of 99%) is, therefore, ensured with the MHD process, and equipment for this is always part of the plant equipment.

Nitrogen Oxides

While it is easy to show that the use of MHD can reduce particulate pollution and virtually eliminate thermal water pollution, it had long been recognized that the situation was not so obvious with respect to the oxides of nitrogen. Therefore, a considerable experimental and analytical effort was addressed to this problem. Here nature has proven kind, and it has been shown that the same high temperature that promotes rapid formation of NO in the MHD burner also promotes rapid decomposition upon subsequent cooling if the cooling rate and the stoichiometry of the flame are suitably controlled. Detailed calculations of the reaction kinetics and experiments carefully designed to simulate the composition temperature-time history of the gas in an MHD plant have yielded NO concentrations well below the standards set by the Environmental Protection Agency (EPA).

Economic Consideration

With regard to all forms of pollution, the advantages ultimately boil down to a question of economics. Any type of power plant can have a clean exhaust if one is willing to pay for it. And the technical requirements for doing so are sufficiently clear that the cost of doing so can be estimated even now with at least ball-park accuracy. We have attempted to do this, using published data where possible, and the results are summarized in Figure 8.

	MHD		CONV. STEAM	NUCLEAR
	EARLY – DEVELOPED		PRESENT	PRESENT
EFFICIENCY %	50	60	40	32
BASIC PLANT COST $/kW	210	150	225	300
BASIC GENERATION COST – MILLS/kWhr				
CAPITAL CHARGES[1]	4.65	3.32	4.98	6.62
FUEL[2]	2.05	1.71	2.56	1.70
OPERATION & MAINTENANCE	0.30	0.30	0.30	0.35
SEED	0.05	0.02		
BASIC GENERATION COST	7.05	5.35	7.84	8.67
ADDITIONAL GENERATION COST – MILLS/kWhr				
DRY COOLING TOWERS	0.50 [3]	NONE [4]	0.80 [5]	1.20
SO_2 – REMOVAL	0	0	0.90 [6]	– –
TOTAL ADD. GENERATION COST	0.50	– –	1.70	1.20

(0) BASIS 1971 U.S. DOLLARS
(1) 15.5% AND 80% CAPACITY FACTOR
(2) COAL COST AT 30 CENTS/10^6 Btu
(3) MHD – STEAM CYCLE, COST PROP. TO STEAM PLANT

(4) MHD – GAS TURBINE CYCLE
(5) OFFICE OF SCIENCE AND TECHNOLOGY REPORT "CONSIDERATIONS AFFECTING STEAM POWER PLANT SITE SELECTION"
(6) BASED ON $40/kW

Fig. 8. Comparative Estimated Costs of Coal Burning 1000 MW (Nominal) Power Plants[0].

Savings of up to 2.15 mills per kilowatt hour in pollution control cost are projected for MHD relative to conventional plants, and if these are even reasonably accurate, the annual savings in pollution control costs alone would very quickly pay the total development cost of MHD.

NUCLEAR APPLICATIONS

MHD conversion combined with a high temperature fission or fusion reactor could prove to be the ultimate power plant of the future in terms of compactness, low maintenance, efficiency, and minimum thermal pollution. Since, at present, only the Nerva rocket reactor produces temperatures high enough to be effective with MHD, one can only speculate what form a commercially acceptable nuclear-MHD plant might take. However, one can list a number of interesting facts suggestive of the future possibilities. They are:

1. Electron beam ionization techniques (developed originally for high power lasers) show promise of leading to an efficient MHD plant using temperatures somewhat higher than the Peach Botton and Fort St. Virain HTGR's but significantly lower than Nerva.

2. MHD is well adapted to a hot loop. The most expensive part of the generator, the magnet, is completely isolated from the working gas or vapor. The relatively inexpensive MHD duct could be regarded as a throw-away item.

3. A recent study suggests that a gaseous core cavity reactor would have an exceptionally good breeding ratio. Because such a reactor has almost no top temperature limit, it would also be an ideal heat source for MHD.

4. Thermonuclear fusion reactors, by virtue of much different and much less severe requirements for the containment of radioactivity, should have far less difficulty producing high temperature.

A relatively straightforward but efficient method of coupling MHD to a fusion reactor is illustrated schematically in Figure 9.

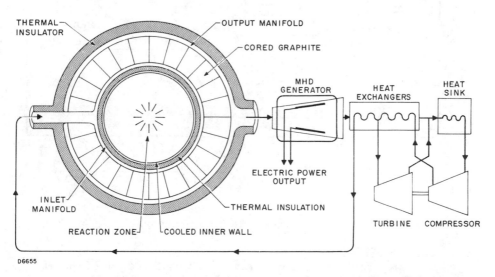

Fig. 9. "Conventional" MHD Conversion System Coupled to a Fusion Reaction by Means of a Graphite Blanket.

It employs a graphite blanket around the reaction zone to absorb neutron and x-ray energy and heat a suitable working gas, for example, helium, for the MHD cycle. Since graphite maintains good structural properties up to 2800°K and since, in this application, it is not heavily stressed, is not called upon to contain fission products as is the graphite used in the HTGR, and is not subject to corrosive atmospheres as is the graphite used in Nerva, it should be possible to operate it at high temperature for long periods of time. A helium outlet temperature of 2500°K seems reasonable, and Figure 10 shows that an overall conversion cycle efficiency of

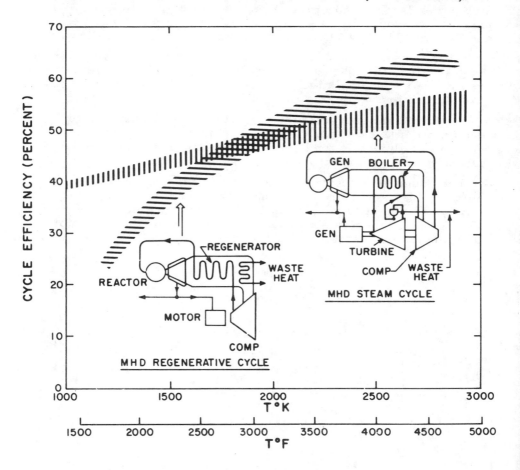

Fig. 10. Top Cycle Temperature.

the order of 60% results.

If some reaction energy remains within the core of the reactor it might either be used in one of the several possible direct-coupled schemes for conversion of charged-particle energy or else used for further heating of the helium in the MHD cycle in order to obtain a temperature above 2500°K and a correspondingly higher cycle efficiency. However, the feasibility of such a refinement remains to be determined.

Given the present state of development of the fusion reactor, a significant practical advantage of this "graphite blanket plus conventional MHD" approach is that its feasibility and major characteristics are to a relatively high degree independent of how the fusion reactor works. That is to say, the design and performance of the reactor does not strongly affect the design and performance of the MHD system and vice versa.

Overall plant efficiency will be maximized by maximizing the fraction of the energy entering the blanket that is deposited at high temperature in the graphite while minimizing that deposited at low temperature in less-refractory material, for example, structure. If it is necessary to carry out tritium breeding in a separate lithium blanket, an especially severe limit is placed upon the fraction of heat releasable in the graphite. A recent study[1] indicates that this fraction is about thirty percent in the case of a purely D-T reactor. Therefore, it is desirable to seek some method of incorporating lithium or lithium compounds within the high temperature portion of the blanket.

Of course, the need for tritium breeding would be eliminated if a reactor using the catalyzed D-D reaction was developed. In this case, the combined use of blanket-coupled MHD for conversion of neutron energy and a direct-coupled scheme for conversion of charged particle energy might well yield the maximum overall plant efficiency.

Because it allows the megavolt temperatures of the reaction zone to be degraded to a mere fraction of a volt, the blanket-coupled scheme may seem rather inelegant. The lack is probably more conceptual than consequential however, since an overall thermal cycle efficiency beyond the 60 to 70 percent possible from this blanket-coupled MHD system is unlikely, especially after all of the cooling requirements of the reactor, both high grade and low, have been met.

In any case, it seems likely that any practical fusion reactor will deposit a substantial fraction of its energy within its blanket and one should seek to make the best possible use of it.

Given a commercially acceptable reactor capable of temperatures above $2000^{\circ}K$, there seems little reason to doubt that MHD would be the preferred way of converting the reactor heat to bulk electricity. In a nuclear closed loop, in which an inert gas would be the probable choice of working fluid, materials problems would be minimal and unrivaled efficiency would result.

STATUS OF MHD TECHNOLOGY

The development of MHD technology has spanned a period of almost 15 years. This development was pioneered in the United States, but the major effort is now occurring in the U.S.S.R. There the construction of a 75 MW MHD-steam pilot power plant incorporating a 25 MW MHD generator has just been completed.

The development work has naturally centered on the MHD generator itself, but major efforts have also involved important plant components such as superconducting magnets, combustion systems, air heaters, seeding and seed recovery systems and methods for pollution control of sulfur and nitrogen oxides.

The study of the MHD generator itself has proceeded along two complementary paths. The first has included construction and operation of relatively large experimental generators designed for short term operation. Parallel to this, studies have been conducted with small scale installations to establish the integrity and performance of electrodes, insulators and channel mechanical design features for continuous long term operation.

In the United States two large experimental MHD generators were built and successfully operated in the middle sixties. These two generators represent perhaps the highlights of generator development in the United States. They were both supported by the Department of Defense, were designed and built by the Avco Everett Research Laboratory, Inc., and were intended for special purpose short-duration applications. One of these generators called the MK V was self-excited and produced a gross power output of 32 MW. This generator is pictured in Figure 11. The other generator

Fig. 11. MK V Self-Excited MHD Generator with 32 MW Gross Electrical Output.

called the LORHO generator (Figure 12) located at the Arnold Engineering Development Center, Tullahoma, Tennessee, delivered an electrical output of 18 MW. Experience gained with these

Fig. 12.   18 MW LORHO Pilot MHD Generator.

larger installations has provided a good understanding of the MHD
process and an experimental basis for predictions of MHD generator performance.

In addition to these larger short-duration experimental
generators considerable work has been conducted with smaller
fossil-fueled generators at much lower power levels of a few kilowatts. This work has mainly been carried out at Avco Everett
Research Laboratory, Inc., University of Tennessee, and Stanford
and has provided design data and information regarding long-term
operation of MHD generators under conditions appropriate to commercial power applications. Experimental work has been encouraging and indicates that the electrical performance and integrity of
electrodes and insulators can be maintained by using the proper
level of axial electrical field and proper operating temperatures of
electrodes and insulators. Work presently underway is aimed at
extending long term channel operating experience up to power levels
of 500 kW in the facility shown in Figure 13.

In parallel with the development of the MHD generator itself, efforts have been devoted to the development of superconducting magnets, combustion systems, high temperature air heaters,
seed recovery systems and pollution control systems. The

Fig. 13. MK VI Long-Duration Facility.

construction and testing of a large 4 Tesla model superconducting
coil magnet in the middle sixties was one highlight of this effort.
        Present work in the United States is funded both by
private industry and government. This includes a number of leading electric utilities,[*] Edison Electric Institute, American Public
Power Association, Avco Corporation, the Office of Coal Research
of the Interior Department, and the Department of Defense. The
work is being carried out principally at the Avco Everett Research
Laboratory, Inc., M.I.T., Stanford, and the University of
Tennessee. Also involved are Westinghouse, G.E., STD Corporation, the Bureau of Mines, and the Arnold Engineering Development
Center.
        By far the most impressive MHD effort exists in the
Soviet Union with the strong support of the State Committee on
Science and Technology and the joint sponsorship of the Academy
of Sciences and the Ministry of Electrification. The Soviet program
is broadly based and involves many institutes of which the chief ones
are the Institute for High Temperatures, the Krzhizhanovsky Power

---

[*] Baltimore Gas and Electric, Boston Edison Co., Consolidated
Edison Co. of New York City, Inc., NEGEA Service Corp., New
England Power Co., Northeast Utilities Service Corp., TVA.

Institute in Moscow and also the Research Institute of the Ukranian Academy of Sciences in Kiev. The most significant accomplishment to date has been the design, construction and preliminary operation of a natural gas-fired pilot plant with a designed MHD output of 25 megawatts. Pictures of this plant are shown on Figure 14 and Figure 15. The plant is located in the north of

Fig. 14. 25-Megawatt MHD Generator in U-25 Plant.

Moscow on a power station site adjacent to the Moscow ring road and is a complete pilot plant in every way, including an inverter system for feeding generated power directly into the Moscow grid system. As such, it constitutes the first commercial pilot plant in the world for network service. Basic development for this plant has been carried on, and continues to be undertaken, in the U-02 installation located in the center of Moscow. This experimental installation is shown in Figure 16. It is a unique 5-megawatts (thermal) experimental facility in which all the elements for an MHD system can be tested for endurance. These two installations are the direct responsibility of the Institute for High Temperatures. The Krzhizhanovsky Power Institute operates a generator test facility, the ENIN-II which may later be converted for experiments with coal firing. The fourth major Soviet installation is located at Kiev where an old power station has been converted for long-duration

Fig. 15. Air Preheaters for U-25 Installation.

MHD tests. Thus, the U.S.S.R. possesses four facilities which can generate a great amount of experience in all aspects of MHD technology and it is expected that runs of up to 5000 hours will be demonstrated within the next few years. In parallel with this effort, work is progressing on the design of the first commerical demonstration unit, and this has variously been reported to have an MHD generator of between 200 and 600 megawatts electrical output.

The MHD effort in the Soviet Union involves several thousand scientists, engineers and technicians, according to estimates from professional people who have been there in recent

Fig. 16. General View of U-02 Installation Showing Preheater in
Foreground.

years. The cost of the U-25 installation has been estimated in
dollar equivalent to be in the 100-150 million dollar range. All of
this indicates the determination and vigor of the MHD development
program in the U.S.S.R.

For the last 10 years, Japan has also undertaken a very
comprehensive program in MHD power generation. All components
of the total system are being studied and extensive system and

economic evaluations are also being conducted. The general character and scale of the Japanese facilities are in many ways similar to those which have been built and tested in the United States except that there is no equivalent to the Mark V or LORHO. On the other hand, a superconducting magnet has been combined with an MHD generator, and the construction of a larger super-conducting magnet for a 1000 kW MHD generator test facility is now almost completed. A small pilot plant has reached an advanced stage of construction and operation is expected during 1973.

Within the past three years, efforts have been initiated in the Federal Republic of Germany. The Bergbau-Forschung Institute at Essen and the KFA Nuclear Institute at Julich are cooperating on the testing of a small propane-fired long-duration experimental facility. The Institut fur Plasmaphysik in Garching together with an industrial group (MAN) has built a 1-megawatt short-duration facility. A 3-megawatt thermal natural-gas-fired long-duration installation is being operated by the Nuclear Research Institute at Swierk in Poland.

REFERENCES

1. A. Pant and R.J. Rosa, "The Feasibility of an MHD-Fusion Interface," Avco MHD Report No. 85, (Feb. 1973).

GENERAL REFERENCE

Symposia on the "Engineering Aspects of MHD" are held yearly in the United States and an international symposium on MHD is held every second year. The proceedings of these meetings are a good starting point for those interested in going deeper into the subject. Basic texts include G.W. Sutton and A. Sherman, "Engineering MHD," McGraw-Hill, N.Y., (1965) and R. Rosa, "MHD Energy Conversion," McGraw-Hill, N.Y. (1968). The re-sults of the short-lived but comprehensive British effort in this area are summarized in J.B. Heywood and G.J. Womack, Ed., "Open Cycle MHD Power Generation," Pergamon Press, Oxford, Eng., (1969).

SOLAR RESEARCH AND TECHNOLOGY*

LLOYD O. HERWIG, DIRECTOR

Advanced Solar Energy Research and Technology
Research Applied to National Needs
National Science Foundation
Washington, D. C. 20550

Presented at the Annual Meeting of the American
Physical Society on February 7, 1974

INTRODUCTION

It is a pleasure to appear before this audience at the Annual
Meeting of the American Physical Society. In this time of rising
concerns over energy, I will try to share with you the enthusiasm
and current plans of the National Science Foundation to make every-
body's lucky star--the sun--an even greater source of life and
energy for the U.S.A. and the world.

In my presentation I will describe the Federal role in solar
energy research and technology for terrestrial applications, the
solar energy technologies and plans in the National Solar Energy
Program, and, in a general way, the research and technology problems that must be solved before practical systems become possible.

THE CHARACTERISTICS OF THE SOLAR ENERGY RESOURCE

The amount of solar energy falling on the contiguous United
States in a year's time is about 700 times our annual rate of
energy consumption. Put another way, on the average, the daily
solar energy arriving on about 5000 square miles of United
States territory is about equal to our total daily energy use.
Our daily total energy use, of course, requires conversion of
many types of fuels for a wide variety of needs. As we proceed,
I will point out that technically feasible solar energy conversion methods can produce many forms of power and energy--
for example, high and low temperature thermal energy,
electricity, and gaseous, liquid, and solid fuels.

*The remarks and information in this paper are the responsibility
of the author and do not represent official positions of the Federal
Government or the National Science Foundation.

It is significant to note that the solar energy falling in a year on a square foot of average land in the Continental United States has a value of about one dollar based upon energy unit value of two dollars per million Btu's. This is based upon average United States insolation of 1480 Btu's per square foot per day, (37 K cal per square meter per day). As you know, however, the value of one million Btu's (252,000 K cal) is very much a function of the temperature at which it can be delivered, the purposes for which it is to be used, and where it is delivered.

Consider the numerous advantages of solar energy. The sun is a large and continuing source of domestically available energy. It is widely distributed, and its use does not add to the earth's overall heat inventory. The wide distribution of solar energy over the United States makes it possible to consider systems providing thermal energy or power at the point of use without recourse to extensive distribution networks and central power stations.

Next, consider two major disadvantages of solar energy that pose challenges to innovators in research and technology. Sunlight is a dilute source of energy--that is, its intensity is relatively low, presenting a technological challenge to achieve its economic conversion to more useful forms of energy. Solar energy is intermittent and variable due to diurnal, seasonal, and environmental obscuration effects; thus the energy must either be used as it becomes available or in conjunction with storage and backup systems.

Because of these two disadvantages, there are requirements for large collection areas and some energy storage. This in turn gives rise to higher initial capital costs than many competing technologies. In considering the cost competitiveness of solar energy systems, however, one needs to take into account life cycle costs including fuel and impact costs, in addition to initial capital costs.

As a result of research and technology projects underway and planned, we believe that by the early 1980's some solar energy systems can meet the challenge of producing substantial quantities of energy at acceptable commercial costs based upon cost accounting that takes a system's life cycle into consideration. Further, other systems can be implemented in the same way by the late 1980's. The general characteristics of the solar energy systems being developed in the National Solar Energy Program are as follows:

no insurmountable technical-barriers to their implementation, numerous conversion methods, and promise of cost competitiveness. Solar energy systems, moreover , would conserve domestic fossil fuels, create new exportable technology products, and thus improve the balance of trade picture. Additionally, solar energy has minimal environmental impacts.

FEDERAL ROLE AND COORDINATION

In April, 1973, the National Science Foundation (NSF) was designated by the President's Office to be the lead Federal agency in planning and coordinating the broad area of solar energy research and technology. Five year program plans for research in each of six selected solar energy technologies have been developed, and are reviewed and revised periodically.

The plans and policies of the National Solar Energy Program are focused on the general objective of developing, at the earliest feasible time, those applications of solar energy that can be made economically attractive and environmentally acceptable as alternative energy sources. The program plans are adjusted to reflect continuing critical reviews based upon new information and understanding, assessments of priorities, identifications of research and technology needs, and levels of funding and research capabilities. An integrated research plan recognizing the multidisciplinary nature of solar research is being implemented with an estimated NSF funding of $13.2M in FY 1974.

PROGRAM ORGANIZATION AND FUNDING

The general objectives of the solar energy program are stated in this way:

> To provide the research and technology base required for economic terrestrial applications of solar energy; to foster the implementation of practical systems to the state required for commercial utilization; and, to provide a firm technical, environmental, social, and economic basis for evaluating the role of solar energy utilization in U. S. energy planning.

The research and technology activities for the National Solar Energy Program are organized under the following six program areas:

. Heating and Cooling of Buildings

. Solar Thermal Energy Conversion

. Photovoltaic Conversion

. Biomass Production and Conversion

. Wind Energy Conversion

. Ocean Thermal Energy Conversion

Federal Funding

The NSF budgets for solar energy applications in past years and for the current year are shown in Figure 1 for each of the six program areas. The budgets are given in terms of Federal fiscal years; e.g., FY 1975 covers the period from July 1, 1974 to June 30, 1975. The NSF solar energy budget shows $1.2M in FY 1971; $1.66M in FY 1972; $3.96M in FY 1973; $13.2M (estimated) in FY 1974; and, $50M (estimated) in FY 1975. Solar energy funding from other federal agencies shows $30K in FY 1971; $50K in FY 1972; $243K in FY 1973; and, $1.7M (estimated) in FY 1974.

The relatively large percentage increases in total funding are very apparent over the period from FY 1971 to the the present. In particular, the increases of more than a factor of three from FY 1973 to FY 1974 (estimated) and FY 1974 (estimated) to FY 1975 (estimated) show the growing Federal interest in exploring the solar energy alternatives.

The total FY 1974 Federal funding of solar energy research and technology performed outside of Federal laboratories is estimated at $15.1M. Five Federal agencies are considering some funding for FY 1974, including the NSF (estimate $13.2M), NASA (estimate $0.9M), the AEC (estimate $0.6M), the DoD (estimate $0.2M), and The United States Postal Service (estimate $0.2M). In addition, there are Federal inhouse research and technology projects that add to the total Federal funding.

Phased Project Planning

The NSF planning for implementing solar energy applications emphasizes a phased project planning approach embodying integrated programs of multidisciplinary research, analysis, experiments, and system studies. The more important steps in phased project planning leading to a new application are shown in Figure 2. The Research Phase can include basic and applied research on advanced approaches to solar energy conversion; feasibility studies; research and analysis on innovative ideas, materials,

NSF/RANN SOLAR ENERGY BUDGET

(MILLIONS OF DOLLARS)

	FY 1971 (ACTUAL)	FY 1972 (ACTUAL)	FY 1973 (ACTUAL)	FY 1974 (ESTIMATE)	FY 1975 (ESTIMATE)
SOLAR ENERGY FOR BUILDINGS	$ 0.54	$ 0.10	$ 0.40	$ 5.9	$ 17.0
SOLAR THERMAL CONVERSION	0.06	0.55	1.43	2.2	10.0
PHOTOVOLTAIC CONVERSION		0.33	0.79	2.4	8.0
BIOCONVERSION FOR FUELS	0.60	0.35	0.65	1.0	5.0
WIND CONVERSION			0.20	1.0	7.0
OCEAN THERMAL DIFFERENCE CONVERSION		0.14	0.23	0.7	3.0
WORKSHOPS AND PROGRAM ASSISTANCE		0.19	0.26		
	$ 1.20	$ 1.66	$ 3.96	$ 13.2	$ 50.0

Fig. 1

STEPS IN PHASED PROJECT PLANNING TO DEVELOP A NEW APPLICATION

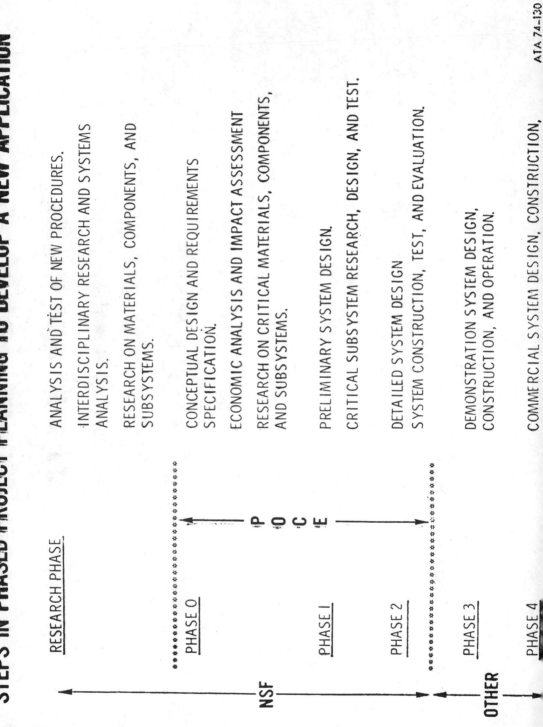

RESEARCH PHASE

ANALYSIS AND TEST OF NEW PROCEDURES.

INTERDISCIPLINARY RESEARCH AND SYSTEMS ANALYSIS.

RESEARCH ON MATERIALS, COMPONENTS, AND SUBSYSTEMS.

PHASE 0

CONCEPTUAL DESIGN AND REQUIREMENTS SPECIFICATION.

ECONOMIC ANALYSIS AND IMPACT ASSESSMENT

RESEARCH ON CRITICAL MATERIALS, COMPONENTS, AND SUBSYSTEMS.

PHASE I

PRELIMINARY SYSTEM DESIGN.

CRITICAL SUBSYSTEM RESEARCH, DESIGN, AND TEST.

PHASE 2

DETAILED SYSTEM DESIGN

SYSTEM CONSTRUCTION, TEST, AND EVALUATION.

PHASE 3

DEMONSTRATION SYSTEM DESIGN, CONSTRUCTION, AND OPERATION.

PHASE 4

COMMERCIAL SYSTEM DESIGN, CONSTRUCTION,

PPCE

NSF

OTHER

AIA 74-130

components, and subsystems; and basic data required for analysis. Proof-of-concept experiments (POCE) are major milestones in the program plan. After a successful proof-of-concept experiment, the plan continues with demonstration systems and commercial systems.

Proof-of-Concept Experiments (POCE)

Proof-of-concept experiments are major milestones in the present five-year program plans.

A system proof-of-concept experiment is undertaken to prove that the full technology base is available to enable a user community to move into the design and development of economically viable systems. In each program area, these experiments are scheduled by FY 1978 to be in either the construction, test, and evaluation phase (2); the preliminary system design phase (1); or the system analysis and selection phase (0)--if results in succeeding phases justify continuing. Subsystem proof-of-concept experiments are programmed as soon as possible to verify the performance, lifetime, and operational and environmental responses associated with materials, components, and subsystems making up parts of a full system.

DESCRIPTION AND SUMMARY OF PROGRAM AREAS

The limited time does not permit a detailed description of either the application areas or five-year programs for all six areas of solar energy applications. I will simply highlight the current efforts with particular emphasis on the highest priority area--solar energy systems for buildings.

In Figure 3, you see our projections for advancement in each of the areas of applications of solar energy. System proof-of-concept experiments will be completed into Phase 2 within five years for three areas--heating and cooling of buildings, wind energy conversion , and biomass production and conversion. In the other areas--solar thermal conversion, photovoltaic conversion, and ocean thermal conversion--component research and subsystem proof-of-concept experiments will be completed, and Phase 0 and Phase 1 will be initiated in system proof-of-concept experiments.

TERRESTRIAL SOLAR ENERGY PROGRAM

TASKS	FISCAL YEAR			
	73	74	75	76
HEATING AND COOLING OF BUILDINGS				
SOLAR THERMAL ENERGY CONVERSION				
PHOTOVOLTAIC				
OCEAN THERMAL CONVERSION				
WIND ENERGY CONVERSION				
BIOMASS PRODUCTION & CONVERSION				

☐ PHASE 0

▨ PHASE 1

■ PHASE 2

ATA 74—129
10-3-73

It should be emphasized that no technological breakthroughs are
required to obtain useful energy and power from early conceptual-
izations of these solar energy systems. In general, each of the
technologies to be described has been demonstrated in the past, in
some cases more than fifty years ago, e. g., the 50 horsepower
steam engine powered by solar energy in an experimental plant
near Meadi, Egypt, in 1913.

The major problem in each application area is to develop systems
that are economically acceptable to the public and commercial sectors
in the United States. To do this will require innovative engineering
as well as new knowledge and understanding of nature's laws; new
and improved approaches to collection and conversion of solar radiation,
and to energy storage, transport, and conversion; new system approaches;
and, perhaps most importantly, new and cheaper materials to increase
system performance, reliability, and economic acceptability. Important problems must also be solved dealing with other factors--social,
legal, political, regulatory, environmental, economic, for example--
which are identified with widespread implementation of solar energy
systems.

A summary of the objectives, technologies, and present
program for each of the six application areas will be described
in the following sections.

Solar Energy for Buildings

The general objective of this program area is to establish the widespread utilization of systems using solar energy for heating, cooling,
and supplying the hot water needs for buildings in the United States.
This can be done, of course, only to the degree that the system
applications are economically viable, technically feasible, and
socially acceptable. More specifically, the five-year objectives are
to obtain increased performance and new options for components,
subsystems, and systems; and, to complete proof-of-concept experiments through test of optimized experimental systems for a number
of economically viable applications.

The application of heating and cooling to buildings has the highest
priority for funding in the current year's budget (about 45 percent
of the total, or about $5.9M) because it is the solar energy area in
the most advanced state of technology and economic viability. It
offers, moreover, an excellent opportunity to make an early impact
on national energy requirements. At present, commercial and
residential building uses account for approximately 25 percent of
the energy consumed in the United States, at an annual cost of
about $18 billion.

Solar energy systems for space and water heating in buildings have received moderate experimental testing in the past, while combined heating and cooling systems have received very little experimental testing, even in laboratory settings. Modest performance data is available for fewer than 25 solar heating systems over the past 30 years in the entire world.

A conventional approach to a solar energy system providing heating and cooling services for buildings will now be described. The sun's radiation is converted to heat by means of an absorbing surface incorporated in a flat-plate collector assembly. The absorber surface has characteristics in common with many ordinary materials that get very warm in the sun, but it is usually mounted in a structure that is designed to reduce direct heat losses by conduction, convection, and radiation. A heat transfer fluid is passed through channels in contact with or integral to the heated absorber surface. The fluid is circulated to a heat storage unit or to the heating and cooling service systems as required.

A common collector configuration consists of a coated black metal surface, including heat transfer channels, sandwiched between heat insulating structures. The sun side is insulated by a transparent structure consisting of one to three layers of infrared-opaque glass, each spaced about one half inch apart. This transparent structure is held about one half inch above the absorbing surface to reduce heat losses by convection and radiation . The back side of the absorbing surface is covered by a thick layer of thermal insulation to minimize heat losses from the bottom of the collector. The collector performance may be improved by using selective coatings to maintain or increase the absorption of solar radiation at the absorbing surface while reducing the loss of energy by re-radiation. Also, transparent coatings may be placed on the glass surfaces to reduce reflections of solar radiation at air-glass interfaces.

Variations in collector configurations and designs are being studied to increase energy collection efficiency; increase the temperature of collection; reduce costs through reduced material, fabrication, and operating costs; and increase performance and lifetime. Collector design is one of the very important subsystems in optimizing the cost of energy from a solar energy system.

In September, an award of $238,000 was made to Colorado State University to design, construct, test, and evaluate an optimized solar cooling system in a fully instrumented experimental house near Fort Collins, Colorado. It will be the first full-scale heating and cooling system test to provide 50 to 75 percent of the needs for climate conditioning and hot water heating. This project will provide important information on system performance, reliability of components and materials, system operational characteristics and maintenance problems, and accuracy of modeling calculations for these types of systems.

Three independent studies by industry/university teams were initiated by NSF in October as the initial phase in system proof-of-concept experiments. They will be completed in FY 1974. These studies will identify the most viable applications of solar energy systems for all types of buildings in all climatic regions. Preliminary system designs of the applications selected under the initial phase will be carried out in FY 1975. The final phase will be performed during FY 1976 and FY 1977 and will include final design, construction, test and evaluation of the selected experiments.

A mobile solar research laboratory is under construction with NSF support to test advanced equipment for the solar heating and cooling of buildings. This laboratory, housed in two trailer vehicles, will be moved to various locations in the United States to collect experimental data on system performance under a wide variety of climatic and sun conditions, which will also be measured with installed equipment. A second major purpose of the laboratory is to acquaint designers, architects, building contractors, zoning and building code officials, mortgage lenders, and others, with the characteristics and capabilities of solar energy heating and cooling systems for buildings, and to learn about local problems facing such applications. Advanced components and subsystems to be tested include a flat-plate collector and two solar energy air-conditioning systems. The project is being managed and partially funded by Honeywell Corporation.

In January, the NSF initiated four projects, called "Solar Energy-School Heating Augmentation Experiments", to advance the systems technology for using solar energy for space heating and hot water needs of buildings. A high school, two junior high schools, and an elementary school in different geographic areas and institutional settings are expected to augment their regular heating systems with different experimental solar energy systems to be installed by March. Four large industrial companies are undertaking these projects to permit the gathering of some data on these solar heating systems during the current heating season. In general, the projects are expected to provide important information on the problems of retrofitting solar heating systems on existing buildings and on the degree such systems can be made economically justifiable and socially acceptable. Results of these projects are expected to be extendable to augmentation of heating systems for factories, warehouses, shopping centers, other low-rise buildings, trailer parks, and crop drying.

In parallel with and in support of the proof-of-concept experiments, demonstrations of applications will be carried out by other organizations. The Department of the Army, The General Services Administration, the National Aeronautics and Space Administration, and the United States Postal Service have initiated phased projects that can lead to solar heating and cooling systems in their facilities. These programs, if completed, will become part of the program for proof-of-concept experiments and will provide important data points in the understanding of the problems of utilizing solar systems.

As the proof-of-concept experiments proceed, advanced and supporting research on new and improved components and systems will continue. Research projects conducted in the past have brought the state-of-the-art to the point that the present experiments can be done. The supporting research and technology will be programmed for the immediate needs of the proof-of-concept experiments. The advanced research and technology program will be conducted on innovative systems, subsystems, and components to obtain new options for increasing performance and reducing the cost of systems. It will produce the technology for second and third generation systems which will, if all goes well, reduce or eliminate the reliance of these systems on auxiliary power sources.

There are now 27 active research and study projects supported by about $5M in Federal funds to nonfederal institutions working on solar energy for buildings. These projects are supported by NSF (22), HUD (1), NASA-Lewis (1), AEC (1), Department of the Army (1), and U. S. Postal Service (1). Another 20 or more projects will be initiated in FY 1974.

Research areas of greatest potential impact include improved collection of solar radiation (higher temperatures and higher efficiencies), heat transfer and transport systems, heat storage materials and systems, space-cooling systems (higher efficiencies at practical solar system temperature), materials (cheaper and better performing), and combinations of functions. In each area of research need, it is imperative that costs of producing, installing, and maintaining the systems be reduced while maintaining or improving the system performance.

Solar Thermal Conversion

The general objective is to prove the technical and economic feasibility of solar thermal conversion systems providing electrical or combined electrical and thermal service. More specifically, the five-year objectives are to complete system studies to select

sites and systems for proof-of-concept experiments and to complete
subsystem proof-of-concept experiments on solar concentrators and
collectors, thermal collection and transfer, energy storage, heat
exchangers, and organic Rankine energy conversion cycles.

Federal funding in this research area is estimated to be about
$2.4M for FY 1974, including $2.2M from NSF and $200K from NASA.
System studies and analytic modeling to optimize system performance
and power costs are underway along with collector-concentrator experimental studies, selective coating experimental studies, mission
analysis, and design and experiments on other system components.

Solar thermal conversion systems collect solar radiation and
convert it to relatively high temperature heat that can be applied
to a boiler in a conventional thermodynamic cycle to produce
electricity or mechanical shaft power. The steam raised by applying
heat to the boiler can also be used for   process   steam or water
for manufacturing or heating needs. For the sake of higher conversion efficiency in thermodynamic cycles, it is desirable to collect
heat at as high a temperature as possible.

Several different system approaches are being studied for concentration of solar energy. In the first approach, large parabolic
cylindrical mirrors track the sun along a single axis and focus
the radiation on a line focus. The intense radiation along the focal
line is absorbed by a selective optical coating on the outside surface
of a cylindrical metal heat transfer pipe placed inside an evacuated
transparent pipe. The heat is removed by a heat transfer fluid to
a storage unit or to a heat exchanger connected with a power conversion unit. In some system concepts, a heat pipe serves as the
heat transfer device from the line focus of the radiation to a local
heat exchanger and storage unit. For a large power plant the heat
and energy must be transported relatively long distances by hot
fluids in pipes to a central storage unit or power conversion unit,
with resulting heat losses. Temperatures in the range of 300 to 600 F
(150-315C) can be obtained at total collection efficiencies of about
50 percent of the intercepted radiation.

In a second approach, a field of two-axis mirrors track the sun
and deflect radiation to a fixed absorber-boiler unit mounted on top
of a high tower. Thousands of mirrors can be deployed to concentrate direct sunlight on the absorber-boiler unit. The high temperature heat can be used to produce steam or other vapors, to operate
a turbo-electric generator. Temperatures greater than 1,000 F
(534 C) can be readily obtained at relatively high total collection
efficiency.

Still another approach is an adaption of flat-plate collectors to higher temperature collection. Temperatures in the range of 300 F(150 C) may be realizable at acceptable collection efficiencies. The advantage of flat-plate collection over the radiation-concentrating techniques is that the system can be used in diffuse sunlight as well as direct sunlight.

The technical feasibility of solar thermal power systems has been established by a number of experimental facilities. As long ago as 1913 in Meadi, Egypt, an array of cylindrical collectors totaling 13,269 square feet was used to concentrate solar energy to produce steam for operation of an engine to pump water. This power system produced peak power of more than 50 horsepower. It operated for about three years. More recently several solar furnaces have been constructed in which mirrors with two-axis tracking are used to concentrate radiation at a central point for high-temperature experiments on materials. The largest of these is located at Odeillo, France, in the mountains near the border with Spain. This solar furnace generates one megawatt of thermal power within a two-foot (61 cm) diameter to create temperatures of the order of 7000 F (4000 C). This heat could be captured in an absorber-boiler used to generate a hot gas, and converted to electricity in a turbine-generator.

At the present time the Federal government is supporting 10 outside projects with a total value of about $2.5M. Of these projects, nine are supported by NSF and one by NASA-Lewis Laboratory. There is some work being done in Federal laboratories operated by NASA and AEC.

Major research problems remain in the areas of radiation collection and concentration, heat transfer and storage, heat exchangers and boilers, and materials performance and lifetime. Special problems include high temperature selective optical coatings, reflector designs and optical coatings, thermal insulation, thermal transients and material fatigue, and materials compatibilities.

Photovoltaic Conversion

The general objective is to develop low-cost, long-lived,
reliable photovoltaic conversion systems to be commercially
available for a variety of terrestrial applications. More specific
objectives in the five-year plan are to reduce the cost of solar
cell arrays made from single-crystal silicon wafers by a factor of
more than 10; to provide the research base for alternate solar cell
technologies--for example, cadmium sulfide, gallium arsenide,
thin film polycrystalline silicon--showing low-cost potential; to
conduct systems and applications studies for low-cost fabrication
of cells and arrays; and, to identify system proof-of-concept
experiments projecting power costs a factor of 10 lower.

The Federal funding for photovoltaic research and technology
for terrestrial applications is estimated to be about $3M in FY
1974.

The use of semiconductor photovoltaic cells, or solar cells,
for direct conversion of solar radiation to electricity was demonstrated in 1954. To obtain the photovoltaic effect, combinations of transparent semiconductor materials, or semiconductor
and thin film metal materials, are placed in intimate contact to
form junctions. These junctions introduce internal fields that,
in the presence of light to produce electrons and ions, give rise
to a potential difference and electric current, if an external
circuit is closed. As long as there is a source of light of appropriate wavelength, the device is effectively a small battery, or
generator, delivering direct current electric power.

Photovoltaic power systems are inherently very attractive
because of the direct production of electricity, the absence of
moving mechanical parts, the large response to diffuse as well
as direct sunlight, and the fact that a large heat rejection
system may not be required.

The most highly developed and best understood photovoltaic
device is the silicon single crystal solar cell that is
used extensively in space power systems. These cells have
been demonstrated to be highly reliable and useful in a wide
variety of space power applications. They are equally reliable
and useful for producing electric power on earth. Efficiencies

greater than 10 percent in the conversion of solar energy to electric energy can be readily obtained at high device yield with present technology. Efficiencies of 15 to 20 percent have been obtained for advanced cells.

Single crystal silicon solar cells are also, unfortunately, very expensive to fabricate:about $30 per watt for moderate quantities of cells with at least 10 percent conversion efficiency for sunlight at the surface of the Earth. For a large power system, with the devices placed in arrays, power would now cost considerably more than $30,000 per installed kilowatt of electricity, much more than for conventional power plants.

There are other photovoltaic materials that can be considered as alternatives to the use of silicon. These include thin film hetero-junction materials such as cadmium sulfide-copper sulfide, homo-junction materials such as gallium arsenide, tertiary materials such as aluminimum-gallium-arsenide, metal semi-conductor junctions such as Schottky barrier diodes, and others. Each of the alternatives has important problems that must be evaluated and solved. In most cases, their advantage lies in cheaper fabrication costs if the other problems can be solved. Solar cells using cadmium sulfide/copper sulfide seem to present the most interest-ing alternative approach to silicon solar cells at this point in time.

Silicon solar cell arrays are used extensively as space power supplies including a 20 kilowatt unit aboard the Skylab manned space station. Also, silicon solar cell arrays are used extensively on earth as relatively small remote power units for communications and warning signals; and, in a few larger installations to pump water.

At the present time the Federal government is supporting 15 outside projects with a total value of about $1.8M. Of these projects, 10 are supported by NSF and five by NASA laboratories. There is considerable additional work being done in laboratories operated by NASA and AEC.

Research and technology in photovoltaic conversion appears to be the application area requiring the highest technology. In particular, solid state and materials research and technology will play an important role.

Major research areas include new approaches to fabrication of single crystal silicon wafers and solar cells, fabrication and characterization of polycrystalline silicon cells and cadmium sulfide/copper sulphide cells, investigations of stabilities and effects in heterojunction and homojunction cells, investigation of crystal defects and impurities on cell characteristics, and characterization of new combinations of photovoltaic materials.

Biomass Production and Conversion

The general objective of this program area is to prove the economic feasibility for large-scale conversion of organic wastes, cultivated organic materials, and water to gaseous, liquid, and solid fuels using biological processes. More specifically, the five-year objectives are to provide an improved technology base for anaerobic conversion of organic materials to methane gas; to show the technical feasibility of producing hydrogen from water by photosynthesizing biological organisms; to identify cultivated crops and associated technology and systems to produce fuel resources; and, to complete a system proof-of-concept experiment through the test and evaluation phase to study production of methane gase from urban organic wastes.

The Federal funding of research in this area is estimated at $1M for FY 1974. At the present time there are nine projects with a total value of $1.2M being sponsored by the Federal government in outside laboratories. Of these projects, seven are sponsored by NSF, one by NASA, and one by the Department of Agriculture.

Wind Energy Conversion

The general object is to develop reliable, cost-competitive wind energy conversion systems capable of rapid commercial expansion. Specific objectives in the five-year plan include increased performance and new options for components, subsystems, and systems up to about 10 Mwe systems; and, completed system proof-of-concept experiments through test and evaluation for 100 Kwe and 0.5 to 2.0 Mwe systems.

Windmills have been in use for many decades to produce electrical power and for centuries to pump water. In fact, the largest windmill electrical power system in the world with a peak electrical output of 1.2 Mwe was.operated near Rutland, Vermont, in the early 1940's in conjunction with a public utility network. Though the system operated successfully for a number of years, it was dismantled in the mid 1940's after a materials defect in the rotor led to a structural failure that was not repaired.

At the present time, the Federal government is supporting four outside projects with a total value of about $0.3M, provided by the NSF. The Federal budget for research on wind energy conversion is estimated at $1M for FY 1974.

Ocean Thermal Conversion

The general objective for research in this program area is to establish system reliability and economic viability of large-scale power plants converting ocean thermal energy into electricity. Specific objectives in the five-year plan are to establish the design of components and subsystems and to obtain performance data; to conduct system studies and subsystem experiments; and, to identify a system proof-of-concept experiment projecting a reliable, practical system.

Another large source of energy, which is adjacent to the coasts of the United States, is the thermal differences that result from solar energy absorption on the surface of the tropical oceans and the vast nearby cold water at 1000 meter depth that has flowed from the polar regions. A 22 Kwe plant was operated for about 11 days in 1929 as the first relatively successful demonstration. A temperature differential of about 35 F (20C) was used to drive a specially designed turbine.

Due to the relatively small temperature differences of the thermal source, the practical engine efficiencies are only a few percent so that very large quantities of water must be moved through a system per unit of power produced. This factor results in systems requiring very large components to handle the large cold and warm water volumes. Fortunately, extremely large replenishable volumes of both warm and cold water are available.

At the present time, the Federal government is supporting two outside projects with a total value of about $0.3M provided by the NSF. The Federal budget for research in ocean thermal conversion is estimated at $0.7M for FY 1974.

NEEDS FOR NEW IDEAS AND TRANSFER OF RESEARCH RESULTS

Solar energy research and technology, as encompassed by the National Solar Energy Program, requires a particularly broad base of research disciplines ranging through the physical and biological sciences, engineering, and the social sciences. In addition, most of the work on applications requires multidisciplinary approaches to particular problems, and also requires frequent assessment of the needs of the system being developed or optimized. The impact

of research directions and results on the cost of delivering useful
energy must receive careful attention. Social, environmental, and
political issues also must be fully addressed to assess the prospects
and problems of technology implementation.

The NSF Research Directorate has sponsored basic research in
universities since 1950. NSF/RANN is one of the recent approaches
to focusing research results and research capabilities on the solution of national problems or needs. In undertaking application
areas, such as solar energy and others, an applied research program
quickly begins to identify gaps in research and technology that are
critical to the design and economic feasibility of a system. These
gaps and critical problems can arise even though the technical
feasibility of the system is considered to be established; that is,
a system has or can be constructed and operated to accomplish the
process--though not necessarily economically, reliably, or for
long periods of time.

For example, production of electric power can be done by collection of solar radiation at high temperatures, transfer of the heat to
water as a working fluid, and conversion to electricity in a steam
turbine-generator. However, careful examination of system designs
and economics raises many critical technical issues, among other
issues. Frequently the critical technical issues are those that require basic-research-type information; such as, optical parameters
and degradation characteristics of selective optical coatings or phase
characteristics of eutectic materials for thermal storage. In selective optical coatings, the questions concern optical constants for
modeling of stacked thin films of various materials, amount and
effect of interfilm diffusion at high temperatures on optical proper-
erties, and activation energies of thin materials and substrates.
In the use of eutectic materials for thermal storage, questions
arise concerning super-cooling mechanisms and perturbations,
surface and bulk corrosion mechanisms with container materials,
and stability of phase transitions.

There is a continuing need for further technical knowledge often
associated with basic research work but required relatively soon.
This type of focused information need is not readily met through the
usual mechanisms for support of basic research. Often it must be
developed through the applied research project. In that context,
there is a research gap that can be met by "directed basic" research--
research for basic information that is recognized as needed to improve the design and economics of a system.

The role of basic research in solar energy applications is very important. First, it provides the research information base to proceed to the desired application; and, secondly, it provides new knowledge and methodologies to generate a broader base for future applications, as they can be recognized. If the research information base is not sufficient in an application area, some "directed" basic research needs to be done.

Who in the research community can provide assistance in solar energy applications? First of all it is recognized that some of the research community may not be interested in or in a position to contribute to solar energy research.

In any case, the solar energy program will benefit from a broad participation of interested professional persons who know and understand the problems in solar application areas and who can propose new or improved approaches to limiting factors holding up practical systems. In fact, the broad spectrum of participation by professional persons has been growing rapidly for most of the application areas, perhaps at least as fast as the available funds for research support.

Each person who understands the problems well enough to propose solutions or who believes he has new approaches to utilizing the sun's energy should consider participating in the program.

How can interested technical persons contribute to the solar energy program? There are many degrees of participation that persons can consider--all of which require some knowledge of the energy source, needs and problems in the application area, and on-going work. Contributions can range from initiating full-time research activities to transferring of ideas or basic research results to others who would explore the suggestions in detail. A few examples of what has occurred will illustrate ranges of participation or interest.

A most interesting example to our program, and possibly to you, involves the transfer of basic research results from high energy physics to solar energy applications. The research results came from an instrumentation project to obtain the most efficient coupling between Cerenkov radiation from accelerated charged particles to a quantum detector. A very generalized approach was made to the problem in the form of a solution to a Hamiltonian equation representing the optical problem. This solution generated an optical conical reflecting surface that maximized the collection and concentration of radiation from the source area. Recently the investigator and his associates were challenged to consider technology applications particularly relevant to energy problems. As a result, the idea was developed to apply this methodology and results to solar collection problems. A project on the "Winston Collector" is now being pursued in a number of new directions with potentially important consequences to solar energy applications.

Another example involves the application of basic research ideas and methodology to the production of hydrogen from water using photosynthesizing organisms and enzyme reactions. The interests of the basic research team became focused on the potential of direct production of hydrogen by biological methods at theoretical efficiencies calculated to be as high as five or more percent. At this time the team has succeeded in its initial goals, and the prospects for longer-range success seem very attractive.

Other degrees of participation by researchers in the solar energy program can be through the transfer of information and methodology from one application area to another, such as in computer modeling and optimization; or through research in areas of concern that are common to many applications, such as materials problems. The former transfer between different application areas is also exemplified in an approach to high capacity thermal storage using recyclable supercooled fluids that have been developed for long shelf-life at room temperature for medical therapy applications. These fluids can have interesting application to problems of thermal storage for solar energy systems.

Applied research workers must be able to assess the information needed and then to have access to it, if it is in the knowledge warehouse. A further desirable goal is to have trained persons intimately familiar with sections of the knowledge warehouse also familiar with problems in technology applications. Bringing these types of persons together is a highly desirable goal.

It is a fact of existence that many ideas and much information are in the knowledge base but are not recognized for their potential contribution to the solution of an important problem. It is worth considerable effort and funds to catalyze the serendipitous application of such ideas and information to practical solutions for problems.

Where "directed" basic research needs to be done, questions arise over where the required information will be obtained and where the funds will come from--basic research funds or applied funds. Often, there is a considerable gap to be filled. Should funds be added to the basic research support or to the applied research support--or should a third type of research support funding be developed. The usual solution has appeared to favor increasing applied research funding to provide the required data. In the energy area, the five-year energy budget and plan appear to provide additional funds for basic research earmarked in some manner for energy-related research. The suitability of the latter approach will be subjected to some testing in the near future. If it is successful it can help close the gap between NSF basic research goals and applied research needs for more information.

CONCLUSION

In conclusion, let me reiterate that the U. S. solar energy plan is based upon the following conclusions:

.   that technical feasibility has been shown for each application area;

.   that support of research and technology in each area--along with phased project planning; consideration of socio-economic, environmental, legal, etc., issues; and‚attention to implementation and utilization issues--can lead to reliable and economically viable systems;

.   that each application area can make a substantial contribution to domestically available U. S. energy resources; and,

.   that each area should be developed to practical systems at the earliest feasible time.

Research is underway and more is being initiated in six principal areas of applications. Research and technology are identifying new approaches to solar energy systems and are focusing on new and improved materials, components, subsystems, and systems in each principal area. System proof-of-concept experiments in each of the application areas are being programmed at the earliest feasible time to show projected performance and economic viability.

Innovative ideas and existing research results must be applied effectively by marshaling the potential contributions of interested professional persons across the broad discipline areas covered by solar energy applications. Additional research results must be obtained through directed efforts as the needs for such results are recognized.

It is the belief of many of us as scientists and engineers that some applications, namely heating and cooling of buildings, wind energy conversion, and biomass production and conversion, can have impact on U. S. energy requirements by the early 1980's Other solar energy applications can become viable economic systems later in the 1980's. The Federal Government is moving ahead strongly to prove solar energy systems as practical, important, alternative sources for the Nation's energy.

GEOTHERMAL POWER

Morton C. Smith*
Los Alamos Scientific Laboratory, The University of California
Los Alamos, New Mexico 87544

ABSTRACT

In its broadest sense, geothermal energy is all of the heat in
the earth's interior. Most of this heat, unfortunately, is not now
useful to man because (1) it is too deeply buried to be reached economically from the earth's surface; (2) potentially useful geothermal
reservoirs at accessible depths have not been recognized or evaluated;
(3) means for transporting the heat to the surface from such reservoirs do not exist; or (4) problems associated with using this heat
at the surface have not been solved. As these difficulties are progressively overcome, the power produced from geothermal sources should
increase to the point at which it can satisfy a significant fraction
of man's total energy needs.
General approaches to overcoming most of these difficulties can
already be identified, including (1) new drilling and fracturing
technology, to permit deeper openings to be made in hotter rocks at
higher rates and lower costs; (2) improved reservoir models and geophysical and geochemical prospecting and evaluation techniques;
(3) new methods of stimulating fluid production from presently unpro-
ductive hydrothermal reservoirs, and of creating fluid circulation
systems in dry ones; (4) solution of the chemical, materials, environmental, and mechanical problems associated with utilization of
mineralized geothermal fluids and of relatively low-temperature geothermal heat sources.

INTRODUCTION

Except for a relatively thin crust at its surface, the entire
interior of the earth is believed to be at temperatures within or
above the melting ranges of the mineral mixtures and metallic alloys
that compose it. Since the earth's volume is about 10^{12} km^3, this
represents an almost incredibly large energy supply, and it is an
inherently clean one--because the energy in it already exists as heat,
and neither combustion nor fission products need be released to make
this heat available. It is, unfortunately, not so far a generally
useful energy resource, because only a very small fraction of it is
immediately accessible to man; because much of its energy is at too
low a temperature to be currently interesting for such purposes as
generating electricity; and because a large proportion of the rest of
this heat exists in forms that man has not yet learned to recover and
use economically. With a little imagination and a great deal of hard
scientific and engineering work, this situation can undoubtedly be

*Includes work done under the auspices of the U. S. Atomic Energy
  Commission.

402

improved, and geothermal heat can become one of the major sources of
the clean energy that man so urgently needs.

In small ways--for bathing, "medicinal" purposes, and spaceheat-
ing--the geothermal energy of natural hot springs has been used by
man for at least thousands of years. Its larger-scale commercial use
began in 1904 when natural steam from wells in the Larderello region
of Italy was first piped to low-pressure turbines which were used to
drive small electrical generators.[1] Since then the uses of geothermal energy both directly as heat and for generating electricity have
expanded continuously, but for several reasons the expansion has been
slow. Now the world's power and pollution problems have caused new
attention to be focused on this oldest of energy supplies. One result is a widely held belief that within a few years the major problems of geothermal energy can be solved, so that it can begin to
contribute significantly to the future of both man and his environment.

THE PROBLEMS OF GEOTHERMAL POWER

The physical nature of our planet is now widely enough understood so that it is probably safe just to state categorically that,
simply by drilling a deep enough hole, it is possible from any point
on the earth's surface to reach rock that is hot enough to be potentially useful as an energy source. There are many well-known methods
by which energy can be extracted from the hot rock at the bottom of
such a hole and transported to the earth's surface, and by which the
energy can then be used beneficially. The problems of geothermal
energy are, therefore, derived not from questions concerning the
possibility of recovering and using heat from geothermal reservoirs,
but rather are primarily those of simple economics. How can we recover and use this heat at a price that we can afford?

To this point in time, geothermal energy has been used commercially only where nature has created a geologic situation in which
heat from the earth's interior is transported essentially to the
earth's surface by the convective circulation of steam or very hot
water. A hydrothermal reservoir of this type is commonly somewhat
leaky, so that there are usually obvious surface expressions of its
presence--in the forms of fumaroles, or geysers, or hot springs.
Where these indications are found holes are drilled and steam or hot
water is recovered through the holes for use at the surface.

Dry Steam

When the hot fluid in a geothermal reservoir is superheated
("dry") steam, it need only be permitted to issue from the well,
passed through a centrifugal separator to remove any entrained particles, and fed into a turbogenerator to produce electrical power.
The two largest producers of geothermal power in the world (at
Larderello, Italy, and The Geysers in northern California) utilize
natural dry steam in this way, and are relatively clean and very
economical power sources. The principal problems associated with
dry-steam geothermal reservoirs are:

1. They appear to be extremely rare. "Vapor-dominated" geothermal reservoirs capable of producing commercially useful quantities of dry steam are now known to exist in less than a dozen places
in the world. The largest of these, when fully developed, will probably produce no more than about 3000 MW (electrical),[2] which is certainly enough to be useful locally but represents only a very small
fraction of the Nation's total energy needs. At this point in time,
there is no evidence that there is enough natural dry steam in the
United States or the entire world to contribute significantly to
solution of our total energy problems.

2. It is, of course, possible that there are many undiscovered
dry-steam reservoirs which are so tightly contained by the rock a-
round them that there are no surface manifestations to reveal their
subterranean presence. Many qualified people believe that this is
the case, and so there is now much prospecting for hidden reservoirs
of natural steam. The prospecting techniques used are largely geochemical and geophysical, and include magnetic, gravity, electrical-
resistivity and seismic measurements, magnetotelluric and micro-
seismic surveys, and a variety of other methods. So far there is
little correlation among the indications from the various geophysical
methods, and they are used principally to identify anomalous locations
at which it appears worthwhile to drill holes in order to find out
what is really there. The task of developing remote-sensing systems
that will identify, map, and evaluate geothermal reservoirs is a very
challenging one for the geophysicist. His work will be supplemented
by that of the geochemist who, however, has difficult problems of his
own in attempting to relate chemical and isotopic water analyses to
the existence and probable temperature of a hydrothermal reservoir.

3. The drilling operation is an essential part of prospecting
for, evaluating, developing, and producing steam from vapor-dominated
geothermal reservoirs. in the hot porous formations typical of such
a reservoir drilling is a difficult, dangerous, and expensive operation, especially if a high-velocity flow of a geothermal fluid is in
fact finally encountered. Improvements in drilling technology are
urgently needed, and revolutionary rather than evolutionary advances
appear to be required. There are, for example, active programs to
develop methods of creating voids in rock by melting or vaporizing
it, or by cracking it with thermal stresses or high explosives or
fluid pressure or cavitation, or of abrading it away with fluid jets
or suspensions of hard particles. Perhaps one or more of these methods will prove important to the development of geothermal energy, or
perhaps some entirely different principle can be applied even more
advantageously. The problem of creating large completely connected
voids in a nonmetallic, composite, crystalline solid--that is, of
drilling a hole in a rock--is not a trivial one, and it deserves much
more attention from the scientific community than it has ever had.
If there were no other reason, one could justify this attention on
the basis that drilling is a billion-dollar-a-year industry in which
there has been only incremental progress for at least a generation.

4. It has not yet been established, physically or legally,
whether a dry-steam reservoir is a wasting asset or a self-renewing
one. The mechanics of a geothermal reservoir are exceedingly complex,

involving heat flow, fluid flow, mass transport, and the interactions of fluid pressures, overburden and tectonic stresses, and the stress gradients induced by changes in temperature and pressure. Aside from the legal and financial implications of reservoir mechanics, their understanding is essential throughout the sequence of prospecting, evaluation, development, and exploitation of geothermal reservoirs, and the fluid reinjection subsequently involved in controlling subsidence and disposing of wastes. Development of the reservoir models that represent understanding and permit prediction will involve sophisticated computer studies based on solid engineering data, or--where that is lacking--on sound estimates that can subsequently be improved. Good people are working to develop such models, but a great deal remains to be done in this area.

5. The chemical problems associated with the production and use of dry natural steam are still troublesome, but are of course largely in the domain of the chemist rather than the physicist. They include particularly the mass transport in the steam of mineral species such as silica which tend to plate out in the plumbing and on turbine blades; of volatile substances such as boron and arsenic compounds, whose discharge into surface waters represents a significant pollution hazard; and of noncondensible gases such as hydrogen sulfide, whose release into the atmosphere presents a particularly unpleasant environmental problem. Corrosion problems with dry unaereated steam have generally been minor, but could become serious if chlorides or sulfates were encountered in significant concentrations.

In the small number of places where commercial amounts of natural dry steam have so far been discovered, the geothermal fields have been developed successfully and produced profitably by industrial concerns. It is therefore generally assumed that industry can and will handle the remaining problems of geothermal steam. To a degree this is undoubtedly true, but the engineering and environmental solutions developed by industry may not be the best ones, and they may not come quickly. There are important and difficult questions related to dry natural steam that still deserve the attention of scientists outside of the industrial laboratories.

Superheated Water

Natural hydrothermal systems in which the reservoir fluid is hot water are much more numerous than are those in which it is steam, as is suggested by the relatively common occurrence of hot springs in many parts of the world. The amount of thermal energy contained in subterranean hot water is enormous, and is potentially capable of satisfying a major fraction of the world's total energy needs. However, the problems associated with producing geothermal water from the reservoir, extracting and using its heat at the surface, and disposing of the by-products of these operations in environmentally acceptable ways, are also very great. Large-scale development of "liquid-dominated" geothermal systems has so far been undertaken for generating electricity only at Wairakei, New Zealand, and Cerro Prieto, Mexico, and for space-heating in Iceland, Hungary, and the Soviety Union (although smaller developments are in progress or in being in several other parts of the world).

The problems associated with the use of natural hot water result primarily from the fact that the solubilities of most minerals increase with the temperature and pressure of the water in the geothermal reservoir. Natural water hot enough to be interesting for commercial use is usually so highly mineralized that it is extremely corrosive and tends also to deposit large quantities of silica, carbonates, and other compounds as its temperature and pressure are reduced. These problems are magnified when--as in the case of the evaporates of the Imperial Valley of California--the porous rocks constituting the geothermal reservoir happen to contain abnormal concentrations of highly soluble minerals. Then the geothermal fluid may be a relatively concentrated "brine" containing up to about 30% by weight of dissolved solids.

In general, the problems discussed above in connection with dry steam exist also for geothermal water, which has additional problems of its own.

1. Drilling difficulties are greatly increased by the corrosive action of the hot brines, which are very destructive of drill bits, drill pipe, casing, and fluid-circulation equipment.

2. This type of corrosion is of course equally damaging to the casing, pumps, valves, and other plumbing and hardware used to bring the geothermal fluid to the surface and to extract heat from it.

3. Most hydrothermal reservoirs are not sufficiently pressurized to maintain artesian flow, so that the hot brine must normally be pumped to the surface. Corrosion-resistant downhole pumps potentially capable of surviving in this environment are now being developed, but so far the universal practice in liquid-dominated systems is to use a steam-lift to "produce" the geothermal water. This is done by permitting a sufficient pressure drop to occur in the well so that the superheated water boils either in the well itself or in the porous formation around it. The rising steam bubbles and the reduced mean density of the fluid column create and maintain a two-phase flow of mixed steam and water up the pipe to a steam-water separator at the surface. One unfortunate result of this procedure is that the pressure drop required to permit boiling also reduces the solubility of certain minerals, notably carbonates, to the point at which they tend to precipitate in and plug the producing formation adjacent to the well or the well bore itself. In some cases this has made it impossible to maintain usefully high rates of fluid production from the well, and in most other cases it requires that the production casing be reamed out occasionally or the formation reopened periodically by some chemical or mechanical method.

4. At the surface a further pressure reduction permits additional water to flash to steam. The steam is separated from the water in a centrifugal separator, piped to a turbine, and used to generate electricity. The noise level around the flash tank is typically very high, as it is wherever fluids are released from the system, and represents a significant environmental hazard.

5. Flashing of the hydrothermal water in general converts 25% or less of its mass to saturated steam at a temperature significantly lower than that of the original geothermal fluid. Although turbines capable of utilizing wet steam at relatively low temperatures are

now available, their efficiencies are necessarily low and their maintenance costs are generally high. Improved power cycles are needed,
which may involve multiple flashing of the hot brine, exchange of
the heat to a lower-boiling liquid, or use of some type of heat machine other than a conventional turbine--for example, a screw-expan-
der or a bladeless turbine.

6. The boiling water from the steam separator is normally discarded, and with it is wasted up to one-half of the heat initially
present in the geothermal fluid. A practical, economical "bottoming
cycle" to use this heat is urgently needed.

7. Most of the mineral content of the original geothermal water
remains concentrated in the hot brine which was separated from the
steam in the flash tank. Unfortunately, the minerals present in
high concentration in this brine do not in general have sufficient
value to justify a recovery treatment, and so the brines are normally
discarded. Commonly they are simply permitted to flow into the nearest large body of water, where they represent both thermal and chemical pollution. (At Cerro Prieto, however, they are instead retained
in large evaporation ponds from which the minerals present in them
may eventually be recovered.) The disposal plan proposed for most
new developments is to reinject the waste solutions into the producing formation at some distance from the production wells through
which the geothermal fluid was initially withdrawn. This at least
reduces the environmental effects of brine disposal, although the
ultimate effect of reinjection on the behavior of the geothermal
reservoir is still unknown.

8. Withdrawal of large volumes of fluid from a subterranean
reservoir normally causes subsidence of the earth's surface above it,
sometimes with very serious results. Reinjection of waste brines
into the producing formation will undoubtedly reduce subsidence, but
it can probably be prevented completely only if the volume of fluid
injected is substantially equal to that withdrawn. In general this
balance is impractical to maintain and significant subsidence commonly accompanies the production of fluids from liquid-dominated
geothermal fields.[3]

9. The lower-temperature geothermal waters in general are also
lower in dissolved mineral content, so that most of the chemical
problems discussed above are less severe when cooler waters are used.
However, the problem of low efficiency in the power cycle is correspondingly increased, and one problem is exchanged for another. Unless and until relatively low-temperature power cycles of increased
efficiency are devised, the lower-temperature natural waters will
probably be used principally for space-heating--which, however, is
an economical and intelligent type of use that should certainly be
encouraged in order to conserve higher-grade fuels for more demand-
ing applications.

The problems associated with the use of natural superheated
water are primarily chemical and thermodynamic in nature, and so--
except perhaps with regard to the development of efficient low-
temperature power systems--they are probably of little interest to
most physicists. They are, however, sufficient both in number and
in difficulty so that the widespread exploitation of a large and

readily available geothermal resource will evidently be delayed for some years until economical solutions to them have been demonstrated.

Geopressurized Reservoirs

In several parts of the world--notably along the Gulf Coast of the United States, from Mexico to Mississippi--there are deeply buried sediments containing moderately hot water under very high pressure, commonly with significant amounts of natural gas dissolved in the water. In part the water was trapped in these sediments as they were originally deposited, and in part it has since been released by a reaction called "thermal diagenesis" which occurs when certain clay minerals are heated. The volume fraction of water in the sediments is sufficient so that the water itself supports part of the weight of the overburden above it, explaining the high fluid pressure in the porous formation and the description of the reservoir as "geopressured" or "geopressurized."

Such "undercompacted" sediments normally occur at depths greater than about 2 to 3 km.[4] Water temperatures are commonly 150 to 180°C,[4] the elevated temperatures resulting primarily from the fact that these sediments are poor conductors of heat so that, even in regions of only normal heat flow, the temperature gradients through them are relatively high. Fluid pressure at the wellhead is commonly 28 x 10^6 to 41 x 10^6 Pa (4000 to 6000 psi),[4] and the content of natural gas about 0.04 to 0.07 m^3/m^3 (10-16 scf/bbl)[4] of water. The salinity of the water is usually low. Since the area underlain by geopressurized reservoirs is thousands of square miles along the Gulf Coast alone, the total energy represented by this combination of low-grade heat, mechanical energy recoverable from the pressurized water, and natural gas, is enormous. However, the problems of extracting and using this energy economically are also formidable, and serious attempts to develop methods of doing so are just beginning.

Among the problems associated with exploitation of geopressurized hydrothermal reservoirs are the hazards associated with drilling into formations containing large volumes of hot water under very high pressure; the normal inefficiency of using low-temperature heat; the requirement for very high-pressure water turbines to utilize the mechanical energy; and the difficulty of separating and drying the gas. However, the magnitude of the resource obviously justifies serious efforts to develop it.

Dry Hot Rock

Even in a dry-steam or superheated-water reservoir, much more than half of the thermal energy is present in the rock, and of course the heat is there even when the steam or water is entirely absent. At sufficient depth, rock hot enough to be potentially useful as an energy source exists everywhere, and in many places at depths shallow enough to be reached at moderate cost with existing drilling equipment. The overall problem in making this energy resource useful is that of extracting heat from the rock at depth and bringing it to the surface in a usable form, at a usefully high rate, and at an acceptable cost in money and environmental damage. Many methods of accomplishing this have been proposed, the most straightforward of which appears to be to simulate on a small scale the hydrothermal systems that in most places nature has failed to provide. This involves drilling holes into the hot rock through which cool water can be injected and steam or hot water can be recovered; then, if it does not already exist, producing sufficient permeability and heat-transfer surface within the hot rock to permit fluid circulation and heat extraction at a usefully high rate for a usefully long time; and finally creating and maintaining efficient systems for circulating the fluid through the rock and extracting heat from the fluid at the surface.

1. Drilling into dry hot rock is somewhat less difficult and dangerous than is drilling into steam or superheated water because it does not involve the corrosion, erosion, and hazards to crews and equipment associated with the escape of high-velocity flows of geothermal fluids. It is, however, slow and expensive because of high rates of bit wear, low average penetration rates, and the high cost of circulating sufficient fluid to keep the drilling tools at acceptably low operating temperatures. Deep oil and gas wells have been drilled successfully into formations at least as hot as 300°C, but to reduce costs, increase penetration rates, and permit drilling into still hotter rock will require the development of new drilling technology. The rock-melting drill ("Subterrene") now under development at Los Alamos Scientific Laboratory may be particularly useful for this purpose since with it an increase in rock temperature simply reduces the amount of energy that an internally heated penetrator must provide to melt a hole through the rock.

2. In porous formations of high permeability an obvious method of energy extraction is to inject water through one drilled hole, permit it to be heated as it percolates through the rock, and--if rock temperature is high enough and system pressure low enough--recover it as steam through a second hole drilled some distance away. Unless the underground system is somehow isolated by natural geologic boundaries, it will probably be necessary to drill not just one but an array of recovery holes around the injection hole, to avoid excessive loss of fluid into the surrounding formations; and at best the consumption of water in such a system will probably be large. However, by developing an extended array of injection and recovery holes, similar to the arrays used in water-flooding for the secondary recovery of petroleum, a very large system of this type could be produced and a very large amount of energy extracted from it. One problem with a large system of this type is that some subsidence of the earth's surface above it would eventually occur as a result of thermal contraction of the rock as heat was removed from it. Another, at any system size, is that water circulating through and being heated by the rock would dissolve minerals from it and redeposit them at the point at which the water flashed to steam. If this were in or near the recovery well, then plugging of the hole or of the formation around it would be very likely to occur.

Underground precipitation problems could be reduced and rate of energy extraction from the recovery well increased if boiling could be prevented and the fluid brought to the surface as superheated water. This could be done by maintaining throughout the system a fluid pressure greater than the vapor pressure of the hot water, which, however, would in most cases probably drive much of the fluid outward through the permeable formation, away from the recovery hole, where it would be lost. When suitable downhole pumps have been developed, it should be possible to use one of these in the recovery hole to maintain the required overpressure in the upper part of the hot leg of the system, and then to maintain the rest of the system at or below normal hydrostatic pressure. This would at least greatly reduce fluid loss into the surrounding formations.

3. In relatively impermeable formations the problem of retaining the fluid in the system is avoided, but the required permeability and heat-transfer surface must somehow be created. Drilling and the use of chemical explosives are expensive methods of producing new surface in solid rock, and appear unlikely to produce economical dry-rock energy systems. The use of nuclear explosives for this purpose appears practical with regard to both engineering and economics,[5] but involves serious environmental problems associated with ground shock and with the containment of radioactivity.[6] What appears to be an attractive alternative is the use of hydraulic fracturing to produce a thin crack of very large diameter in the hot rock. This fracturing method, which has been widely used for some years to stimulate production of petroleum and natural gas, involves using a high-pressure pump at the surface to produce a fluid pressure at the bottom of a drilled hole sufficient to crack the rock constituting the hole wall and then to extend the crack outward into the surrounding formation. The stress situation in the earth is such that, except at relatively shallow depths, the crack produced is vertical with its azimuthal orientation controlled by the local tectonic stress field. Crack diameters up to at least 2 km have been produced by hydraulic fracturing, representing a heat-transfer area sufficient to support a high rate of energy extraction for a very long time. Cooling of the rock will result in thermal contraction which will create new voids adjacent to the original hydraulic fracture. If these voids form as cracks opened widely enough so that water will circulate through them, then the underground circulation system may extend spontaneously into previously uncooled rock, and its useful life may be very long.

There are many unanswered questions about such a system, and, under sponsorship of the U. S. Atomic Energy Commission, answers to them are now being sought by Los Alamos Scientific Laboratory in experiments in the hot granite underlying the Jemez Plateau of north-central New Mexico. It has already been demonstrated there that drilling and hydraulically fracturing the granite involve no unexpected difficulties, that the fractures produced are vertical and extend outward from the borehole in a specific direction, and that the permeability of the granite--at least locally--is low enough to contain pressurized water without significant loss. However, the fracture dynamics of such a system are not yet well enough understood

to insure that a very large hydraulic fracture can be made to grow
outward and downward, as is desired, instead of preferentially upward. To minimize the impedance of the fracture to fluid flow
through it, it is desirable to hold the crack open with fluid pressure; but the stress situation at the crack tip may be such that the
pressure required to accomplish this will cause the crack to "run
away." If so, at least until there has been enough cooling so that
thermal contraction of the rock will hold the crack open, it may be
necessary to prop it open with particles injected with the fracturing fluid--as is commonly done in petroleum and gas stimulation.
Computer simulations of such systems indicate that convective circulation within the crack can be expected to maintain a high rate of
heat extraction from the crack surfaces. These simulations, however,
are based on very uncertain assumptions concerning crack geometries
and impedances, concerning which little detailed information actually
exists. The geochemical problems associated with dissolution, repre-
cipitation, and alteration of minerals in such a system are expected
to be minor compared with those of natural hydrothermal systems, but
are certain to be troublesome. So long as injection of fluid into
active fault systems is avoided, it appears that creation and operation of such a circulation loop involves no significant risk of
triggering earthquakes. However, until the state of knowledge in
this area is greatly improved, the earthquake hazard cannot be ignored.

There are, then, many problems related to dry, hot rock geothermal energy systems which deserve theoretical attention and experimentation both in the laboratory and the field. Dry hot rock is by
far our largest accessible supply of directly usable energy, and--
because the energy already exists in it as heat--it is a relatively
clean energy source. As drilling and heat-extraction technology
improves, it appears possible that energy from this source may eventually become economically available almost anywhere, and may then
contribute significantly to solution of some of the world's worst
power and pollution problems.

Lavas and Magmas

The most impressive demonstration of the existence of high-
intensity energy in the earth's interior is presented by an active
volcano. There have been many imaginative suggestions for energy
extraction either from the lava pools created by such eruptions or
from the subterranean magma chambers in which the lavas are believed
to originate. Most of the energy systems proposed have involved
either injection of a heat-exchange fluid into the molten rock, which
appears usually to involve a serious large-scale explosion hazard, or
immersion in it of a heat exchanger, which is likely to lose efficiency rapidly as an insulating layer of solid rock freezes around
it. Although accessible lava pools and magma chambers are rare, they
are large reservoirs of high-temperature heat, and the possibilities
and problems of extracting energy from them deserve serious and
imaginative scientific attention.

CONCLUSION

Physicists as a class have, historically, been greatly concerned
with the problems of the microcosm and the megacosm, and in general
have remained largely uninvolved in humanity's day-to-day problems
of eating, drinking, breathing, moving about, and keeping warm. In
a time of world-wide energy and pollution crises, it is heartening
that a large group of physicists is now actively seeking ways in
which their special types of knowledge can contribute to solution of
the immediate problems that beset mankind and threaten his future.

To physicists thus concerned, I recommend the problems of geothermal energy. They relate to one of the largest, cleanest, and
most broadly distributed energy supplies that is accessible to man,
and they exist in all degrees of difficulty and sophistication.
Many of them are the large problems of engineering, which will be
attacked by the engineer himself in large-scale experiments in the
field. However, many more lie in such scientific areas as geophysics,
seismology, fluid dynamics, heat transfer and terrestrial heat flow,
solid-state physics, rock mechanics and fracture dynamics, and the
modeling of complex dynamic systems. Here the engineer needs scientific background information that so far does not exist. I hope
that the physics community will help to satisfy this very urgent need.

REFERENCES

1. E. Barbier and M Fanelli, Overview of Geothermal Exploration and
    Development in the World, Pisa, National Research Council (1973).
2. J. O. Horton, Geothermal Resources, Part I, (U. S. Government
    Printing Office, 1973), p. 43.
3. Ibid, p. 50.
4. W. J. Hickel, Geothermal Energy, College, Alaska (University of
    Alaska, 1973), p. 16.
5. American Oil Shale Corporation, Battelle-Northwest, Westinghouse
    Electric Corporation, U. S. Atomic Energy Commission, Lawrence
    Livermore Laboratory, and Nevada Operations Office-AEC, "A
    Feasibility Study of a Geothermal Power Plant," PNE-1550 (1971).
6. G. M. Sandquist and G. A. Whan, "Environmental Aspects of Nuclear
    Stimulation," in Geothermal Energy (Stanford University Press,
    1973), pp. 293-313.

SOLAR SEA POWER

Clarence Zener
Carnegie-Mellon University, Pittsburgh, PA 15213

ABSTRACT

The high cost of a solar energy collector has presented a stumbling block to the generation of large amounts of useful power from solar radiation. The present paper discusses attempts made to design an economically viable system of power generation using the surface of the tropical oceans as a solar collector, and the deep cold water as a heat sink. No new technology need be developed to make such a system technically feasible. Ingenuity is required, however, in envisioning methods of performing the various necessary functions of such a system at the lowest possible cost.

If indeed such power systems have a high probability of being economically viable, a new energy economy is to be developed. The author indicates how physicists have the proper training to guide the development of such a new energy economy.

§1. INTRODUCTION

Of all the various new sources of energy now being discussed, Solar Sea Power (SSP) enjoys two unique features. First, it is the least publicized. Second, it offers the greatest potential for satisfying the demands of an energy-hungry and food-hungry world.

The lack of publicity is easy to understand. Journalists naturally hesitate to write about what may turn out to be a quack-pot idea. So they consult their scientific and engineering friends. The one thing that all scientists and engineers know about energy is that the conversion efficiency from heat to power is governed by Carnot's law:

$$\text{Efficiency} \leqslant \frac{T_{source} - T_{sink}}{T_{source}}$$

A quick calculation shows them that the 20°C difference between the surface and deep water of the tropical oceans limits the theoretical efficiency to only 6.6%, the practical efficiency to perhaps only 2%. They then express their honest intuitive feeling that such a low efficiency will require so large a plant for a given net power output that the power costs will be astronomical. This conviction has been so strong that an official delegation from MIT told a Congressional committee[1] last year that it will never be practical to take advantage of the temperature differences between the upper and lower layers of the ocean! The first lesson I learned upon changing from physics to engineering was that intuition is a dangerous guide to truth in unfamiliar non-linear systems. A Solar Sea Power Plant (SSPP) is a highly non-linear system, as in fact are most engineering systems.

The high probability that SSP will satisfy the world's needs for energy and for food arises from three fortunately coexisting circumstances. (1) As we shall see later in this talk, the probability is high that the capital cost of a SSPP will not exceed that of a conventional fossil fuel power plant of the same net power output. And of course its fuel cost will be zero. (2) Over half the sun's heat which strikes the earth is absorbed by the surface layers of the tropical oceans. Most of this absorbed heat is then lost through the evaporation of water. If only a small fraction of this heat now used in evaporation were to be passed through a SSPP, the new power generated would supply the projected world population for the year 2000 with the same energy per capita as is now used in the USA. (3) The last fortunate circumstance is that the ocean serves as a very efficient means of transport. In fact, so efficient is the ocean in transporting its own water that a SSPP can suck into its boiler the vast quantity of warm water it needs from many miles away, and expend less than 1% of the gross power generated in so doing. The low cost of shipping bulk cargo by ocean insures that the output of SSPP complexes could be carried readily to all coasts in the world. One of the first important outputs will be ammonia, to be used as a fertilizer. Currently ammonia is synthesized in the Gulf states using as a raw material natural gas. It is then sent via pipeline to all parts of the Midwestern farm region. The skyrocketing price of natural gas is causing a diversion of gas away from ammonia feed stock, and is thereby threatening an agricultural crisis in this country. Now hydrogen is an even more suitable feed stock for ammonia than is natural gas. And hydrogen is a natural product from a SSPP. The electrical power generated is used directly to electrolyze water. I thus visualize SSPP complexes whose product is primarily ammonia. This fertilizer will then be shipped directly to all parts of the world. A continuous 18,000,000 kilowatts would be required to generate the 14,000,000 tons of ammonia now produced in the USA. To supply the whole world with the same per capita use as in the USA today would require 300,000,000 kilowatts. The hydrogen generated by a SSPP could of course be shipped directly to the USA, thereby fueling the much discussed hydrogen economy.

The concept of Solar Sea Power is of course not new. D'Arsonval[2] pointed out as early as 1881 that we may produce electric power from a heat engine which uses the upper warm layer of the tropical oceans as a heat source, the deep cold water as a heat sink. D'Arsonval even suggested the overall features of such a heat engine. Specifically, he suggested that one use as a working medium a fluid with a fairly large vapor pressure at ambient temperatures, and thereby avoid the enormous turbines which would be required if one attempted to use the ocean water itself as the working medium. Nor is the attempt new to actually construct a SSPP. The French engineer Georges Claude[3] actually built a plant on the Cuban coast in 1930. This plant was not, however, economically successful, partly because Claude did not follow the advice of D'Arsonval of using an intermediate working fluid, and partly because the age of high priced fuel had not yet arrived.

What is new is a government program aimed at a large-scale utilization of solar energy using the ocean as the thermal collector. This program is supported by the National Science Foundation, and currently consists of two design projects, one at the University of Massachusetts, one at Carnegie-Mellon University. The University of Massachusetts project is directed toward taking advantage of the current in the Gulf Stream. Our project is aimed at a design which is independent of ambient ocean currents. It is this second project that I shall discuss.

§2. OVERALL VIEW OF SOLAR SEA POWER SYSTEM

The gross features we visualize in a Solar Sea Power Plant are depicted in Figure 1. Warm water above the thermocline flows inwards towards the warm water intake. Density stratification prevents water from below the thermocline from being sucked into the SSPP warm water intake, at least for plants below a given critical power level. McMichael[4] has shown this critical power level to be about 700,000 kilowatts. In passing through the boiler this water is cooled several degrees, and is then ejected below the thermocline, where it spreads outwards away from the SSPP at a level characteristic of its temperature. Similarly, deep cold water flows towards the cold water intake. In passing through the condenser this water is warmed several degrees. It then flows down to a level characteristic of its new temperature, and then flows, radially outwards. Density stratification prevents mixing of the condenser intake and exhaust water.

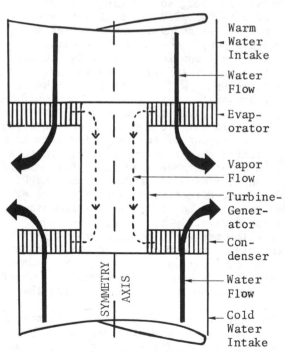

Warm Water Intake

Water Flow

Evaporator

Vapor Flow

Turbine-Generator

Condenser

Water Flow

Cold Water Intake

SYMMETRY AXIS

Fig. 1. Sketch of a Solar Sea Power Plant.

The radial symmetry of the water intake and exhaust flow as depicted in Figure 1 would, of course, be modified by ocean currents. Such currents will of course aid the density stratification in preventing output and input sea water mixing. The forces induced by the ocean currents will however require some sort of anchoring.

Unfortunately, at the high Reynolds number characteristic of our plants, fluids do not readily conform to potential flow when decelerating, as in the warm and cold water exhausts. The kinetic energy of these exhausts therefore represents an unrecoverable loss. Typically these velocities are 4 ft/sec, corresponding to a head loss of one-quarter of a foot. In order to gain a perspective as to how important such losses are, we must know the work which can be extracted from a unit mass of warm water which flows through the boiler. Assuming a typical 2°C drop in water temperature, and a difference of 20°C between the temperature of the heat source and heat sink, a Carnot efficiency would give a work corresponding to a head of 186 ft. Our exhaust losses are therefore truly negligible.

Work must be performed, however, in lifting the cold water from about 3,000 feet to the condenser. When due account is taken of the difference in temperature of the cold water and of the not-so-cold water through which it is being raised, calculation shows that this work typically represents a head loss of 1.5 ft. Again, this loss is not serious.

Of special significance is the head loss arising from the friction of the water rising within the long cold water pipe. The power dissipated by this friction becomes a greater fraction of the gross power developed the smaller is the plant capacity. Thus whereas the capacity of a single SSPP is limited on the high side by the requirement of no mixing of the boiler input and output water, the capacity is limited on the low side by the head loss in the cold water pipe. We have seen the upper limit is about 1,000,000 kilowatts. The lower limit is about 10,000 kilowatts.

§3. PHYSICAL PRINCIPLES OF HEAT EXCHANGERS

Before the working medium (w.m.) can perform any work, it must change phase from a liquid to a vapor phase, absorbing heat in the process. After the w.m. has performed work, the reverse phase change must occur, from vapor to liquid, thereby releasing heat. The equipment for transferring heat to and from the w.m. are called the boiler and condenser, respectively.

Although the concept of boiling and of condensing is readily understood, this heat exchanger equipment constitutes the major cost in conventional power plants. In SSPP's, not only does this heat exchange equipment likewise constitute the major cost of the plant; it also contributes overwhelmingly to the irreversible losses within the plant.

The basic reason for the dominating role which the heat exchangers play in both the total plant cost and in the plant's irreversible losses lies in the very poor molecular thermal conductivity of water. The latent heat within the warm water is convected into the boiler from miles away with essentially no irreversible expenditure of power. Within the boiler the latent heat is convected to within a fraction of a millimeter of the surface of the boiler tubes, but only at the expense of a considerable expenditure of power. The final part of the journey, clear to the tube wall, can only be accomplished by strict thermal conductance across the non-turbulent laminar layer. But a temperature drop is required to drive this thermal flux across the laminar layer. And of course this temperature drop will not be available to help drive the heat engine. Another temperature drop is required to drive this same thermal flux from the heat exchanger wall to the fluid-vapor interface of the vaporizing w.m. Let δT be the total temperature drop consumed in driving this thermal flux. Essentially a similar δT will be required to drive the heat of condensation in the condenser from the condensing vapor-fluid interface to the heat exchanger wall, and then across a laminar water layer into the turbulent convective flow of the cold water flowing through the condenser pipes.

Considerable insight can be obtained by quantifying the above discussion. Let q be the thermal flux per unit area of the boiler (or condenser). Then since q is proportional to δT

$$q = k \, \delta T .$$

But the Carnot efficiency which we can expect from our heat engine is given by

$$\eta_{Carnot} = (\Delta T - 2\delta T)/T$$

where ΔT is the temperature difference between the intake warm water and the intake cold water. The gross power obtained per unit area of boiler surface is therefore

$$k \, \delta T \, (\Delta T - 2\delta T) .$$

This gross power is a maximum where the total temperature drop consumed to induce thermal flow, $2\delta T$, is one-half of the total available temperature drop. Thus one-half the total available work potential is irreversibly lost by thermal flux flowing down a temperature gradient from the warm water into the w.m. in the boiler, and from the w.m. into the cold water in the condenser. This irreversible loss corresponds to a head loss of 93 ft. In addition, the heat exchangers have a further head loss consumed in pushing water through the heat exchanger tubes.

§4. DESIGN STRATEGY

The above brief outline of the function of heat exchangers is, I hope, sufficient to give a taste of the dominant role trade-offs must play in the design of a complete plant. We cannot go to a manufacturer or to a specialist in a given component, such as a heat exchanger, a pump, a turbine, or a generator, and ask for a design of such a component satisfying a given set of performance specifications at least cost to the manufacturer. The optimum set of design specifications of the individual components are themselves not known. Our only fixed specification is that the plant produce a given net power output at a minimum possible initial capital cost, due regard being given to durability. It is clear that we must abandon the concept of assembling a set of components, each designed according to

some preconceived specifications. Rather, our design strategy must
be to write down an estimate of the cost of every part of every component, as well as equations which express the constraints which
proper functioning places upon each part. The ensemble of parameters
describing the various parts are then regarded as the variables of
the plant. These variables are then chosen to minimize the total
cost, subject of course to the various constraint equations.

The above strategy requires that we know in detail the physical
processes taking place in every part of the plant. We have found it
unrealistic to rely completely upon empirical performance, for the
conditions under which our various components operate are frequently
outside the range of past experience. Our struggles with the boiler
design, cited below, furnish a good example of design changes forced
by our requirement of understanding all physical processes.

Conventional boilers are of the pool boiling type. Here tubes
carrying the heating fluid are submerged in the w.m. Superheating
of the w.m. on the tube wall nucleates vapor bubbles. The bubble-
rich w.m. is then convected to a vapor separator which separates the
vapor and the liquid. The formidable task of modeling such a system
forced us to the much simpler system of a "falling film" evaporator.
Here the working fluid forms an enveloping falling film on vertical
tubes through which the heating fluid passes. Evaporation is then
directly into the pure vapor phase, thereby bypassing the intermediate bubble-liquid two-phase system.

Even a large number of system variables cannot encompass all
possible designs. For example, in Figure 1 we place the condenser
just below the evaporator, and bring the cold water up to the condenser. But we could have placed the condenser down deep where the
water is cold, and then bring the working fluid vapor down to the
condenser. A decision between these two topologies must be based
upon comparative costs. But a fair comparison between two topologies
can be made only when the designs for each topology are optimized.
Because of the many possible topologies which one can envision, it is
necessary to have a rapid optimization system.

One unique advantage of our design group is that Carnegie-Mellon
University has two of the three developers of Geometric Programming.[5,6] This is an optimization procedure which, among many other
valuable features, makes a rapid comparison between the relative
costs of different topologies. In fact, in those simple cases where
computers are not necessary, comparative costs of two different optimized topologies can be obtained without finding the optimizing variables themselves! In those cases where a computer must be used, a
new program need not be written for each topology. A program has
been written for all topologies, in fact, for a large class of engineering systems. We are completely dependent upon Geometric Programming for our progress in design. In fact, our original conviction
that Solar Sea Power has a high probability of economic viability was
instilled by the insight given by this optimization procedure.

§5. ECONOMICS

We believe we now have an overall design concept in which we understand fairly well all the detailed physical processes taking place. Our objective is, however, to design a SSPP of minimim cost per net kw output, not just a SSPP that will operate.

The outstanding question is, of course, what will an optimally designed SSPP cost, assuming such plants were in production? As previously mentioned, the major cost of such a plant is for the heat exchangers. We estimate that the evaporator and condenser each will require about 20 ft^2 of tubing per net kw output. If this tubing can be made of plain carbon steel, the cost of a SSPP will certainly be less than that of a conventional fossil fueled power plant. If this tubing must be made of aluminum, the cost of a SSPP will be comparable to that of a conventional fossil fueled plant. If the tubing must be of copper nickle tubing, the cost will certainly be higher. Only actual in situ tests can tell us the lowest cost material that will withstand the ocean environment without suffering degradation by chemical and/or biological fouling.

In the case of a fossil fueled power plant, the initial capital cost is only one part of the cost per kw hour of the generated power. The other major part is the fuel. These are combined in Table I to give the generated cost per kw hour, and are compared with the tentative cost of electric power generated by SSPP's.

Table I. Cost Comparison

	Fossil Fueled	SSPP
Capital Cost ($/kw)	320	320
Mills/kw hr		
Capital*	6.4	6.4
Fuel**	10.0	0
Maintenance	.1	?

* Computed on a Capital Recovery Factor of 2×10^{-5}/hr, corresponding to a 7.5% return on investment after taxes, and to a 13 year life.

** Computed on $1/10^6$ BTU for gas, natural or from coal; $5.80/Barrel for oil.

The economic viability of Solar Sea Power will depend not only upon our ability to generate low cost electric power, but also upon our ability to utilize this low cost power. As mentioned in the Introduction, it appears that the major part of this power will be used to generate hydrogen in situ. This hydrogen will then either be converted into other forms of chemical energy, such as ammonia for fertilizers, or be shipped directly to the industrialized countries. We thus anticipate that Solar Sea Power, rather than nuclear power, will fuel the much-discussed hydrogen economy.

§6. THE FUTURE

I have attempted to point out the futility of component experts being assigned the tasks of designing their own special components. The components cannot be designed independently of one another, but only as parts of an interacting system.

In the same spirit, I contend that the technical design of a SSPP is only one component of mankind's attempt to continue to have an abundant supply of energy without either despoiling his environment by hasty, one-shot strip mining of coal or oil shale, or by reliance upon an all-nuclear economy whose safety requires completely rational behavior in all parts of society, including the governmental sector. The maturing of a Solar Sea Power Age will require many more activities than simply technical design. Can we scientists and engineers leave the planning and implementation of these other components to specialists in business and in government? I believe not. In this area of new energy sources we cannot rely upon the free enterprise system to develop that system which will be most beneficial to society, especially from the environmental standpoint. We certainly cannot rely upon any federal administration to plan beyond its own term in office. If we as scientists and engineers really become convinced of the viability of Solar Sea Power, we must take upon ourselves the task of planning the implementation of the various necessary steps. If we as a technical community can arrive at a consensus at what should be done, I have enough faith in our form of government to believe that it will be done. It is in this spirit that I welcome critical discussion now and at any time in the future.

REFERENCES

1. Committee on Science and Astronautics, House of Representatives, Hearings before the Subcommittee on Science, Research, and Development on Energy Research and Development, May, 1972.

2. J. D'Arsonval, Revue Scientifique, 17 September 1881.

3. G. Claude, Mech. Engr. 52, 1039, (1930).

4. F. McMichael, "Sea Water Inlet-Outlet Hydrodynamics," Semi-Annual Report to the National Science Foundation, NSF Report No. NSF/RANN/SE/GI-39114/PR/74/2, Carnegie-Mellon University, January 1974.

5. R. Duffin, E. Peterson, C. Zener, Geometric Programming, Wiley, New York, (1967).

6. C. Zener, Engineering Design by Geometric Programming, Wiley, New York, (1971).

BIOGRAPHICAL BRIEFS

ANDELIN, JR., JOHN P.

Administrative Assistant to Congressman Mike
McCormack, obtained his Ph.D. in low temperature
physics at California Institute of Technology in
1966. Following positions at the Ford Scientific
Laboratory and Harvard he joined Congressman
McCormack's staff in 1971 as Scientific Advisor.

BRUECKNER, KEITH A.

Executive Vice President of KMS Fusion, Inc.
Professor of Physics at the University of Pennsylvania
from 1956-1959, he moved to the University of California at San Diego and became Dean of Graduate Studies
there in 1965, having served from 1961-1962 as Vice
President and Director of Research of the Institute
of Defense Analysis. He joined KMS Industries in
1968 and took his present position in 1971. His
principal interests have been theoretical nuclear
physics, plasma physics, MHD and fusion.

CAIRNS, ELTON J.

Assistant Department Head of the Electrochemistry
Department of General Motors' Research Laboratories.
Following 7 years' research at the General Electric
Research Laboratory and 7 years at the Argonne
National Laboratory, he joined GM in 1973. His con-
tinuing interests have been in electrochemistry, fuel
cells and related areas.

DAKIN, THOMAS W.

Manager of Electrical Performance Insulation Materials
in the Westinghouse Research Laboratories. Since
coming to Westinghouse in 1946 his principal concern
has been with dielectrics and insulating materials.

DONOVAN, PAUL

Director of the Office of Energy Research and Develop-
ment Policy, National Science Foundation, joined the
Foundation in 1968 after previous positions at the
Brookhaven National Laboratory and Bell Telephone
Laboratories. After 3 years in the Physics Session
he became Division Director of Advanced Technology
Applications, from which position he chaired the
Energy R&D Task Force of NSF.

DYOS, M. M.

Manager of Advanced Systems in the Advanced Reactors
Division of Westinghouse Electric Corporation, is responsible for the design of commercial size liquid
metal fast breeder reactors. Following several years'
experience with gas-cooled reactors in Great Britain
and Australia, he came to the United States in 1965 to
join General Atomic on helium cooled reactors, and
went to Westinghouse in 1968.

FISCHER, TRAUGOTT E.

Became head of surface physics for the Esso Corporate
Research Laboratory in January, 1972 after serving 6
years as Associate Professor in the Department of
Engineering and Applied Science at Yale University,
with previous positions at Bell Telephone Laboratories
and the Federal Institute of Technology in Zurich.
His career-long interest has been in surface physical
properties and phenomena.

FISHER, JOHN C.

Currently Manager of Energy Systems Planning for the
Power Generation Business Group of the General Electric
Company. Following 21 years of research and research
management, he left GE to serve a year as Chief
Scientist for the U.S. Air Force. He returned to GE
in 1969 and came to his present position in January
1972.

FURTH, HAROLD P.

Professor of Astrophysical Studies and Co-Head of the
Experimental Division of the Princeton Plasma Physics
Lab. He has been working in controlled fusion
research since 1956, with special interest in plasma
confinement in toroidal magnetic bottles.

GIBBONS, JOHN H.

Became the first Director of the Office of Energy
Conservation (now in the Federal Energy Office) in
September, 1973. Prior to this he directed the NSF-
sponsored Environmental Program at Oak Ridge National
Laboratory and the Environmental Center of the University of Tennessee. His professional interests have
ranged from nuclear physics to power plant siting to
technology assessment.

GREENWOOD, ALLAN

Chairman of Electric Power Engineering at Rensselaer
Polytechnic Institute, was until 1972 senior con-
sulting engineer in the Power Delivery Group of
General Electric. He is a member of the Advisory
Committee on Energy Research for New York State.

GREGORY, DEREK P.

Director of Energy Systems Research for the Institute
of Gas Technology responsible for research programs
on hydrogen-energy, fuel cells, future automobile
fuels, energy conservation, and methanol manufacture
and use. He is a member of the National Academy of
Engineering Committee on Alternative Aircraft Fuels.

HARDING, JOHN T.

Program Manager - Magnetic Levitation for the Federal
Railroad Administration in the United States Department of Transportation, spent six years in the Jet
Propulsion Laboratory before moving to the Ford
Scientific Research Laboratory in 1965 as a staff
scientist. Since joining DOT in 1969 his principal
interest has been in magnetically levitated high
speed trains.

HARWOOD, JULIUS J.

Research Planning Manager and Acting Director,
Physical Science Laboratory, Ford Scientific Research
Staff, came to Ford in 1960 after 14 years with the
Office of Naval Research, the last 7 as Head of the
Metallurgy Branch, concerned principally with metal-
lurgical and other materials science programs.
He is a past president of the Metallurgical Society.

HERWIG, LLOYD O.

Director, Advanced Solar Energy Research and Technology, National Science Foundation. Following 7
years' work on reactors with Westinghouse, and 3
years with United Aircraft, he came to NSF in 1964.
He became Program Manager of the Advanced Technology
Division in 1971, and went to his present position in
1973.

HEYWOOD, JOHN B.

Became Associate Professor of Mechanical Engineering
at the Massachusetts Institute of Technology in 1968
after spending 3 postdoctoral years with the Central
Electricity Governing Board, England. His interests
have included studies of pollution and of magnetohydrodynamic power generation, in addition to his
current focus on combustion modeling.

KANTROWITZ, ARTHUR

Chairman of the Board and Chief Executive Officer of
Avco Everett Research Laboratory, Inc., founded by
him in 1955. Prior to his association with Avco, he
was Professor of Aeronautical Engineering and Engineering Physics at Cornell University. A past chairman of the Division of Fluid Dynamics of the American
Physical Society, his central technical interest has
been in physical gas dynamics, and in development of
commercial MHD power generation.

KOUTS, HERBERT J. C.

Became Director in 1973 of the Division of Reactor
Safety Research of the United States Atomic Energy
Commission. Previous to this he had been at Brookhaven National Laboratory with principal interests
and responsibilities in reactor physics. He has
been a member and chairman of the AEC's Advisory
Committee on Reactor Safeguards, and was a 1963
recipient of AEC's E.O. Lawrence Memorial Award.

LEVIN, FRANKLYN K.

Research Scientist with the Esso Production Research
Company. His career has been spent almost entirely
in geophysical exploration for petroleum and in related areas, with the Carter Oil Company and Jersey
Production Research Company. He joined Esso Production Research in 1965.

LIIKALA, RONALD C.

Manager, Engineering Systems Analysis, Batelle-
Northwest has been broadly involved in research
and development of nuclear technology with emphasis
on technical and economic feasibility of using
plutonium in power reactors. Most recently his
interest has been in radioactive waste disposal.

MYERS, PHILLIP S.

Professor of Mechanical Engineering at the University
of Wisconsin at Madison, where he has spent his professional career. He has published about 100 papers
on engines and combustion. He was 1969 National
President of the Society of Automotive Engineers, and
has been a recipient of the Horning Memorial Award.

PERRY, HARRY

Consultant to Resources for the Future. He was with
the U.S. Bureau of Mines from 1946 to 1967, concerned
with all aspects of coal, becoming Director of Coal
Research in 1964. In 1970 he became Senior Specialist
on Energy for the Congressional Research Service of the
Library of Congress. He has been a special assistant
to the Director of Licensing for the AEC and a consultant to the Senate Interior and Insular Affairs
Committee.

RIBE, FRED L.

Associate Division Leader for Controlled Thermonuclear
Research, Los Alamos Scientific Laboratory, came to LASL
following his doctorate in 1951. His general interests
have been in nuclear and plasma physics and in the last
several years in problems in thermonuclear reactors.

SMITH, MORTON C.

Group Leader for Physical Metallurgy at Los Alamos
Scientific Laboratory, was Professor of Metallurgical
Engineering at the Colorado School of Mines for 7 years
before coming to LASL in 1954. His recent interests
have included rock-melting penetrators and geothermal
energy.

STARR, CHAUNCEY

President of the Electric Power Research Institute.
Following a 20 year industrial career during which he
served as vice president of Rockwell International
and president of its Atomic International Division,
he became dean in 1967 of the UCLA School of Engineering and Applied Science. He came to his present
position early in 1973.

VINEYARD, GEORGE H.

Director of Brookhaven National Laboratory. He came
to the Laboratory in 1954, became Chairman of the
Physics Department in 1961 and after two intermediate
positions became Director in 1973. Active in a wide
variety of physics activities, publications, and
organizations, he was Chairman last year of the
Division of Solid State Physics of the American
Physical Society.

YONAS, GEROLD

Manager of Plasma and Electron Beam Physics Research
at Sandia Laboratories, heading a major program on
advanced applications of relativistic electron beams.
A 1966 Ph.D. from California Institute of Technology,
he was in charge of electron beam research at Physics
International before joining Sandia.

ZENER, CLARENCE M.

University Professor at Carnegie-Mellon University,
has had a distinguished career in industry and academia.
Director of the Westinghouse Research Laboratories from
1956 to 1962 and director of science 1962-1965, he left
to become Dean of the College of Science at Texas A&M
in 1966 and University Professor in 1968. He cites his
interests as "theoretical physics and engineering".

AIP Conference Proceedings

		L.C. Number	ISBN
No. 1	Feedback and Dynamic Control of Plasmas (Princeton 1970)	70-141596	0-88318-100-2
No. 2	Particles and Fields - 1971 (Rochester)	71-184662	0-88318-101-0
No. 3	Thermal Expansion - 1971 (Corning)	72-76970	0-88318-102-9
No. 4	Superconductivity in d- and f-Band Metals (Rochester 1971)	74-188879	0-88318-103-7
No. 5	Magnetism and Magnetic Materials - 1971 (2 parts) (Chicago)	59-2468	0-88318-104-5
No. 6	Particle Physics (Irvine 1971)	72-81239	0-88318-105-3
No. 7	Exploring the History of Nuclear Physics (Brookline 1967, 1969)	72-81883	0-88318-106-1
No. 8	Experimental Meson Spectroscopy - 1972 (Philadelphia)	72-88226	0-88318-107-X
No. 9	Cyclotrons - 1972 (Vancouver)	72-92798	0-88318-108-8
No.10	Magnetism and Magnetic Materials - 1972 (Denver)	72-623469	0-88318-109-6
No.11	Transport Phenomena - 1973 (Brown University Conference)	73-80682	0-88318-110-X
No.12	Experiments on High Energy Particle Collisions - 1973 (Vanderbilt Conference)	73-81705	0-88318-111-8
No.13	π-π Scattering - 1973 (Tallahassee Conference)	73-81704	0-88318-112-6
No.14	Particles and Fields - 1973 (APS/DPF Berkeley)	73-91923	0-88318-113-4
No.15	High Energy Collisions - 1973 (Stony Brook)	73-92324	0-88318-114-2
No.16	Causality and Physical Theories (Wayne State University, 1973)	73-93420	0-88318-115-0
No.17	Thermal Expansion - 1973 (Lake of the Ozarks)	73-94415	0-88318-116-9
No.18	Magnetism and Magnetic Materials - 1973 (Boston)	59-2468	0-88318-117-7
No.19	Physics and the Energy Problem - 1974 (APS Chicago)	73-94416	0-88318-118-5